The Making of a Special Relationship
The United States and China to 1914

The Making of a Special Relationship
The United States and China to 1914

Michael H. Hunt

Columbia University Press

New York

Publication of this book is supported by a gift in recollection of Helen Boyden Lamb Lamont

Library of Congress Cataloging in Publication Data
Hunt, Michael H.
 The making of a special relationship.

 Includes bibliographies and index.
 1. United States—Foreign relations—China.
 2. China—Foreign relations—United States.
 I. Title.
 E183.8.C5H86 1983 327.73051 82-9753
 ISBN 0-231-05516-1
 ISBN 0-231-05517-1 (pbk.)

Columbia University Press
New York   Chichester, West Sussex

Copyright © 1983 Columbia University Press
All rights reserved
Printed in the United States of America

p 10 9 8 7 6
c 10 9 8 7 6 5 4 3 2 1

Clothbound editions of Columbia University Press books are printed on permanent and durable acid-free paper.

To Paula
For constant help
and much else

Contents

Preface	ix
Part One: First Contacts	1
1. The Rise of the Open Door Constituency, 1784–1860	5
2. The Chinese Discover America, 1784–1879	41
Part Two: Emerging Patterns of Interaction	81
3. The Politics and Diplomacy of Exclusion, 1879–1895	85
4. The United States in Li Hung-chang's Foreign Policy, 1879–1895	115
5. American Policy and Private Interests, 1860–1899	143
Part Three: The Patterns Hold	185
6. China's Defense and the Open Door, 1898–1914	189
7. Exclusion Stands the Test, 1898–1914	227
8. American Reform and Chinese Nationalism, 1900–1914	258
Epilogue: The Special Relationship in Historical Perspective	299
Notes	315
Index	405

Preface

EVERY BOOK HAS a personal story to tell. This one had its inception at the outset of my teaching career. I was intent on putting America's mid-twentieth-century Asian entanglement in historical perspective for a special group of Yale freshmen interested in American–East Asian relations. The road to Pearl Harbor, the fall of China, war in Korea, and recurrent crises in the Taiwan straits were all very much on my mind. But above all, preoccupying instructor as well as students, was the costly and divisive Indochina War then in progress.

My first instinct was to turn to the literature on American–East Asian relations to search out the historical roots of the contemporary crisis. The subsequent discovery that the mainstays of that literature were disappointingly irrelevant to my concerns provided the initial impetus to this work. Tyler Dennett's *Americans in Eastern Asia* (1922), the logical starting point for my investigation, epitomized the problem. Its breadth of scope and coherent interpretative themes had given it enormous staying power. Although now less often read, Dennett's work supplied a conception of American–East Asian relations which historians borrowed and built on for nearly fifty years after its appearance. But it was precisely that conception that struck me as dated, narrow, and one-sided and hence in need of revision. Writing just after World War I, Dennett had sought to demonstrate the necessity of American cooperation with the other powers in shaping a stable, peaceful Asian order. *Americans in Eastern Asia* spoke to the issues of a bygone era. What's more, it emphasized American diplomacy and high policy to the neglect of nonofficial contacts, and it virtually excluded Asian perspectives.[1]

The flaws in the literature soon convinced me that I must find my own way. Relations with China, a topic to which I inclined by virtue of personal interest and special training, seemed a natural place to start. While a broader framework embracing Japan and Korea as well as China would have been the ideal way to remedy the gap in the literature, I was prepared to settle for the more modest goal of dealing in depth with both sides of the Sino-American relationship and to leave it to other historians to reknit the strands of

East Asian international relations. New questions, new sources, and a growing monographic literature would have to serve as my excuse for abandoning the admirable pan-Pacific outlook of Dennett and others.

The penultimate step in devising this project was to temper my present-mindedness so that the world I wanted to recreate would be recognizable to the characters—Chinese and American—who peopled it. I have found it helpful in this regard to keep Garrett Mattingly's injunction before me: "Nor does it matter at all to the dead whether they receive justice at the hands of succeeding generations. But to the living, to do justice, however belatedly, should matter."[2] The welcome end of the Indochina War and its gradual relegation to mere "history" have helped moderate some of the intense concern that brought me to this project in the first place, and I hope thereby make me a fairer judge of events and personalities treated here.

Finally, I needed a workable interpretive framework that would accommodate my concern at the outset with making the past speak to the present while doing no violence to the integrity of the past. What has emerged as the main theme of this volume is the rise of a "special relationship" between the United States and China. Taken in the sense established by long usage, the term underscores the role of American benevolence, Chinese gratitude, and mutual good will in Sino-American contacts. My own conception of the "special relationship" is significantly different. As the substantive chapters that follow demonstrate and as the epilogue makes explicit, a process of cultural, economic, and diplomatic interaction through the nineteenth century among a large and diverse cast of Chinese and Americans—from obscure trans-Pacific travelers to eminent public figures—had by the early twentieth century bound two countries widely separated by culture and geography in a relationship notable for its breadth, complexity, and instability. To the extent that I establish to the reader's satisfaction the formative character of the era treated here and its role in the genesis of the mid-twentieth-century collision and estrangement between a United States ascendant in Asia and an intensely nationalist China, then I will have in substantial measure responded to the concerns that first inspired this work.

A reader can plow straight through this book, taking its three major parts in order. Each part corresponds to a major period in the evolution of Sino-American relations from the late eighteenth to the early twentieth centuries, and each is prefaced by an historical cameo introducing the major issues to be developed in the chapters immediately following. Or a reader can approach this work topically by jumping from chapter to chapter. Chapters 1, 5, and 8 deal with the "open door constituency," my shorthand term for a set of interest groups—American businessmen, missionaries, and diplomats—with a common commitment to penetrating China and propagating at home a paternalistic vision (conventionally associated with the open door) of defending and reforming China. These same chapters also relate the American interest groups to the Chinese, both those who aided them and those who opposed them. In the first half of chapter 2, in chapter 4, and in chapter 6 high level policy, characterized by the desperate ploys of embattled Chinese on the one side and the mounting pretensions of Americans on the other, receives its due. The trials of the Chinese immigrant, the growing force of nativist sentiment in the United States, and the defensive policy of protection taken up by Chinese officials are treated in the second half of chapter 2 as well as in chapters 3 and 7, and stand in thematic counterpoint to the successful American penetration of China.

A note on my research strategy is also appropriate here. My initial plan was to construct the story of bilateral relations chiefly from primary sources on the Chinese side and secondary sources on the far better-studied American side. This two-tiered approach did not work out. Although I have waded through a wide range of Chinese documents, I have to my delight found a good deal of help from the scholarly research of others, much of it completed while this work was in progress. Conversely, as I read on the American side I repeatedly gave in to the urge to go back to original sources to develop neglected points I thought important, to test controversial or doubtful historical generalizations, or simply to get a better feel for the people and period in question. The result is a history that rests exclusively on original research at some points, that relies substantially on standard monographs at others, and that at times mixes the two. Readers thus hold before them neither a conventional research monograph nor a derivative survey but rather a hybrid, a "research survey." Historians of American–East Asian relations, of U.S. foreign relations, and of modern China will, I hope, find here enough in the way of fresh material and perspectives to repay the effort of reading on. At the same time I have tried to keep this a full and accessible introduction for those unversed in Sino-American relations.

A brief comment on the handling of sources in this volume:

frequently cited collections of documents have, in the interest of economy, been abbreviated and an alphabetical listing of those abbreviations provided at the front of the Notes section at the back of the book. Before the notes for each chapter I have supplied a brief essay on the relevant historical literature. These essays indicate my many debts to other historians as well as some of my differences with them. They are also intended as a guide for those who wish to dig deeper. Biographical information in this account is, unless otherwise indicated, taken from standard references.[3] I have omitted from my book citations the names of publishers in order to hold down an already voluminous set of notes.

My personal obligations have piled far higher than I had imagined possible when I set to work on this project a decade ago. John Blum, Dorothy Borg, Sherman Cochran, Warren Cohen, Waldo Heinrichs, Tom Hietala, Jane Hunter, Akira Iriye, and Howard Lamar deserve special thanks for their critical comments on various drafts of this work. My in-laws, Paul and Theresa Schreiter, have on several occasions made scholarship more pleasant by providing a delightful and quiet retreat. *Business History Review* was good enough to sanction the reuse here of material that appeared in volume 51 (Autumn 1977) as "Americans in the China Market: Economic Opportunities and Economic Nationalism, 1890s–1931." I am also grateful to Yale University Press for permission to draw from *Frontier Defense and the Open Door* (1973) in portions of chapter 6 of this work. The National Endowment for the Humanities, the Yale Concilium on International and Area Studies, the Yale Council on the Humanities, and the Colgate Research Council created opportunities which I hope this work justifies. Finally, Ceci Long and Mandy Hollowell, both at the University of North Carolina at Chapel Hill, and the Columbia University Press staff, foremost Bernard Gronert and Jennifer Crewe, deserve recognition for their help in the home stretch.

Chapel Hill
Summer 1981

The Making of a Special Relationship
The United States and China to 1914

Part One

First Contacts

MIDDAY ON A SUNDAY in late September 1821, a six-pound pottery jar fell from the top deck of the *Emily*, an American ship out of Baltimore, at anchor off Canton. Somewhere below at that instant was a boatwoman selling fruit. Her husband's discovery of her body in the water the next day set in motion a month of contention. The Chinese charged that the owner of the jar, a Sicilian sailor named Francesco Terranova, had felt himself shortchanged and hurled the jar in anger, hitting the woman on the side of the head and knocking her unconscious into the water. Terranova for his part denied any connection with the woman's death, and had the sworn testimony of his shipmates to support him.[1]

The Chinese judicial machinery began to turn after Captain Cowpland of the *Emily* refused to buy off the family of the deceased. He would not stoop to this form of "bribery" practiced by the Chinese, particularly to settle a case that appeared to him to involve nothing more serious than accidental homicide. That decision in turn brought in the provincial governor general, Juan Yüan, who was determined that if the foreigners would not "be quiet and observe the laws," then they would have to feel the full weight of China's justice.[2] With the issue unavoidably drawn, an emergency committee made up of the principal American merchants and captains in the port, including such prominent figures as Jonathan Cushing, Samuel Russell, and D. W. C. Olyphant, quickly formed and immediately conceded Chinese jurisdiction in this case and sought in return assurances that Terranova be given "a fair and impartial trial, in which all evidence for his defense, foreign or Chinese, shall be equally and impartially received; the friends of the accused to be present during his trial. . . ." Confident that Chinese practices would meet the standard of a "fair" trial and pleased with the apparent reasonableness of the Americans, the Chinese authorities acceded to the American request that Terranova remain in their custody and, going a step further, decided to convene the trial right on board the *Emily*.

On October 6 the trial began and at once revealed the differences in outlook between the forty unarmed Americans in attendance and the presiding magistrate, who had arrived accompanied by an imposing array of war vessels and a personal escort of a thousand men. The committee of American merchants complained against the inadequate means of translation afforded them and the lack of opportunity to cross-examine Chinese witnesses or to present testimony in Terranova's behalf. Repeated interruptions by the Americans on these points at last roused the magistrate's ire. Trial was not, according to accepted Chinese judicial practices, an adversary proceeding; the responsibility for hearing the evidence and reaching a decision was the magistrate's alone. He abruptly ended the proceedings and returned to his boat. Juan Yüan now ordered all American trade halted and an embargo placed on American vessels until Terranova was surrendered for retrial in the city.

The trade stoppage struck directly at the livelihood of the American merchants, and so on October 24 the emergency committee handed Terranova over. Within the next four days Terranova, now bereft of the support of his fellows, was retried under the eyes of the highest judicial authority in the province, confessed throwing the jug without intending any real harm, was judged guilty of killing in an affray, and at dawn on the fourth day was marched to the execution grounds and strangled. That same day his body was returned to the Americans, and the next day trade resumed.

The first phase of direct Sino-American contacts, opened by the celebrated voyage of the merchantman *Empress of China* from New York to Canton in 1784, has usually been associated with profitable commerce. (Indeed, it was. That venture returned 30 percent on an investment of $120,000.)[3] With the passage of time those early China trade days have also taken on a romantic aura. But to probe beneath the surface is to lay bare tensions and ambiguities inherent in relations between two distinctly different peoples just discovering each other. Americans in Canton did not like having to submit to Chinese control, as their attempts to shield Terranova make clear. They regarded the Chinese legal system as barbarous not so much because they failed to grasp its niceties as because its fundamental moral and social values and its special provisions for disciplining foreigners struck at their own fundamental ideals and interests.[4] But since the Chinese government would not waive its jurisdiction, Americans would for the moment have to accept the status quo and hope for a change for the better (though when and how was far from clear). The Chinese were at the same time coming to terms with the Americans as a distinct people, but of precisely what quality remained undetermined by the time of the Terranova affair and for some time thereafter. For the moment they were content to assume that Americans had the same crafty and deceitful disposition attributed to the other "barbarians."

Distant economic developments were in ways unimagined at the time of the Terranova case to intensify Sino-American interaction in the 1840s and 1850s, and begin to resolve some of the ambiguities on both sides. The growing dependence of the Indian economy and the British China trade on China's consumption of opium finally in the early 1840s forced the British government to overthrow the Canton system. Thrown on the defensive, Chinese officials along the coast now took an urgent interest in the newly arrived Westerners and the divisions among them that might be turned to China's advantage. Some were impressed with the rise of a new power on the North American continent whose sharp and deeply rooted jealousy of the British carried hope of aid for a beleaguered Chinese foreign policy. Americans saw British dealings with China in a different light. While fretting over the British advance, Americans were at no time loath to follow behind, claiming all the rights secured by the British. The result was to intensify American penetration not only by multiplying the points of access along the coast but also by endowing Americans, like other foreigners living under the new treaties, with substantial immunity from Chinese control.

Not long after this major shift in the terms of Sino-American contacts along the China coast, the economic development of the newly opened American West was to set in motion interaction on an entirely new front. In the 1850s the demand for labor there began to divert increasing numbers of Canton delta emigrants from the well-worn routes to Southeast Asia. The resulting movement of Chinese across the Pacific was to raise in a different context the familiar and contentious issues of conflicting cultural values and group interests. These contacts on the opposite shores of the Pacific and the questions they gave rise to were for the moment superficial and marginal to the national life and even the foreign policy of each country, yet they would yield to more clearly defined and in time even fixed patterns of interaction with ultimately serious consequences.

Chapter One

The Rise of the Open Door Constituency, 1784–1860

THE LATE EIGHTEENTH century and the first two-thirds of the nineteenth was the seedtime for American interests in China. Over those eight decades an American community emerged along the coast. The members of that community and, to a degree, their allies at home made up what might appropriately be designated the open door constituency because of their shared commitment to gaining and maintaining access to China on favorable terms. The groups thus united were, however, neither equal in influence nor fully reconciled in their specific goals. Down to the latter part of the nineteenth century, mercantile interests were predominant. The sporadic trading ventures of the 1780s had evolved by the 1850s into a substantial commercial establishment lodged in the treaty ports along the China coast. This mercantile presence overshadowed the nascent mission movement and called into existence a diplomatic service charged with the defense of American commercial interests. Yet neither of these junior members of the open door constituency were satisfied with a subordinate position. Government representatives often took as much interest in promoting American prestige and influence in China as in meeting the wishes or needs of the mercantile community, while missionaries struggled for secure access to the masses of unsaved Chinese against the obstacles imposed by a cautious, commercially oriented American policy no less than by the policy and popular preferences of China.

Mercantile Beginnings in Canton

Americans gained independence from Britain at the expense of losing valuable markets in Europe and the West Indies. But at the same time freedom from British mercantile restrictions opened new markets. China was one of these. No longer constrained to obtain Chinese goods through London at prices set by the monopolistic British East India Company, merchants in Boston, Salem, New York, Philadelphia, and Baltimore were free to establish direct contact with Canton, China's sole foreign trade mart. They set out unfamiliar with the Canton trade, uncertain of their success, and virtually on their own but nonetheless with certain assets that would bring them success. To begin with, they had, in Samuel Eliot Morison's phrase, "a desire to get on in the world." They already knew how to finance international trade through remittances paid out of London, and their contacts in that important commercial hub were extensive. They were also well schooled in the problems of carrying on a global commerce with the attendant uncertainties of predicting the movement of distant markets (Canton was over 15,000 miles away, roughly four months by sail) and the need to evaluate the character and acumen of agents and captains (on whose soundness the success of any enterprise might turn).[1]

Ingenuity in applying the system of mercantile capitalism, by which most of the China trade was financed and conducted until late into the nineteenth century, was perhaps the key ingredient in the success of the new Canton trade. Even a smooth voyage blessed with favorable winds would run only slightly under a year. More time was required for touching at other ports in the Pacific, Latin America, the East Indies, Europe, or the Mediterranean, wherever there arose opportunities for trade complementary to that conducted at Canton. Not only did the long voyage routinely tie up a sizable amount of capital for a year and a half to two years, but the ship and goods it was invested in were subject to a variety of dangers, including uncharted seas, disease on shipboard, storms, pirates, and the depredations of French and British men-of-war between 1793 and 1815. To spread the financial burden and share the risk, trusted friends and family relations would band together, usually on an ad hoc basis, to send a ship to Canton and market the goods it carried.

To increase their flexibility and knowledge of the market, American merchants began to post agents to Canton as early as 1803. A full-time resident would be free of the necessity of completing business in a brief trading season when newly arrived goods would often flood the market. The resident agent could buy and sell anytime during the year, whenever prices were favorable, and with passage of time would gain familiarity with the market. The decision by Perkins and Company of Boston to establish the first agency in Canton reflects that port's rising importance in the estimate of American merchants. (London was the only port that theretofore Perkins had paid such a tribute to.) True to the style of the merchant capitalists, Perkins relied heavily on business associates and kin to protect interests in Canton. The first agent was a twenty-six-year-old who had worked in the firm's Boston counting house. Perkins supplied roughly a third of the needed funds, a dozen other individuals and firms the balance. With the agent went a Perkins relation, also experienced in the counting house.[2]

The Canton trade that Perkins and others conducted was hopelessly lopsided at first. Americans went to China primarily to buy Chinese products prized in the United States and Europe—silks, nankeen cotton cloth, porcelain, and above all tea. (That beverage had become part of the daily fare in New England as early as the 1720s, and by the early 1780s most Americans had acquired the tea-drinking habit.) China traders had difficulty at first finding goods to carry to Canton to offset their purchases there. New England ginseng came easily to hand, but it was regarded by the Chinese as an inferior grade, and importers tended to flood the market. In their global meanderings, trading vessels collected other offerings—sea otter skins (purchased at low cost from Indian trappers), seal pelts (secured by hunting expeditions organized on shipboard), sandalwood (obtained at bargain rates from a monopoly controlled by the Hawaiian monarchy), and other exotic natural products. But even at best these items were never enough, and constant exploitation soon depleted the supply.[3] At least until the 1830s the balance had to be made up in specie, an unprofitable export, for Spanish silver dollars and bullion were difficult to assemble in sufficient quantity, drew no interest in the months they lay idle while in transit, and ran the risks of the voyage.

Americans finally discovered opium as the ideal medium of exchange. Excluded for the most part from the major source of supply in India, they located an alternative source in the Turkish port of Smyrna (the modern Izmir). Though inferior in quality, the Turkish drug could be mixed with a better product or consumed straight by less discriminating smokers. Philadelphia and Baltimore ships had gotten into the trade as early as 1804. Insignificant before the war with Britain and suspended for the duration, trade in Turkish opium exploded after the war, accounting in 1817–1818 for as much as one half of the goods (exclusive of specie) brought by Americans into Canton. By the 1820s Americans may have been carrying in annually as much as 2,000 chests (approximately 133 tons) of "turkey," far more than any merchant had earlier thought China capable of absorbing but even so only one-fifth what the British were bringing in from India. By 1821, in the midst of this expansion of the opium trade, resident agents in Canton had found a way to get around Chinese prohibitions—by stationing a storage ship at Lintin and other points on the outer reaches of the Pearl River. The agent could make his sale in Canton and leave it to Chinese smugglers to carry the opium from storage ship to shore for distribution. Perkins set up the first American storage ship in 1823 and quickly emerged as the dominant dealer in Turkish opium. The rise of this opium trade staunched the flow of silver from the United States to China, so successfully indeed that it may have contributed to the sharp inflation at home in the 1830s.[4]

The period after 1815 was one of consolidation for the American Canton trade. Only those firms with a high volume of trade could afford to keep their own opium storeship off Canton or resident agent in that port. The trade was becoming not only more costly to conduct but also more competitive, particularly for Americans. For over two decades, between 1793 and 1815, Europe had been at war, giving the advantage of neutrality to American ships and hence the American trade. But with the return of peace, the Europeans returned to Canton. The dissolution of the British East India Company monopoly in 1834 further expanded the number of merchants acting as middlemen between Canton and foreign markets but without a commensurate increase in overall demand in either of those markets. Even the dealings in opium, which had increased in the late 1830s to twice what they had been in the previous seven years, suffered. The end of controlled opium exports from India created "a drug on the market" (a phrase that dates from Canton days), and oversupply took its toll among small firms less able to absorb a loss or hold an opium shipment for a better market. One by one the nine American firms (one-sixth of the foreign total) operating from Canton in 1836 began to cut back, perhaps in response to the decline in profits recorded in their ledgers, or out of fear of falling into trouble with Chinese officials or their own consciences over the opium trade, or perhaps for a combination of those reasons. Wetmore and Olyphant stayed with the Canton trade but made a principled decision against handling opium, while some large firms, including the Browns of Providence, Astor of New York, and Girard of Philadelphia, got out altogether.

Out of the pressures of the 1820s and 1830s, Russell and Company emerged as the premier American firm in China. Founded in 1824 by the leading China merchants of Boston and Salem as the successor to Perkins and Company, Russell maintained its commanding position through much of the rest of the nineteenth century, thanks to the skill and experience of its agents together with its capital strength and business contacts in Boston and London. Within less than a decade of its creation, Russell and Company had succeeded to Perkins' old position as the dominant American firm trafficking in the Turkish drug, and after 1834 it entered heavily into the Indian opium trade as well. But its agents handled not just opium but the whole range of Canton goods, buying and selling on the firm's account as well as for others on a commission basis. The senior members of the firm selected their agents from among the able young relatives or business associates (often the two coincided) eager for the chance to accumulate and carry home the small fortune a half-dozen years' work in Canton could bring. A list of Russell and Company staff reads like a *Who's Who* of the early China trade—Perkins, Heard, Low, Hunter, Forbes, Delano, Sturgis, and King.[5]

The final step in constructing a profitable trade relationship was that of finding economic intermediaries to put Americans in touch with the China market, both for export and import items. In the Canton system they found from the outset a ready-made and generally satisfactory solution to their problem. That system had been shaped by a long Chinese experience of dealing with foreign merchants on the South China coast. Local officials supervised foreign trade and taxed it, usually at about one third of value. Foreigners were confined exclusively to Canton and tolerated on the condition that they respect Chinese regulations and keep to their "factories," a row of waterfront buildings partitioned by nationality and serving as a combination of residences, warehouses, and offices. There foreigners passed the trading season spanning the late fall and winter. Contact with Chinese was held to a minimum, and approaching Chinese officials directly was forbidden. With the end of the trading season all foreigners were supposed to leave Canton, and at least until the 1830s most foreigners complied, escaping in Portuguese-governed Macao the heat, confinement, and long hours of work associated with factory life.[6]

The most important of the intermediaries provided by the Canton system was the cohong. It was a corporate body of major merchants of that city selected by the Chinese government to supervise foreigners and deal with their commercial requirements. Thus the cohong bore political responsibility as the price of its profitable position in Sino-foreign trade. Cohong merchants each traded on their own accounts and rented out the "factories" to the foreigners. They carried communications back and forth between local officials and the foreign "guests." Guaranteeing the cohong's commercial integrity was a shared sense of group responsibility and a common fund of capital pooled by the merchants to cover bad debts any of their number might incur. Of these cohong merchants, Howqua (the name borne by Wu Ping-chien and his sons, Wu Yüan-hua and Wu Ch'ung-yüeh) did the most business with Americans, especially with Perkins and, later, Russell. This Chinese mercantile family based its business relationships with Americans not on sentiment but on a clear and immediately profitable basis. In the main it sold tea to the Americans for specie, thus dispensing with the import goods the British brought in to balance their trade. As Americans began to rely increasingly on opium in place of specie, Wu Ch'ung-yüeh kept his distance from the drug, perhaps because (as he argued) of the official prohibition (his brother had gone to prison in 1831 for violating it), perhaps no less because of the added difficulty of disposing of that import before seeing a profit. The Americans also benefited the Wu family by offering a cover for putting some of its fortune (estimated in 1824 at $26 million) to work abroad—in international trade and in enterprises in the United States.[7]

For the convenience and reliability of the trade mechanisms in Canton, Americans like other merchants paid a price. Their trade was geographically circumscribed and thus limited to the interior market accessible from Canton. Foreigners were supposed to trade only through one of the designated merchants, whose prices sometimes smelled of collusion. And of course foreigners were ultimately subject to Chinese laws and exactions (though in the main they were left to tend to their own affairs in their physical isolation while the economic intermediaries tended to the regularized corruption that was part of Canton trade). While private British merchants grew increasingly restive under these conditions, American merchants in Canton were less so, in part because their trade prospered, in part because they had no hope that their government would intervene to redefine the terms of trade. "The American Government," declared "an American" in the Canton press in October 1830, "requires of us to submit peaceably to the laws of the country we may visit; hence we consider ourselves bound to obey the laws of China. Other foreigners may take a different view of their obli-

11   Rise of the Open Door Constituency

gations and their governments may uphold their resistance. We do not question the propriety of their conduct. We well know the terms on which we are admitted to trade."[8]

General acceptance, however, did not mean complete passivity. American traders and their British counterparts sometimes tried to get around the system. In the 1820s they sought partially to evade the cohong monopoly by effecting better deals with "shopmen," small Canton merchants who did not have to pay the exactions that went with official supervision and were less likely to coordinate pricing with the cohong. Although the local governor general in Canton rejected a petition when the foreign community sought to regularize this practice, "shopmen" continued to handle certain specified categories of goods.[9] Some foreigners sought to evade the Canton system altogether by trading along the coast, a development given impetus by the growing tide of opium imports and mounting official opposition. With Canton too dangerous and the cohong obliged to oppose the opium trade, merchants found substitute intermediaries in Chinese smugglers.

The American merchants' success at coming to terms with the Canton market was reflected in their tendency to engross an increasing percentage of China's growing foreign trade (though it should be remembered Americans never challenged British predominance).[10] But even more important were the handsome profits that sprang from this trade. The money made early by the *Empress of China* was no fluke. One investor in the Canton trade in the latter part of the 1780s boasted a 126 percent return. The *Neptune* out of New Haven in 1796 slaughtered 80,000 seals in the South Pacific to trade in Canton and returned in 1799 with a clear profit of $220,000. John Perkins Cushing, the young assistant to Perkins and Company's agent, had gone out to Canton in 1803 and taken charge of business at the age of sixteen following the death of the agent. A man of demonstrated business acumen, he returned home in 1831 with nearly a million dollars to show for his long exile. Rufus King, an eminent New York Federalist, invested $20,000 in various Canton trade ventures between 1808 and 1813 and somehow, despite Republican trade restrictions and British command of the sea in 1812 and 1813, managed to more than double his money. John Jacob Astor accumulated much of the half-million that he eventually invested in Manhattan real estate from the Canton trade between 1800 and 1809. The Browns of Providence made 35 percent on their first venture to Canton in 1788–1789. Nine of their voyages to the

Pacific between 1815 and 1821 yielded an average of 63 percent. Russell and Company's net profit in 1834 was about $90,000. For the following two years it came to $310,000.[11]

Diplomacy and Commercial Interests

Merchants had from the very outset of the Canton trade appealed to political leaders to lend their support to "an increasing and profitable branch of our commerce." In response, prominent figures in the Federalist party—Alexander Hamilton, John Adams, and John Jay—were ready to go as far as the limited resources of the government would allow. They had been quick to recognize the value of the China trade in the 1780s, and in the following decade they pressed within the Federalist administrations for measures favorable to it. Canton and India together, Hamilton informed the House in 1791, offered "an additional and extensive field for the enterprize of our merchants and mariners [and] an additional outlet for the commodities of the country." The results were modest: the establishment of favorable tariff treatment, maintenance of an honorary consul at Canton, and attempts at prying open British India to Americans in search of goods to sell in China.[12] Yet even this was more than the Republican administrations of Thomas Jefferson and James Madison were willing to do. Neither would devote government resources to an enterprise about which they were ambivalent. Both men recognized the practical importance of foreign trade to the national livelihood but regarded warily the attendant entanglements and indeed were prepared to sacrifice it in waging economic warfare against Britain and France from 1807 onward.[13]

With little to work with, these early American leaders seized on the idea of cultivating Chinese good will as an economical way to advance American commercial interests and establish an advantage against European rivals. This impulse, to persist in the American approach to China, had its roots in early merchant reports. Those who had made the trip in the 1780s had reported that the Chinese were "very indulgent towards us" and "highly pleased at the prospect of so considerable a market for the productions of their own empire." John Adams had already picked up the refrain in November 1785: "Much will depend upon the behavior of our people who may go into those countries [of the East]. If they endeavor, by an irreproachable integrity, humanity, and civility to conciliate the esteem of the natives, they may easily become the most favored nation; for the conduct of European nations in general, heretofore, has given us a great advantage." Jefferson, in a curious episode in the summer of 1808, was guided by the same sentiment. A Chinese— ostensibly a "high mandarin" named Punqua Wingchong, who (Jefferson was told) had come to the United States to collect debts and attend to business—wanted permission to pass through the embargo for the ship he had engaged to take him and his property home. Jefferson at once recognized an opportunity for "making our nation known advantageously at the source of power in China, to which it is otherwise difficult to convey information." He calculated that by using this dignitary to inform the Chinese of "our nation, our circumstances and character, and of letting that government understand at length the differences between us and the English, and separate us in its policy," he could "bring lasting advantage to our merchants and commerce with that country." [14]

In fact the pressing problems of establishing true independence and security for the new nation virtually ruled out any meaningful official support for the Canton traders for the first four and a half decades. American consuls, in reality merchants who attended to official business in their spare time, enjoyed neither salary nor standing before the Chinese government, and their ships went without naval protection against the British, the French, and Chinese pirates. They were limited to petitioning in the accepted manner to Chinese officials over minor grievances and carping in private over more serious ones, but always carefully avoiding any major frontal challenge to the principles of the Canton system. Their response to the first visit of an American warship to Canton in 1818 illustrates their sense of isolation and vulnerability. The captain of the frigate *Constitution*, then on a two-year cruise in Asian waters, had resolved to use his time in Canton to cultivate the friendship of the Chinese. His first appearance evoked nothing but the usual studied Chinese indifference, and finding the welcome no warmer on his return, he sailed into local waters up toward Canton in violation of well-publicized regulations. Alarmed by this pointless bravado, American merchants publicly disassociated themselves from the proceedings so that their own position would not be compromised. In 1821 Americans in Canton demonstrated that same accommodating spirit when they surrendered Terranova for trial. As late as 1830 at least some felt "bound to obey the laws of China." [15]

The Opium War (1839–1842) and the collapse of the Canton

system after two decades of steady erosion were finally to arouse the American community in Canton and its allies at home and force Washington to be more attentive. Opium smuggling had eaten away at the system. Faced with routine foreign defiance of official prohibitions and a spreading addiction to opium, officials throughout the empire debated methods of stamping out these related evils. A local antiopium campaign in 1839 drew the issue by conditioning the continuance of any foreign trade on a cessation of all dealings in opium and the surrender of all opium stocks. Regarding the new regulations as intolerable, the British government resorted to force and after a series of victories overturned the restrictions of the Canton system in favor of less fettered commercial relations to be defined by treaty. The first of these, the milestone Anglo-Chinese agreement of 1842, freed trade from the cohong monopoly and official exaction. Foreigners were thenceforth free to reside and trade in any of five ports along the coast (Shanghai, Ningpo, Amoy, and Foochow as well as Canton), beyond official Chinese supervision and jurisdiction. The treaty also required that the Chinese henceforth accept foreign envoys and deal with them as at least nominal equals.

For the following two decades U.S. China policy would develop within the context of Anglo-Chinese relations. Britain, the leading Western power, represented the impatient and aggressive half of the main equation through the 1840s and 1850s. The Chinese half of the equation was characterized by a peculiar mix of vulnerability and obstinacy. The Chinese government was unwilling to surrender its claim to control foreigners and foreign trade or to lie down in the face of an aggressive military policy. But resistance proved increasingly difficult through the 1840s and 1850s. The Opium War confirmed foreign suspicions of Chinese military and naval inferiority, and subsequent clashes of arms did nothing to rehabilitate China's reputation. The spread of rebellions throughout the empire in the 1850s compounded the problem by diverting Peking from foreign affairs while sapping imperial strength. Against the greatest of those rebellions, the Taiping, the dynasty mobilized its entire resources and only after two decades of conflict, with deaths running perhaps to 20 million, did it restore its control. In the meantime Peking sought to temporize in its efforts to contain foreign encroachments, but its vulnerability to coercion remained manifest to all foreign observers.

Against the background of this gathering crisis in Anglo-Chinese relations, the Americans in Canton began calling for more official support. In 1830 they had for the first time not only welcomed a naval visit but also expressed their conviction that a naval presence would raise their prestige in Chinese eyes and secure for them prompt redress of grievances. In 1839, as the opium crisis came to a head, Americans in Canton became more insistent on Washington's protection than ever before, calling again for a warship in port and petitioning Congress to send a commercial agent to China to negotiate a treaty that would secure their position. Merchants in Boston and Salem represented by Thomas H. Perkins suggested that same year to Congress that a "respectable national force" be dispatched to Chinese waters during the coming clash between China and Britain, and some urged Congress to take the still bolder step of sending a minister to Peking.[16]

Throughout this period Washington became increasingly responsive. Visits to China by naval vessels, though still spasmodic and brief, became more frequent through the 1830s.[17] At home mercantile pressure, reenforced by news of the British treaty ending the Opium War, created the first serious interest in a diplomatic mission to China. John Quincy Adams, chairman of the House Committee on Foreign Affairs and always solicitous of the maritime interests of his Boston and Salem constituents, strongly supported it, as did Secretary of State Daniel Webster, after doing his own survey of merchant opinion. Finally, in 1842 President John Tyler recommended the mission to Congress. His message (prepared by Webster) concluded hopefully that American exports, which had doubled over the last ten years, would continue to grow as new ports opened along the China coast. The mission, led by Caleb Cushing, set off in 1843 with instructions from Webster to make known American friendship and to keep abreast of the gains won by Britain in those "very important marts of commerce" recently opened. Diplomacy was to augment American exports and serve "the commercial and manufacturing, as well as the agricultural and mining interests of the United States."[18]

This decision to establish direct diplomatic ties with China gave rise at mid-century to a foreign service in that country that was (as elsewhere) above all else amateur and unstable. Party loyalties counted heavily in the selection process. Of the first nine ministers to China between 1844 and 1860, seven were deserving politicians. They served on the average but a year each, leaving chargés to fill the gap on no less than nine different occasions totaling some seven years. Congressional parsimony kept salaries low, and together with the political patronage system impeded the development of a professional staff. Lacking a permanent residence or direct access to the central government in Peking, these early emissaries lugged their official papers and personal effects up and down the 2,000 mile-long China coast, dealing now with one, now with another set of provincial officials and always dependent on a sometimes grudging navy for transport and on the hospitality of prominent merchants, above all Russell and Company. The creation of a full-time consular service in 1854 only had the effect of partially displacing part-time merchant consuls (in Canton and Shanghai these were usually partners of Russell and Company because they alone could afford the time and money to fulfill the social and official duties of the office) with underpaid and short-tenured political appointees.

Behind the foreign service, and in some senses a part of it, was the navy. The Asiatic Squadron, officially organized in 1835, began in 1842 to maintain a constant presence off the China coast (with the sole exception of a two-year hiatus between 1846 and 1848, during the war with Mexico). Its force, initially two or three vessels but grown by 1860 to thirty-one, had to protect American lives and commerce not only there but also in the rest of maritime East and South Asia. To these older duties it added new ones as an auxiliary to diplomacy. Warships transported diplomats as they moved up and down the China coast. Their guns, even when silent, helped to underline to the Chinese the seriousness of diplomatic demands. Naval commanders sometimes became the diplomat's alter ego; they even substituted for the minister on four occasions.

The dispatch of diplomats and warships reflected Washington's desire to keep informed of the China crisis—and by timely action to stay abreast of the British—but Washington at no time intended to become actively involved in the Anglo-Chinese struggle. A foreign policy directed primarily at extending and consolidating control over a substantial portion of the North American continent forced China well down the policy agenda until the late 1840s, and thereafter the national divisions over the future of slavery, sharpened by the very success of this expansionist policy, kept China on the periphery of official concerns. For policy makers China was distant, and their own influence there one of manifest weakness.

This weakness helped keep alive among policy makers the old strategy of cultivating the Chinese as one easy way to advance American commercial interests. But the willingness of the British to share in the fruits of their victories suggested a complementary and no less economical way of safeguarding those interests. The

United States had only to follow behind British diplomats and gunboats and lay claim to the concessions they would wring from Peking. As long as the British held to the principle of free trade and the Chinese acquiesced in the magnanimous policy, the United States could collect the dividends of gunboat diplomacy while maintaining the forms of amity toward China. Webster's instructions for the first mission to China had embodied that two-pronged strategy; subsequent administrations kept it alive.

Washington's approach was a recurrent source of discontent among American diplomats sent to China. Their criticism of Washington's policy was a compound of impatience over Chinese attempts to check the advance of civilization and anxiety over the unchecked British tide in Asia. Cultivating a people like the Chinese who only respected force was futile, and passively relying on Britain ("jackal diplomacy") left that old arch foe of American security and commerce dangerously free to seize exclusive advantages detrimental to the United States. The alternative, advanced by such dissenters in the foreign service, was to take up the more energetic line of advancing alongside rather than behind the British. By sharing the risks and costs, the United States would be in a better position as a partner (admittedly a junior one) to claim a share of the gains and to restrain British avarice if necessary. A variant, no less energetic in conception, was to imitate Britain in her use of force and acquisitions of territory along the coastal periphery but to keep a free hand, thus avoiding entanglement in her imperial ambitions. Thus might national dignity and honor be preserved, the national interest in commerce advanced and, incidentally, the careers of the worthy politicians occupying the China post promoted. Slow communications—it usually took six months for a query to reach Washington and a reply to get back to China—provided envoys the time to develop their own divergent strategies.

The tendency for the government's agents on the scene to adopt forceful methods first became apparent in the conduct of naval commanders. Schooled in the ways of the sea and sensitive to questions of national honor, they were frequent advocates and willing instruments of gunboat diplomacy. For example, when in 1839 the Chinese cordoned off the foreign factories to force the surrender of opium stores, the first instinct of the commander of a visiting American warship was to sail up to the city and liberate the foreign community by force if necessary. (He was at last dissuaded on the grounds that his actions might set off mob violence, possibly with serious consequences to Americans in the factories.) In 1842, on the next appearance of an American naval vessel in Chinese waters, the commander took his ship right up into the Canton anchorage to force a settlement of outstanding grievances by several American merchants. It worked, and the next year on a second visit he repeated the performance—with equal success. To Washington he communicated the lesson learned: force was the key to dealing with the Chinese. "The presence of a fleet of United States ships appearing here would do more to obtain a favorable treaty than any other measure."[19]

The tension between the preferred strategies of Washington and of its envoys also became evident in the first formal diplomatic mission to China. Its head, Caleb Cushing, was a political ally of the embattled Tyler administration and a strong friend of the China trade. (Indeed, J. P. Cushing, head of Perkins and Company, was his cousin, and Massachusetts his political base.) Cushing's proclivity for an aggressive China policy had been evident as early as December 1842, when he had warned the President of the threat that Britain posed to American commerce in the Pacific. The British fleet, having just squeezed commercial advantages from China, might now attempt to open Japan to trade or even threaten Hawaii. The task of securing for Americans equal commercial footing with the British seemed at first easy enough. The British, Cushing then already knew, would not put obstacles in the way of American trade in China, while word from Americans in Canton confirmed Cushing's and Webster's own view that the Chinese would welcome an equipoise to the threatening British, whose conduct (in Cushing's words) was "outrageous," characterized by "base cupidity and violence and high-handed infraction of all law, human and divine." But by 1843 Cushing had begun to think in terms of outstripping rather than just equaling the British. Before a Boston audience he spread out a vision of American ascendance in East Asian affairs. He himself would be the emissary from the rising new center of world civilization to a once vital but now stagnant Asia. "We have become the teachers of our teachers."[20]

Once in China, Cushing decided to gamble on demands that might raise American prestige while boosting his own political fortunes. He wanted nothing less than access to Peking to deliver a letter to the Emperor from the President. Webster had suggested such a course in his instructions. But he had also provided as an alternative that the President's letter might also be sent to the Emperor, stressing that Cushing go to China only as "a messenger of peace . . . to cultivate the friendly dispositions of the government and people, by manifesting a proper respect for their institutions and manners, and avoiding, as far as possible, the giving of offence either to their pride or their prejudices." In this manner, and by emphasizing the former antagonism between the United States and Britain, Cushing was to make clear the differences between the policy the United States would pursue in China and that of Britain.[21]

Cushing himself in 1842 had stressed the need for flexibility and informality in meeting Chinese sensibilities. However, on arrival in China in February 1844 Cushing made his demand to travel to Peking formal and categorical. Further, he stood by it. When in April of that year he concluded that the Chinese were putting him off, he backed his demand with veiled references to war, and he dispatched a frigate up river toward Canton to intimidate the Chinese.[22] Only after receiving notice from his Chinese counterpart that he would not get the treaty he wanted until he renounced the journey to the capital did Cushing backpedal. In short order (on July 3, 1844) the two sat down in a temple of the Goddess of Mercy in the village of Wanghsia outside Macao to sign the document inaugurating official relations between the United States and China. Within two months Cushing headed home to see his treaty receive unanimous Senate approval. It gave American commerce and Americans in China all the advantages the British had won and promised by its most-favored-nation clause that any new treaty gains by Britain or others would automatically accrue to American interests in China.

Good fortune and prudence saved Cushing from falling out with the policymakers he served. His successors who suffered from the same tendency to assume a more forceful stance were not all so fortunate.[23] James Buchanan, Secretary of State in the Polk administration, sent out Alexander Everett as Cushing's successor in 1845 with instructions to take every opportunity "to cultivate the good will of the Chinese Government and people," to give them advice and information conducive to their national security and "their progress in arts and in arms," and thus to "promote the commercial prosperity of your own country."[24] However, once in China, Everett panicked over British efforts to force their way into Canton (a prelude, he feared, to the seizure of Chinese territory) and recommended that the United States join with St. Petersburg and Paris to head the British off. Everett's death several months later allowed Washington to turn a blind eye to the idea. The Fillmore administration, unable to get anyone to accept the low-paying China post, left matters in the hands of the missionary Peter Parker. During a two-and-a-half-year stint as chargé early in the 1850s, Parker showed even more than Everett how strong were the temptations to set aside restraint. However, rather than defending the intransigent Chinese against the British as his predecessor seemed inclined to do, Parker wanted to act with Britain to secure new advantages for all and to impress the Chinese with American naval power. Washington was no more interested in Parker's ideas than it had been in Everett's.

From 1853 to 1857, in the context of renewed Anglo-Chinese tensions and growing rebellion in China, American diplomats were time and again afflicted with an itch for action. Humphrey Marshall, a rough and ready Kentuckian who served through the single year of 1853, carried to China an intense Anglophobia, soon joined by a distaste for the Chinese. The Taiping rebels no less than their imperial opponents were "impotent, ignorant, conceited . . . and superlatively corrupt." With China rendered vulnerable by civil war, Marshall feared that Britain and Russia would make a grab for territory that could seal "the fate of Asia" and "nullify the [Pacific] projects of the United States." Marshall recommended to Washington a policy of intervention, guided (befitting the exceptional American role in the world) by a humane and charitable desire to bring peace to China, protect her against ruthless European aggressors, and "elevate her people . . . by a proper and peaceful, but scientific employment of their natural energies." Intervention on China's behalf would also, Marshall was careful to stress, promote Chinese interest in American cotton goods so that one day they would eclipse their Manchester competition.[25]

The Pierce administration rejected out of hand Marshall's bold proposal, and in the fall of 1853 decided on his recall. But Pierce had trouble finding a sufficiently docile replacement. The appointment of Robert J. Walker, a major figure in the Polk administration earlier and an ardent and influential expansionist, fell through, apparently over Walker's demand for adequate naval support and the freedom to use it as he saw fit to extend American trade. The China post then went in 1854 to Robert McLane, a Maryland congressman who, initially at least, pursued a modest policy closer to Washington's preferences. He liquidated disputes that Marshall had stirred up over control of Shanghai and the customs and kept in touch with the British in the first of several expeditions to the North China coast calculated to scare the Chinese into treaty revision. But finding the Chinese unbending, McLane proposed a "strong policy" of joining the other treaty powers in the use of force, if necessary, in order to open China and thus "offer to the American manufacturers a market more valuable than all the other markets of the world to which they yet had access."[26]

Poor health forced McLane home and saved the Pierce administration another imbroglio over China policy. But Peter Parker, serving again, first as chargé (1855) and subsequently as minister (1856–1857), revived his earlier plan of alliance with the British and accordingly proposed an increase in the number of American warships in Chinese waters. As a further step toward forcing the Chinese into line and opening the way to Peking, Parker wanted to seize Taiwan in order to anticipate Britain's taking strategic island bases in the East, to secure for the United States commercial and strategic advantages, and even to demonstrate the regenerating influence the United States could exert in Asia. Washington not surprisingly remained skeptical about the efficacy, not to mention the necessity, of displays of force, and it was no more amenable then to taking Taiwan than it had been earlier when proposed by Commodore Matthew Perry during his celebrated voyages to Japan in 1853–1854 or promoted by such merchants as Gideon Nye, Jr. In 1857 Washington once more had to recall one of its overeager diplomats.[27]

The Buchanan administration thereafter managed with fair success to hold its agents to a policy of advancing American commercial interests through treaty revision accomplished without coercion. The President, facing serious sectional divisions at home, was in no mood to tolerate obstreperous envoys, a type he had already encountered as Polk's Secretary of State. Buchanan's first envoy, William B. Reed, suffered one lapse when he asked for more latitude to use force to compel the Chinese to negotiate with him. But otherwise he placidly followed the Anglo-French expedition north in 1858, while keeping in touch with the Chinese. After the guns of the expedition had forced the Chinese to grant foreign diplomats the right of residence in Peking, to open up the Yangtze, and to promise the protection of foreign missionaries and their converts, Reed was there to share fully in the spoils.[28] When supplementary treaty negotiations began in September 1858, the American chargé replacing Reed, S. Wells Williams, carefully left the British negotiator, Lord Elgin, to strike an agreement; then in November he stepped forward to claim similar terms for the United States. The task of finalizing these bargains fell to Buchanan's second appointee, John E. Ward. While the second Anglo-French expedition prepared to fight its way to Peking, the Ward party made its way to that city peaceably to exchange copies of the 1858 treaties.

In 1860 the scene of an American envoy (Ward again) quietly trailing north behind an Anglo-French expeditionary force was repeated for the third time in as many years. Humiliated in their previous assault against Chinese coastal defenses, the British with their French allies quickly broke through this time and marched inland to seize Peking and humble Chinese arrogance. The flight of the court brought an end to diehard Chinese resistance and set firmly in place the new era of treaty relations. Though Americans had contributed virtually nothing to its realization, they would nonetheless share fully in the privileged position Britain had carved out for foreigners in China. It was as if Washington had discovered (and diplomats on the scene overlooked) the appropriateness, for the moment at least, of the old Taoist dictum, "do nothing and all things shall be accomplished."

The new treaty system created new opportunities and greater safety for foreign trade. American traders after the Opium War had the run of five ports, and the major firms set up their main offices in Shanghai, not Canton. Once a backwater mart, Shanghai by the early 1850s had risen to preeminence on the basis of its easy access to the populous Yangtze Valley and to the major tea and silk producing districts. American merchants were no longer bound to the cohong but, still in need of economic intermediaries, they continued to deal with many of the same merchants who had formerly constituted it.[29] The foreign takeover of the customs service rounds out the list of economic changes effected by the diplomacy of this era. The creation in 1854 of the Imperial Maritime Chinese Customs (as it was grandly called) insulated foreign trade from the corrupt and arbitrary methods of Chinese officialdom and in return assured Peking a large and steady flow of revenue.[30]

Under the new dispensation inaugurated by the treaties of the early 1840s, American trade expanded, from $9.5 million in 1845 to $22.5 million in 1860. Exports quadrupled, though the overall balance of trade still ran against the United States. Opium, now tolerated perforce by local Chinese officials and finally accepted even by Peking in 1858, retained its place as a major American import into China. It was joined by a growing volume of cotton goods (including not only cloth but raw cotton, yarn, and thread) produced in the United States. Americans, once themselves purchasers of Chinese cloth, had already begun to turn the tables. Sales reached a high of $2.8 million in 1853, just before unrest in the interior dealt the cotton trade a heavy blow from which it would not recover until later in the century. Teas and to a much smaller extent silk continued in demand in the United States. Some firms, breaking out of the passive approach that had theretofore characterized their handling of Chinese exports, began to put themselves in closer touch with their sources of supply in the China market. In 1853 Russell and Company sent a representative into the Bohea tea district (then under Taiping control) to secure its own exports, thus setting an example for others. In another case American buyers, dissatisfied with the poor quality of export silk, tried (though with little success) to upgrade sericulture and reeling techniques. American shippers jumped into the only new export trade—the transport of coolies (a term properly applied only to laborers sent out under sharply restrictive contracts and often under conditions of coercion or fraud) to Latin America, particularly Cuba and Peru.[31]

While trade expanded, it also continued to consolidate into fewer hands, a process evident since 1815. New York emerged as the unchallenged distribution center for China imports on the East Coast, leaving its old rival Boston as well as Baltimore far behind. Philadelphia and Salem simply bid farewell to China trade days. New York's only close competitor in the export trade was San Francisco. In the diminishing field of American firms involved in the China trade, Russell and Company retained the lead. Augustine Heard and Company, founded in 1840 by two former Russell partners, ran a distant second. Like its rival, Russell and Company, Heard began as a commission business and later branched out into coastal shipping, foreign exchange, and insurance. Only these larger firms had the capital needed to diversify into new fields, to introduce the clipper ship in the early 1840s and, soon thereafter, to operate the first commercial American steam vessels in Chinese waters.[32]

Traders that held on through the 1840s and 1850s had to weather considerable commercial turbulence stirred up by diplomatic and military crises and the Taiping rebels, who for a time endangered the treaty ports, disrupted production and transport within the coastal economy, and dried up purchasing power. But even in these adverse circumstances, merchants proved themselves adept at turning a profit. A case in point arose in 1839 when the Canton authorities as part of their antiopium campaign had demanded the surrender of all foreign-held opium. At that time the market was glutted. By readily complying with these demands,

American merchants not only satisfied the Chinese but also gained a hefty profit by claiming the compensation earlier promised by the British government for stores of opium surrendered under duress. When the opium crisis forced the British to leave Canton, the Americans stayed on, not only maintaining their own legitimate business but also handling the British trade on a commission basis until the restoration of peace in August 1842. Similarly, in the latter part of the 1850s American merchants made money out of the very turmoil they complained about to Washington. Internal trade obstructed by fighting moved to the coast and into the hands of foreign shippers, Americans included. To both imperial forces and Taiping, Americans sold ships, provisions, weapons, and ammunition, often at spectacular prices. Taiping purchasing agents had to pay five times the old market price for ammunition, while powder that once fetched three silver dollars commanded twenty-five to twenty-six dollars in late 1853.[33]

Fragmentary evidence gives some sense of the rewards extracted from the China trade, at least by the larger commission houses. Between 1844 and 1860 Augustine Heard's share of his firm's profit amounted to $250,000, roughly $16,000 a year. The profits for the firm as a whole were around $50,000 in the late 1840s and between $180,000 and $200,000 per annum in the 1850s. Russell and Company seems, in the years for which information is available, to have done as well: for example, earning a net profit of $220,000 in 1849 and $104,000 in 1854. Robert Bennett Forbes of Russell and Company anticipated making $150,000 over two business seasons. But there were also warning signs that the days of the commission houses were numbered. The return on time and capital invested in China was no longer as attractive as it once had been. In the early 1840s John Murray Forbes of Russell and Company estimated the return on the China trade at not over 6 percent. At the very least investment of capital in the China trade seems not to have been as attractive as investing at home. Continuing a trend of disinvestment already evident in the 1830s, fortunes earned in China were applied in the 1840s and 1850s to domestic enterprise, especially in the West and in railroads.[34]

The Missionary Presence

The American missionary enterprise in China began modestly enough, a faint trans-Pacific reverberation of the great religious revival experienced in the northeastern United States in the early nineteenth century. Preachers called from the pulpit for a revitalized Christianity, particularly among the youth, to meet the challenge of a secularizing society. The emphasis on evangelism among the unsaved resulted in the establishment in rapid succession of five mission boards—the Philadelphia Bible Society (1808), the American Board of Commissioners for Foreign Missions (1810), the American Baptist Board of Foreign Missions (1814), and Episcopal and Methodist Missions (both 1820). It was from this institutional base that the early overseas missionary endeavor—in the Middle East, Hawaii, and India as well as China—would develop. In 1830 the American Board sent out the first American missionary to China, Elijah C. Bridgman, an Amherst graduate in his thirties. By 1839, on the eve of the Opium War, he had six American coworkers (both Congregationalists and Baptists), but they were still in Bridgman's words only "a feeble band" in the face of millions upon millions of unsaved humanity.[35]

The early China missionaries were in the main a solemn and noncosmopolitan lot, typically born and raised in rural areas where pietism was strongest. After graduation from college in New England or New York, they would embrace mission work out of a deep conviction of mankind's sinfulness and the imperative of saving heathen from eternal damnation. There was little room for humor or compromise in their outlook. Viewing themselves as the instruments of God's will, they took as their task one of the most difficult imaginable—effecting and sustaining the conversion of a strange people thought to be sunk in sin. As a divine instrument, missionaries felt obliged to cultivate a habit of critical introspection, necessary to determine worthiness for the high calling they were embarked on and to steel themselves for the trials and sacrifices of mission life. They operated on slim resources (usually on a subsistence salary). Family separations were common, as were mental breakdowns. Disease was a constant threat. And the unsaved were trying in their obduracy. The certitude of serving some providential purpose helped the missionary to accept with a degree of fatalism these frustrations and sacrifices. Even the ultimate adversity of death, missionaries ritualistically observed to one another, offered eternal joy and release from the pains of this world.[36]

Confined initially to Canton and Macao, missionaries saw opportunities widen considerably in the 1840s and 1850s. The treaties of 1842–1844 opened up for them no less than for the merchants new points of lodgment on the South China coast, with Shanghai emerging as the main base. The treaties also legitimized mission activities for the first time. At Canton in the 1830s Chinese regulations had forbidden open proselytizing, studying with Chinese teachers, or buying Chinese books. The treaties not only gave positive sanction to these activities within the portions of the treaty ports open to foreign residence, but also provided for the renting of land for cemeteries and the construction of churches. Imperial edicts issued in 1844 and 1845 added to missionary opportunities by extending official toleration to Chinese converts to Christianity. The next round of treaty-making in 1858–1860 incorporated the earlier imperial concessions of toleration and added to missionary prerogatives the right to travel outside the treaty port sanctuaries into the interior (but no farther than could be covered in one day's journey). The American Board responded by closing down its Southeast Asia operations in favor of concentrating on China and dispatched new volunteers from home. By 1850 the American missionary community, which now also included Methodists, Presbyterians, and Episcopalians, had expanded dramatically—to a total of eighty-eight.[37]

But missionaries found it difficult to make a perceptible dent in China's vast and resistant population.[38] At first, under the Canton system, official restrictions had prevented an open appeal to the Chinese. Bridgman and others had to perform religious services for the Chinese "with closed and, not unfrequently [sic], with locked doors." Evangelical efforts made then and later were greeted often by hostility and invariably by disbelief at the presumption of barbarians who brought with them opium yet promised a better life if only the Chinese would abandon wholesale the accepted practices and views of their culture. Chinese had to give up polygamy and ancestor worship. Gambling, drinking, and other vices were intolerable. A convert would have to resist the tendency to syncretism in religious patterns and accept the foreigners' conception of man's sinful nature, a stark contrast to the Mencian conception of man's essential goodness. A missionary-educator working in Canton candidly complained of the barrier these many points of difference put between him and the Chinese. "We meet them day after day, but . . . there is little or no play of sympathies between us. Our intercourse is much like that of two untaught mutes, that meet . . . and then part again in utter ignorance of each other's spiritual being." To these cultural obstacles to conversion was added, once British power came into full play along the China coast, the taint of profiting from gunboat diplomacy. As one missionary lamented, "Our preaching is listened to by few, laughed at by many, and disregarded by the most."

Successes were so scant relative to the investment of time and money that had missionaries been businessmen measuring their returns in hard cash, not immortal souls, they would have long since gone home. It took the American missionaries in Canton seven years after Bridgman's arrival to win their first convert—and he soon strayed. Methodists laboring at other points along the China coast put in ten years of work before claiming their first convert in 1857. The Methodist Episcopal Church (South) invested four years of labor in Shanghai before winning over its first converts.

The disappointments associated with the first thirty years of operations along China's periphery impelled the mission movement toward a search for new, more efficacious methods. That search, an essential feature of the Protestant mission movement in China through the end of the nineteenth century and into the twentieth, began in the Canton enclave. The most pressing task evangelists faced was to devise ways to communicate with potential converts. For that, literary work was essential in laying a foundation for the future. By surreptitious study of the Chinese language in Canton or by going to Chinese communities in Southeast Asia to study, the first generation tackled the problem of communication. Once they had acquired rudimentary skills, the Canton group impatiently set to work. Using his command of the language, Bridgman (aided by a Chinese assistant) prepared gospel tracts and translations of world history and geography intended to reveal the potency and richness of Western religion and culture and shake the Chinese sense of superiority. S. Wells Williams, a young man from upstate New York with a scholarly mind and with skills as a printer, directed the production of missionary works from Macao, while from his pen issued pioneering dictionaries, handbooks, and guides invaluable to his colleagues. He authored *The Middle Kingdom,* a two-volume work which appeared in 1848 and established him as an authority on China. Along with Bridgman he served as editor and a major contributor to the *Chinese Repository.* From its inception in 1832 to its closure in 1851 that periodical offered missionaries as well as others a unique window on Chinese affairs.[39]

Canton in the 1830s also saw the birth of what was to develop into a major humanitarian effort by Americans in China. It ran along two lines—one exploiting "superior" Western medical techniques, the other involving education. Nineteenth-century missionaries saw their charitable work not as an end in itself but as a means to awaken the spirit of the Chinese to the Christian message. In 1835 the Canton ophthalmic hospital launched the effort to turn medical skills to Christian purposes. Its founder, Peter Parker, a Yale graduate in his thirties, was a trained doctor, plagued by recurring personal doubts and psychosomatic disorders. He drew scores upon scores of patients (over two thousand in the first three months), but in an experience to be shared by later missionary doctors the gratitude of many translated into the conversion of a very few. The educational alternative got its start in 1839 when the Anglo-American missionary community in Canton collected donations from merchants for a school for Chinese, located in Macao. With a staff recruited largely from among Yale students, the school tried to attract promising young Chinese and give them a Christian education, but it met with indifferent success down to its closing ten years later.[40]

Those who wished to place greater emphasis on literary, medical, and educational work as a supplement to evangelism found that their views were generally out of favor with sponsors at home. Rufus Anderson, head of the dominant American Board, rejected the view that the missionary was a reformer, scholar, or man of affairs. His task was simply preaching the gospel and thereby "*reconciling immortal souls to God*"; education had a place only in training a native ministry. Unable to point to converts won by often costly nonevangelical activities, Parker, Williams, and others of their persuasion were thrown on the defensive and were under mounting pressure from home to subordinate or even abandon their work in favor of evangelism.[41]

Yet the equal fruitlessness of a strictly evangelical approach left missionaries no alternative but to continue experimenting.[42] The mission experience in Foochow illustrates the important place nonevangelical techniques continued to occupy even in the face of board disapproval. The first missionaries arrived in 1847, attracted by the concentration of "over half a million of immortal souls, spellbound by idolatry or atheism, in the capital of one of the largest provinces of the empire. . . ." With evangelism uppermost in mind, newcomers set to learning the local dialect, a task at which they made at least some progress due to the substantial time most missionaries stayed at one station. (For example, the seventy-two American missionaries representing Methodists and the American Board between 1847 and 1880 stayed in Foochow an average of eight years each, nearly four times the average length of service for American ministers and chargés over those same years.) Fluency made it possible to carry the Word to Chinese who came into the missionaries' homes—the tutors and servants and their families, all of whom were expected to attend regular household services as part of the terms of employment. Fluency also made it possible to reach out to a broader audience by street preaching, by operating chapels, and by distributing tracts prepared in missionary printing establishments in Macao, Hong Kong, and Shanghai. These tracts, sometimes sold at a nominal price to ensure that the Holy Word would be read and not turned to some unworthy use (such as kindling, toilet paper, or insulation), were distributed to the curious attracted into chapel, to students come to take the official exam, or to country folk within a day's travel of the city.

But when preaching alone proved ineffectual in winning converts, Foochow missions began to diversify into education, care for fondlings, and medical assistance. Much effort went into schooling. Missionaries tried to attract students by promise of instruction in Chinese as well as Christian subjects and also by such material inducements to their overwhelmingly poor applicant pool as free meals and school materials and a cash allowance. When attendance at day schools proved irregular and brief, boarding schools were set up with even greater material inducements, but on the condition that parents enroll their children for a period of four to six years.

Missionaries also recruited cultural intermediaries in the persons of tutors and native helpers, who served a role comparable to that occupied by cohong merchants, opium smugglers, and compradors relative to American businessmen. Initially, these were meant to help overcome the Chinese language barrier, for a missionary sermon, replete with crude phraseology and incorrect pronunciation, produced as much confusion and humor as it did divine light for a Chinese audience. In Foochow the first intermediaries were tutors who also functioned as interpreters, translators, go-betweens, and informants. They sometimes became the first converts. Later, missions recruited colporteurs to distribute tracts in the city and out into the countryside, and began to assemble a team of Chinese aides who could win and superintend converts in areas where missionaries could venture only briefly, if at all. As it became apparent that these auxiliaries could multiply the missionary effort at a much lower cost than it would take to support an exclusively foreign force, missionaries began to spend an increasing amount of time training and supervising their Chinese aides. A Methodist magazine described the approach in the military metaphors common to the mission movement, "We must furnish the officers for the great enterprise, and the Chinese must furnish the rank and file."[43]

The results of all this activity in Foochow, as elsewhere, before

First Contacts

1860 were dismaying. Two American and one British mission could not among them count one convert in the first nine years of work in that city, and they were understandably discouraged, depressed, irritable, and defensive in making their accounts to the home boards. One missionary in his second year in Foochow in 1849 moaned, "Oh! there is a desolateness in life here which nothing but the grace of God can render tolerable."[44]

In their impatience with the failure of the Chinese to respond either to direct evangelism or even to acts of service, some missionaries began to take an interest in diplomacy as a means of advancing the Christian cause. No sooner were the treaties of 1842–1844 in place than missionaries began to push for the grant of new treaty rights, the most liberal possible construction of the existing ones, and in general a more tolerant policy on the part of the Chinese government. Whatever the rhetoric used by the mission boards about keeping divine and secular interests distinct, the China missionary had no doubt that the success, perhaps survival, of his cause depended on state power. And he had no doubt that as an agent of civilization and American influence he was as much deserving of support as the merchant. Working through their boards, missionaries sought the establishment of nearby consulates to ensure diplomatic protection. They welcomed the presence of American and British gunboats for the chastening effect it was thought to have on Chinese arrogance and hostility and ultimately for clearing "the way for His kingdom."[45]

Some missionaries took the next logical step—becoming personally involved in the making of China policy. In one sense, taking up diplomacy was merely an extension of the search begun in Canton in the 1830s for ways to better penetrate and convert China. Diplomats drawn from American political life and consuls recruited from the mercantile community were sympathetic in small matters (for example, the need for consular presence beneficial to all Americans in the treaty ports), but in the main their preoccupation with commerce and political advancement inclined them to give spiritual matters short shrift and to urge caution and compromise on the missionaries. Diplomats drawn from missionary ranks might be able to change that orientation. At the same time that the American Board was pouring cold water on nonevangelical activities, the government's desperate need for China experts and for some continuity in the legation provided a new opening for the disgruntled or rejected missionary with Sinological skills.

Missionary entanglements in diplomacy went back to the 1830s.

When Edmund Roberts—a Portsmouth merchant who had promoted himself into an appointment as the Jackson administration's envoy to look after Asian commerce—visited Macao in 1832, Bridgman stepped forward to become his chief escort and informant, while Bridgman's *Chinese Repository* served as the authority he turned to for information on China. Later on, at the time of the Opium War, Bridgman and Parker made their services available to American authorities in Canton and their views known in Washington. In 1841 Parker, on extended home leave, met with the members of the outgoing Van Buren administration, with President Harrison (and after his death his successor John Tyler), and then with Secretary of State Daniel Webster, Henry Clay, and John Quincy Adams in order to express his views on the need for a mission to China. Once the mission was approved, Cushing sought the assistance of both Parker and Bridgman. (The American Board yielded only grudgingly, observing that it thought a claim to two missionaries excessive and warning that missionary involvement in "secular embassies" took valuable men away from "their more appropriate labors," and might provoke "the suspicions of the natives, that after all missionaries are agents of the governments of their own countries.") Bridgman and Parker served Cushing as translators and interpreters and, no less important, shaped Cushing's negotiating strategy by assuring him that intimidation might work as long as the Chinese "have no desire or intention of comparing strength with a foreign government. . . ." The two advisers also successfully looked after missionary interests in the final treaty. Cushing left China lauding his two aides as "praiseworthy and meritorious" advisers with "long and exact knowledge of China." In need of such expertise as well as continuity in the legation, Washington turned to Parker to head it some six times between 1846 and 1857.[46]

Eventually, S. Wells Williams became Parker's successor as a mainstay of the legation. Williams turned to diplomacy for want of support by the American Board for his literary and printing projects and for want of personal interest in evangelism. Having deferred and then altogether decided against ordination, Williams committed himself to a diplomatic career that allowed him to stay in a China that had become "home" to him and to carry on his literary labors in his spare time and under less pinched personal circumstances. His help to Perry's Japan expedition brought him favorable notice in Washington, and thereafter, for the twenty years between 1855 and 1876, he was steadily on the government payroll as interpreter and translator but also as temporary head of the China mission on eight occasions (for a total of five years). Williams' chance to serve the missionary cause came in 1858 when he, assisted by another missionary Sinologue, W. A. P. Martin, accompanied Minister Reed up to the North China coast to negotiate treaty revisions. The two missionaries succeeded in including in the treaty an explicit clause on toleration, even though Reed, indifferent on this issue, had not made Chinese acceptance of their demands a prerequisite for bringing the talks to a conclusion. In 1859 Williams and Martin once again accompanied a new minister, Ward, on a second trek up the coast and inland to Peking.[47]

By 1860 missionaries were hardly closer to winning China for Christ than in 1830 despite the infusion of funds and personnel, despite the persistence of evangelistic efforts, despite the experimentation with new mission methods, despite the recourse to diplomacy. In the face of these efforts the Chinese had shown some active resistance but even more profound indifference. Even so, disappointed home boards and frustrated, doubt-ridden missionaries continued stubbornly along, driven as at the beginning by a spiritual imperative.

Early Images of China

Before the *Empress of China*'s voyage and possibly for some decades after, what little Americans knew of China was derived from European literary sources and the decorative arts. The resulting outlook blended curiosity about China together with a sense of China as a curious place.

Among early Americans, intellectuals took the greatest interest in China. Benjamin Franklin epitomized the tendency to approach China as a source of inspiration and innovation. He as well as others in the Philadelphia circle of intellectuals had read the French Jesuits who had lived in the East and the philosophes who had turned Jesuit travel reports to their own end, and they had accepted the Enlightenment view of China as a benevolent despotism, a harmonious society ruled by officials chosen on merit and guided by the wise, secular code of Confucianism, a country where the arts and philosophy flourished while the peasantry in simple tranquility tilled the soil. Franklin, ever the practical scientist, looked to

China for new ideas and methods—on such varied topics as census-taking, production of silk, windmills, and heating systems. Another Philadelphian, Charles Thomson (secretary of the Continental Congress and later of the U.S. Congress), emphasized the practical advantage of learning from "the industry of the Chinese, their arts of living and improvements in husbandry, as well as their native plants."[48] Others outside Philadelphia also looked to China as a model. James Madison took an interest in the intensive agricultural techniques of the Chinese. Jefferson regarded their political and commercial isolation with envy. If the United States could assume such a position relative to Europe, he observed wishfully in 1785, "we should thus avoid wars, and all our citizens would be husbandmen." That same year James Monroe noted, perhaps with the same sense of envy, that China was "seperated [sic] and perfectly independent" of the world powers.[49]

The knowledge on which these admiring views were based was thin—indeed they thrived on distance from the idealized object of study. A collection of random facts from books and a chance interview with an occasional British or American traveler resulted in at best an imperfect conception, even among the curious Philadelphians. Franklin labored under the delusion that China, although the most populous of countries, still "clothes its Inhabitants with Silk, while it feeds them plentifully." He thought Chinese census methods "effective," admiring what he assumed to be their goal of providing a "forewarning of scarcity from overpopulation." (In fact, they provided but crude population estimates which were used primarily for taxation.) In 1788 Franklin foisted on the public a description of Chinese prisons as clean, airy, and comfortable places where inmates were kept profitably employed (a model far indeed from the real thing). The learned Dr. Benjamin Rush erroneously believed China's large population lived within a territory "not larger than one of our Colonies." George Washington long believed that "the Chinese, tho' droll in shape and appearance, were yet white" until a correspondent disabused him of that misconception in 1785.[50]

At the popular level conceptions of China, where they existed, were an amalgam of the curious and the fantastic. Chinese motifs on a variety of goods brought back from Canton—cottons, wallpaper, furniture, toys, antiques, and above all the inexpensive and widely used chinaware which ships used as ballast on the return voyage—offered a window on a fairy-tale land: hazy, timeless, peaceful. The quality of the goods themselves suggested a nation of skilled artisans. The curios that the well-to-do collected carried the same message. When the rich built houses and gardens along Chinese lines they further popularized China's standing as an exotic land. A child's reader on China published in 1837 built on this theme, piling up without rhyme or reason one curiosity after another—the Chinese diet of dogs, rats, cats, snakes, and frogs; the elephants and rhinoceroses to be seen in the wild; and the queer-looking letters that "the people of China can read just as easily as you can read A.B.C." An interest in Chinese exotica also drew the public to exhibits of Chinese artifacts in Philadelphia, Boston, and Salem. Between 1839 and 1842 no less than 100,000 spectators in Philadelphia paid to see a "Chinese museum" containing more than 1,200 arts, crafts, and natural history specimens as well as life-sized figures all assembled by the merchant Nathan Dunn. The catalogs from that exhibit are supposed to have sold 50,000 copies, each offering a balanced and often admiring view of China and even expressing sympathy for the Chinese struggle then under way against the opium trade.[51]

A weak countercurrent of condescension, even antipathy, existed alongside the more tolerant and sometimes inquisitive view of China. A 1784 geography characterized the Chinese as "the most dishonest, low, thieving people in the world." Foreshadowing what was to develop into a fixed American preoccupation, Franklin remarked on the difficulty of written Chinese and attributed it to "the obstinate Adherence of that People to old Customs. . . ." Jefferson agreed in an 1818 comment that Chinese characters were "so complicated, so voluminous and inadequate" that they would block progress and make the "expression of ideas . . . very imperfect." In 1839 John Quincy Adams launched what was probably the first major broadside against the Chinese by a public figure in the United States. He attacked the Chinese bias against unrestricted commerce ("among the natural rights and duties of men"), their baseless pretensions to superiority, and the flaws in their culture, including idolatry, pervasive despotism, polygamy, and infanticide, and he praised Britain's resort to force to reform China. But Adams appears to have been out of step with his prominent contemporaries. None rose to second his views, and the editors of the *North American Review* judged them too intemperate to publish. It was finally left to the *Chinese Repository*, that mouthpiece of the China missionaries, to put Adams in touch with a smaller but more approving public.[52]

Americans who set off to China may have been influenced by one or another of these currents of thought. But after their arrival Americans began to develop their own distinct views shaped by personal needs and experience and by those of the group to which they belonged. It hardly needs to be said that Americans in China tended to look down on the Chinese. Bridgman concluded in the 1830s that "darkness covers the land and gross darkness the people"; Parker found China a "moral wilderness"; and Williams wondered at "the torpor of mind" of the Chinese, and under the force of years of daily contact assumed in private an increasingly deprecatory view of them as "grown children" "with such dirty bodies, speaking forth their foul language and vile natures, and exhibiting every evidence of their depravity." One pioneer diplomat described the life of the common people as mean, brutal, and stagnant; their rulers were weak, inept, and despotic. "Of all uncouth figures, that strut their little hour upon the stage of life, a China-man is surely the most grotesque animal."[53]

Moreover, it is equally obvious that the concrete interest of the open door constituency in access to China, whether for trade, evangelism or diplomacy, made them impatient in the face of the stubborn Chinese rejection of "modern" forms of international relations—impatient to the point at times of favoring the use of force against the recalcitrant. Every discussion held on the China coast on how to deal with a people of this character must have repeated the generalizations that time and experience seemed to make ever more compelling. "Disinterestedness is all very well," Augustine Heard, Jr., observed in a formulation already become cliché, "but the first element to impress the oriental is power. Fear is more potent than love, and consideration is more likely to be imputed to weakness than any higher motive." A group of Boston merchants warned Washington in 1843 on the eve of the Cushing mission that the Chinese would treat only "through fear of armed compulsion, or through a politic desire to offer us voluntarily what has been *forced* upon them by others. . . . If they find that we recede from any position once taken, they learn our weakness and will take advantage of it to the utmost. . . ." The missionary community carried the same message to diplomats and policy makers—sometimes directly, sometimes through their boards. S. Wells Williams, with his special insight on the role of fear and force in "God's plan of mercy," was "sure that the Chinese need harsh measures to bring them out of their ignorance, conceit, and idolatry." "Nothing short of the Society for the Diffusion of Cannon Balls will give them the useful knowledge they now require to realize their own helplessness."[54]

However, within this general sense of superiority and impatience, views of China and the Chinese frequently reflected the special interests peculiar to merchant, missionary, or diplomat. Occasionally, these interests proved to be compatible.[55] The drama of the Taiping Rebellion offers a good example of the way in which compatible interests evoked similar responses from each of the groups. In reaction to this political development promising "progressive" change, a broad range of Americans in China initially entertained high expectations only to be disappointed on closer examination.

Early reports on the Taiping Rebellion reaching missionaries along the coast indicated that the rebels were sincere, if ill-informed, Christians who would open the way to China's salvation.[56] They opposed the evils of idolatry, slavery, footbinding, prostitution, and infanticide and the use of opium, liquor, and tobacco. Taiping administration, Bridgman concluded after a tour, deserved credit for "remarkable energy, order and devotion." But a closer look made possible after 1853, when the Taiping fought their way into the lower Yangtze Valley, revealed that they were not genuinely Christian, that they were violent and despotic, and that they displayed that supremely irritating Chinese sense of superiority. The early, nearly unanimous approbation had dissolved by late 1854. Thereafter only a vocal minority including Roberts, Parker (when not restrained by his official role), and W. A. P. Martin consistently argued the Taiping case, and even they argued with diminishing effect and energy. In 1860 hopes for the Taiping briefly revived following news of reforms in Nanking and a major military victory, but the rebels still could not stand up under scrutiny. In a widely circulated report that year one missionary who had set off optimistically for Taiping-controlled Nanking lamented that he had found "nothing of Christianity but its names, falsely applied."[57]

The merchant view of the Taiping was no less rooted in self-interest. Initially, in 1853, merchants commented positively on a movement that seemed favorable to trade, but the next year they reversed themselves as the rebels lost ground and as contact disabused any illusions about expansion of trade under Taiping auspices. By 1860 one of the Heards reflected the merchants' desire to see an end to civil conflict when he opined that "it would be better for the benefit of China and humanity, to say nothing of cotton goods [he could have added tea], to exterminate the whole [Taiping] party."[58]

Some diplomats and naval officers had also at first expected some good from the Taiping—in this case in advancing American ideals and interests in China. Matthew Perry, while serving off the China coast, saw the civil war as a conflict "between a despotic government . . . and an organized revolutionary army gallantly fighting for a more liberal enlightened religious and political position. . . ." Out of a Taiping victory, Perry hoped, would come new concessions for Americans. Townsend Harris, then a government diplomatic agent in Hong Kong, agreed with Perry. But their view was not to prevail. Marshall, an agnostic, held no brief for Christianity—Taiping or otherwise—and reports coming out of the Taiping capital in Nanking had convinced him by mid-1853 that the rebels were no better than the imperialists. Diplomatic opinion turned definitively against the Taiping in 1854 when Minister McLane during a personal visit to Nanking concluded that nothing was to be gained by support of a group no less arrogant than the Ch'ing.[59]

But there were also divergent interests within the open door constituency that gave rise to divergent, even antagonistic, points of view. Missionaries, moralistic in outlook and self-sacrificing, were sharply critical of merchants for their high living in safe, comfortable enclaves; for their indifference to China's idolatry; for their consuming materialism; and for their participation in the opium trade. Merchants, on the other hand, tended to view their missionary neighbors as zealots whose intolerance would create trouble for trade.[60] Both merchant and missionary in turn harbored doubts about the politician-diplomats that Washington sent out ostensibly to support and protect them.[61] Established American treaty port residents—with their financial links to London, their close social and sometimes professional ties to British merchants and missionaries, and their dependence on British power for their privileges and safety—did their best to counteract the Anglophobia the diplomats sometimes brought with them, and to tutor these newcomers in the locally perceived "realities" of China. But as students, diplomats were held back by their greater interest in the political careers to which they would return after an interlude abroad.

These variations in fundamental outlook within the open door constituency were reflected in differing reactions to central features of the China scene. For example, with the Canton system most merchants were relatively well content. They had come to China, as they had gone to Europe, the East Indies, and Latin America, primarily in search of profit. Their dominant concerns had to do with swings in the tea and opium markets, credit and exchange problems, shipping schedules, and—once business was done—food, sport, and light entertainment to fill the tedium of their exile. Isolated as they were within their foreign community, they gave little attention, at least in their laconic jottings on Canton life, to the Chinese around them. When they occasionally allotted praise or blame, it was usually in the context of their business activity: they looked down on the masses of Cantonese as an unruly and cheating rabble, while they respected the cohong merchants for their honesty and dependability. For many an American merchant, the Canton trade could offer an incomparably "high average of fair dealing" as well as the chance to make a small fortune.[62]

By contrast the Canton system denied missionaries what they most wanted—free access to the Chinese. Their tense impatience with its constraints is reflected in Bridgman's 1835 assault on the idea "that nations have a right to manage their own affairs in their own way. . . ." China's exclusive system of foreign relations was "replete with evils," worst of all its obstruction of religious and other civilizing work. To extirpate these evils, Bridgman saw it as the duty of Britain, aided by France and the United States, to take "strong and determined measures."[63] Not until late in the 1830s, as the opium crisis brought down the Canton system, disrupting trade and hence profits, did merchants come to agree with the missionaries on the urgent need for reform, but even then merchants were far from accepting as a prescription for action Bridgman's sweeping cultural critique. Merchants wanted to reform trade, not China.

The views of the only diplomat personally exposed to the Canton system fell in with those of the missionaries. Edmund Roberts, who looked to Bridgman and the *Chinese Repository* for orientation, left after his first visit in 1832 with a contemptuous view of China desperately in need of the "intellectual, moral and religious improvement" that much wider Sino-foreign contact, in place of the prevailing restrictions, would bring. W. S. W. Ruschenberger, a physician who accompanied Roberts on his second visit in 1836, was, on the other hand, an amused and tolerant observer. He admired the way in which Chinese isolation had secured the country from foreign turmoil; foreigners dissatisfied with the prevailing system, he advised, should leave. Openly critical of the views of the *Chinese Repository*, Ruschenberger pronounced the Chinese "civilized" and "industrious," though he identified serious flaws in family life, the military, politics, and the court that belied Chinese claims to superiority over foreigners.[64]

Chinese consumption of imported opium similarly divided Americans in China. Missionaries uniformly condemned opium use as "a great barrier to all our plans" (in Williams' words) since a people degraded by opium could not respond to the gospel. But good could come out of evil if God so willed. So the opium trade became in the end providential when it precipitated a blow against Chinese arrogance and brought about a restructuring of relations with China in a way beneficial to mission interests. By contrast, merchants had no moral qualms about the trade and marshaled an array of justifications. Opium was innocuous or at least no worse than the use of alcohol in the West. The Chinese wanted it. It was not the business of traders to question Chinese taste—or to leave the profit to others. Chinese officials, though formally hostile to the trade, revealed their true attitude, it was argued, by actually allowing it to continue out of sight. On the opium issue, Roberts joined the missionaries in deploring the use of the drug, and Ruschenberger sympathized with Chinese attempts to stop its import.[65]

Between the views articulated by the disparate members of the open door constituency and the perceptions of China which were held in the United States down to the 1860s, it is difficult to demonstrate a clear and direct relationship. The rise of a small American community in China certainly helped close the distance between China and observers in the United States and, if nothing else, helped to correct some of the older, blatant misconceptions that Franklin and others had labored under. But just as the groups within that community occasionally held differing views, so too did they express them with varying degrees of confidence, frequency, and persuasiveness. Merchants did not have the impact that their numbers and long experience might have given them. Their minimal contact with the Chinese and their isolation in treaty port enclaves limited the range and depth of their information, and few tried to address an audience beyond the mercantile community at home and its political allies. Diplomats (aside from missionary recruits) had even less impact. They stayed only briefly in China, hardly long enough to build a claim to special insights, and then returned home to resume an active role in domestic politics.

It was the missionaries with their long periods of residence and unmatched exposure to the culture who developed the most impressive and widely recognized claim to expertise. More than other Americans in China they were eager to put their views before the public, a task in which the mission boards assisted them. Williams, for example, gave over one hundred lectures during a home leave in 1845–1846 and then worked them into *The Middle Kingdom* in order to "produce more sympathy in behalf of the moral life" of China.[66] The *Chinese Repository,* learned accounts, letters and reports retailed in mission periodicals, and talks and interviews while on home leave provided the public a steady diet of missionary opinion. Still, the missionary message was on the whole self-serving and out of touch with the growing secular sense of national mission, and so it must have seemed to those Americans at home who took an active interest in China.

Even well into the nineteenth century China, so central in the mental map of missionaries, less so for merchants and diplomats, must have been distant from the daily concerns of most Americans. When Americans bothered to think about China, it seems to have been in terms no longer as unrealistically respectful or admiring as they once had been. But even this shift probably had less to do with the influence of the open door constituency than with changes at home and in China. The firming up of national self-confidence following the War of 1812 and the gathering pace of technological invention and industrial development in the United States made China appear more backward and less deserving of study for utilitarian reasons. The domestic and foreign crises into which China was plunged from the 1840s onward would have overwhelmed any lingering traces of the old romantic and admiring view of China— whether or not Americans had been on the spot to observe and report. The most that could be claimed for the open door constituency was that it had probably expanded the space China occupied on the fringes of American consciousness. But it would take the arrival of Chinese immigrants in large numbers, a development in the offing in the 1850s, before the Chinese could claim a more central place in the American imagination.

Chapter Two

The Chinese Discover America, 1784–1879

BY THE MIDDLE of the nineteenth century the United States had come to hold a special attraction to some Chinese, just as China had come to hold a special promise to a select group of Americans. The United States first came onto the Chinese horizon through the work of scholar-officials seeking to understand the Western powers whose intrusions had begun to unsettle their world. As problems of foreign policy along the China coast became more acute from the 1830s onward, investigations into the origins and interests of the Americans assumed a pressing, practical character. In the 1850s, at the very time the United States was becoming for the first time an important point of reference in Chinese foreign policy, the trans-Pacific odysseys of the men of the Canton delta, drawn to the "mountain of gold" by economic opportunities, introduced the United States to Chinese in a second way. These Chinese emigrants did not go as permanent residents intent on assimilating into American society, but as sojourners. They would in time set off a "native" opposition on the Pacific slope that bears at points a striking similarity to the hostility Americans encountered in China. By the 1870s this conflict between Chinese emigrants and their American opponents had grown to such proportions that it occupied a central place in Sino-American relations.

The American Infatuation

With the arrival of the *Empress of China* in Canton in 1784, the United States as a separate country began to intrude on the Chinese consciousness, but even then Americans occupied a decidedly marginal place in thinking on maritime affairs, itself a topic of only peripheral concern to Chinese intellectuals and officials until the

1830s. During those early decades the dominant Chinese tendency, well established before the first direct Sino-American contact in the 1780s, was to lump Westerners ("ocean barbarians") together as a single undifferentiated tribe. The contempt for outside cultures, which sustained the simple vision and which militated against exploring its diversity, was itself deeply rooted in cultural tradition. "If he be not of our kin," the *Tso-chuan* reminded generation after generation of exam-bound Chinese students, "he is sure to have a different mind."[1] These foreigners, who came from across the "western ocean," were thought to have animal-like natures, more like sheep and dogs than men. The very written characters selected to name these people reflected this sharp distinction drawn between civilized Chinese and the wild barbarian. For example, the general term for those who arrived by sea, the Americans included, was represented by the character for sheep with the sign for water added alongside. In general, foreign names, both national and personal, were transliterated or translated well into the nineteenth century in ways intended to demean and give unsavory connotations.

The dim early perceptions of the Americans derived in large measure from Chinese observation of the dominant British community in Canton. Since the Americans spoke the same language, sought out the company of the British, and shared with them a seemingly unrestrained devotion to trade, it was natural for some Chinese to classify the Americans as merely a subgroup of the English tribe. Others, less sophisticated, simply took the Americans as another kind of "sea-going barbarian," a type whose essential traits had been previously defined by watching the British. Time and again reports coming out of Canton intoned, "The sole object of the foreigners is to trade. . . ."[2] They warned that their minds were essentially "inscrutable," and that their daily behavior revealed a great craftiness in pursuit of profit. Some few observers took note of the mechanical ingenuity of these newcomers and their essentially frivolous creations. With the onset of Sino-British tensions in the 1830s the image of the foreigner in Canton began to take on darker coloring. As Chinese authorities sought to "rein in," the foreigners became even more unreasonable. The British as the quintessential foreigners then impressed Chinese more and more with their violent side, the fearful destructiveness of their weaponry, and their ability to draw Chinese traitors to their side.

Chinese geographers knew of the existence of the North American continent from Jesuit works of the late seventeenth century.

And in 1787, four years after independence, Americans as a people distinct from the English appeared for the first time in the Chinese record. Even this brief delay may have been more attributable to the desire of the first American traders in Canton to be counted as English to avoid various exactions than to any indifference on the Chinese side. Then and in subsequent years the Chinese would refer to the United States as the "flowery flag country" (*Hua-ch'i kuo*) after the picturesque national banner whose stars appeared to the Chinese as flowers, by various literal translations of "United States" (e.g., *Lien-pang kuo*), by a considerable range of transliterations of "America," or even some combination of these, before ultimately settling on *Mei-kuo* ("beautiful country") as the standard term. The surviving early references are, however, few and terse, suggesting miniscule interest and no knowledge of the United States beyond the fact that it was a trading nation at Canton with "numerous" ships. By the 1810s the conception of the United States had expanded to a nation at odds with the British. The two have, the governor general in Canton noted in 1814, "engaged in plundering each other's goods and money." Three years later that same official further observed that the American barbarians were "most respectful and submissive to us." They had, strange though it seemed, no king, and instead selected a chief by lots to serve a four-year term. Two published works from the early 1820s added little to this picture. A geography from that time identified the United States as a somewhat narrow and isolated island, about ten days to the west of England, which had once controlled it. Americans followed customs similar to those of the English (including monogamy), and traveled about in steamships which they were adept at building. Governor general Juan Yüan included an entry on the United States in one of his many scholarly compilations, a work on the "outer barbarians" completed in 1822 (his fifth year in Canton and the year after disposing of the Terranova affair). It simply indicated that the Americans, whose ships in Canton had become as numerous as those of the English, occupied a part of the "vast territory" of the North American continent. Seven years later Juan's successor claimed that the Americans, "although not docile and submissive," were seldom as "rascally" as the British (a distinction that some officials writing at about the same time openly questioned).[3]

From the late 1830s onward Chinese observers labored to bring the United States into sharper focus. An aggressive British policy made imperative a fuller understanding of the foreign troublemakers and helped bring the United States out of the haze. Ch'i-ying, who was to negotiate the American treaty with Cushing, observed succinctly in an 1842 memorial to Peking that "to control barbarians, one must first know their nature."[4]

Americans in Canton, both by their actions and by the information they conveyed, also played a pivotal role in this growing Chinese appreciation of the United States. Merchants left their mark chiefly by demonstrating the national devotion to trade. The far less numerous and late-arriving American diplomats sent confusing signals. On the one hand, they stressed the fiercely independent attitude of the United States toward Britain and a deep concern over equitable commercial opportunity, whereas on the other hand their behavior often suggested dependence on—if not outright collusion with—the British. The missionaries exercised an influence all out of proportion to their numbers by using their Sinological skills to celebrate the achievements of the West generally and the United States in particular. As a byproduct of their effort to shake Chinese arrogance and resistance to the Christian message, missionary informants, preachers, and translators imparted much of the basic cultural and historical information that gave nuance and depth to early Chinese commentary on the United States. The most important of these was without question Elijah C. Bridgman, whose full Chinese language account, "A brief guide to the United States of America" (*Mei-li-ko ho-sheng kuo chih-lüeh*) appeared in Singapore in 1838 and subsequently circulated along the China coast.[5]

This search for more information on the United States centered naturally enough in Canton, where both the crisis and the Americans were to be found. Liang T'ing-nan, a native scholar and counselor to a succession of governors general, was one of a growing number of Chinese guided by the premise that understanding the foreign danger was the precondition for meeting it. Liang's first published notice of the United States in 1838 was no more than a reproduction of Juan's simple sketch of 1822. But several more years of investigation and the discovery of Bridgman's study enabled him to describe for his readers the course of English settlement of the thirteen colonies and the dispute over taxation that led to conflict and American independence (in 1788!) under the leadership of George Washington. Liang also touched on such diverse topics as the topography of the United States, its climate, social and political institutions, and commerce. The pioneering works by Liang as well as Bridgman were in turn to shape the other notable study touching on the United States that dates from this period, Lin Tse-hsü's "Guide to the four continents" (*Ssu-chou chih*). Lin, who arrived in Canton in 1839 as the imperially appointed commissioner charged with closing down the opium trade, hastily embarked on gathering as much information as was available on the barbarians he had been sent to manage. The resulting compilation provided detailed treatment of the establishment of the American colonies and states, the structure and operations of the government, the topography, and the population (including prominently blacks and Indians). This highly descriptive work, weighted down with dates and statistics, nonetheless had a clear message for the attentive reader: the United States, after only sixty years as a nation, had already become a "major power" and a rival to the English. Lin's account held up for particular praise the extensive system of schooling, the unmatched skill of the Americans at putting the steam engine to practical use, and the responsiveness of the government to the people ("no different than the rule of worthy emperors").[6]

In the relatively placid period between the end of the Opium War and the turmoil of the 1850s two works appeared which finally supplied the first coherent picture of the United States and struck the central themes that were to resonate in later works. The first of these works was Wei Yüan's "Treatise on maritime kingdoms" (*Hai-kuo t'u-chih*) of 1844. Wei, born in 1794 in the inland province of Hunan, had already established himself as a widely respected scholar before developing an interest in the late 1830s in the problem of "maritime defense." That interest grew in intensity as a result of firsthand exposure to the British threat along the coast and an encounter in 1841 with his friend Lin Tse-hsü, then on his way from Canton into exile after his attempt to bring the British opium traffic under control had miscarried. Lin left behind a copy of his "Guide to the four continents," and Wei set to editing and supplementing it. The first results of his research appeared in print in 1842 under the title "Chronicle of imperial military campaigns" (*Sheng-wu chi*), a work already in progress before his meeting with Lin. Two years later the full scope of his investigations became public in the "Treatise on maritime kingdoms." This latter, full-scale study of the foreign problem was guided by starkly utilitarian concerns. Wei offered his work as a textbook on how to handle the unprecedented barbarian threat confronting China. He wanted to correct misconceptions and supply fuller information, the better "to use barbarians to attack barbarians, to use barbarians to soothe barbarians, and to study the barbarians' superior techniques to master barbarians."[7] While urging a better understanding of these foreigners, Wei maintained an unshaken sense of Chinese cultural superiority evident in his consistent use throughout his writing of the condescending term "barbarian" (i) and his belief that China was obliged to show generosity and solicitude for those respectful and obedient outsiders who came to China to make their livelihood.

Out of these preoccupations and biases came a surprisingly sympathetic portrait of the United States. Wei's account traced the emergence of a new country into a rapacious world of international rivalry. The struggle for the East Coast of North America ended with the aggressive British in control and the French deeply embittered over the loss of their settlements there. Britain's bullying was, however, to prove her undoing. "The British barbarians levied numerous and heavy taxes which caused the thirteen parts of America to start a righteous revolt to drive them out. At the same time the Americans asked France to help them. . . . The Americans cut the British supply lines. The British soldiers were in hunger and distress, and the British ceded territory and asked for peace." In gaining their independence and "recovering [sic] the twenty-seven parts [states] of their original land," Americans demonstrated martial virtues. In drawing on France to defeat Britain, they showed themselves clever. They subsequently constructed a sound political system and made their country rich, so that by the 1830s the United States had come to stand, as Wei understood it, alongside France as one of the "powerful countries of the West." Yet the United States, despite its wealth and power, "did not bully small countries and did not behave arrogantly toward China." Those Americans who came to Canton to trade were, along with the French, "the most amicable and obedient" of the foreigners, and they naturally resented the British, "the most fierce and arrogant of people," for attempting to dominate China's trade. Old rivalries of the West, Wei's account told his readers, lived on, even along the China coast.[8]

Wei's "Treatise" remained the definitive Chinese account of the United States for only a few years. It was overtaken in 1849 by a more detailed and cosmopolitan work, "Brief survey of the maritime circuit" (*Ying-huan chih-lüeh*), by Hsü Chi-yü, an official one year Wei's junior and one of a small but growing group of Chinese experts on the West. Hsü's "Brief survey" had much in common with Wei's "Treatise," both in genesis and themes. Like Wei, Hsü hailed from the interior (in his case, landlocked Shansi in the north). He had not become involved in the barbarian crisis until 1840, the beginning for him of a decade of service in a variety of posts in the southern coastal provinces of Kwangtung and Fukien (where he ultimately held the post of governor). After one local defeat during the Opium War, Hsü in his agitation over the English peril had complained, "I can neither eat nor sleep, trying to think of ways to help."[9] Hsü, now aroused, set himself the same task Wei had undertaken—to search out new information that might shed light on these "intractable" and seemingly "unfathomable" intruders and thereby suggest solutions to China's foreign crisis.

In the five years between 1843 and 1848, a relatively tranquil period in Sino-foreign relations, Hsü devoted to the project what time he could spare from his official duties. He built on Wei's "Treatise" just as Wei had drawn from Lin's "Guide." But directed by genuine curiosity as well as his utilitarian concerns, Hsü also gleaned information from Chinese travelers and barbarian experts, foreign informants, and foreign printed sources to add substantially to and amend the work of his predecessors. For his account of the United States he consulted David Abeel, an American missionary in Amoy. (Abeel found to his disappointment that his interlocutor was "far more anxious to learn the state of kingdoms of this world, than the truths of the kingdom of heaven.")[10] Hsü also consulted Bridgman's account of the United States. Finally, he talked with an American doctor in Amoy, and he may have sampled a periodical published in Chinese by the Anglo-American missionary organization, the Society for the Diffusion of Useful Knowledge. The result, consistent with the spirit and breadth of Hsü's inquiry, was a study of genuine intellectual sophistication whether compared to extant Chinese accounts of the United States or even to guides to China prepared by contemporary Americans. Hsü in effect invited his countrymen to think of foreigners in a less stereotyped, more openminded way by avoiding the more blatant forms of condescension (such as using the term "barbarian") that had marred the work of his predecessors and by suggesting that the world was diverse and that the Chinese could learn from it.

On close examination the Americans revealed themselves to Hsü, as they had to Wei, as an admirable people—"docile, goodnatured, mild, and honest"—yet also possessed of "wealth and power."[11] Hsü was particularly intrigued by the success of the Americans at repelling the domineering British and at making their country strong and rich in a short time. Because of the obvious relevance of such achievements to China's contemporary plight, Hsü gave more attention in the "Treatise" to the United States than to any other single country, even more than to the formidable and overbearing British. Hsü told part of the American success story in terms of the rapid development of unsettled land. Americans, who once hugged the Atlantic Coast, managed in two hundred years to possess the heartland of North America where "the favorable climate and fertile land are almost as good as China's," and to achieve "a prosperity which has overflowed into the rest of the world." With scarcely concealed admiration, Hsü traced the process: "Cities were laid out along the coast. . . . Trade flourished and gradually became very abundant. Because of this, wealth and power were achieved."

Hsü recognized that development continued even as he wrote. Americans had extended their landholdings across the continent, ousting the "aborigines." He was particularly impressed by the way Americans applied their technical skills to integrate a once wild frontier into the national political and economic system. They built canals linking many small rivers into a transportation network. "They also build fire-wheel carts [steam locomotives], using rocks with melted iron poured on them for the road in order to facilitate their movement. In one day they can travel over three hundred *li* [about a total of one hundred miles]. Fire-wheel ships [steamers] are very numerous. They move back and forth on the rivers and seas like shuttles. . . ." Here was a pattern of settlement, without counterpart in China's own frontier experience, in which technology imparted a marvelous speed and prosperity and consolidated political loyalties.

Hsü depicted as no less a marvel the social and political order of this newly developed land. The marvel in this case was not (as might be anticipated) its foreignness but rather its startling similarity at many points to Chinese ideals. The American society, as Hsü described it, was divided into categories familiar to his readers. Scholars (including clergy, doctors, and lawyers) were at the top, followed by peasants (meaning farmers), laborers, and merchants. Hsü took the parallel between the two countries a step farther by picturing Americans as lovers of education and learning. They were also, like the Chinese, a homogeneous people. Although somewhat less than half came from non-English stock, all were Christians and English-speaking. Even his account of the American political system, described as "a wonder" and unprecedented either "in ancient or modern times," emphasized patterns of behavior that strikingly paralleled Confucian ideals of civic virtue.

Hsü largely credited American political achievements to George Washington. Compared by Hsü to the mythic sages of Chinese antiquity, the American leader was to become through this account and others influenced by it a familiar figure to Chinese interested in foreign learning and an especially important one to a subsequent generation of Chinese nationalists. Hsü's Washington fit the archetype of the dedicated national leader devoted to the people and the state rather than personal or family power. Sketched out according to the conventions of Chinese biography, Hsü's Washington emerges more Chinese than foreign. Raised by his widowed mother, he displayed as a child a natural talent for civil and military affairs and from an early age had great ambitions. "His bravery and eminence surpassed all others." Although the British rulers withheld the recognition he deserved, not so "the people of his native place." When they rebelled against oppressive British taxation, they called Washington out of retirement, and with his "patriotic zeal" he led them through eight years of bloody war to victory. Though Washington then "desired to return to his fields," he could not set aside the popular mandate. He thereafter "governed his state with reverence and respected good customs. He did not esteem military achievements; he was very different from [the rulers] of other states." Hsü was particularly impressed by Washington's encouragment of agriculture and commerce and his belief that "it was selfish to take a state and pass it on to one's descendants; he said it was better to choose a person of virtue for the responsibility of governing. . . ."

Washington's political legacy thus was a state operating in accord with Chinese political ideals in which the educated and virtuous ruled while remaining attuned to the needs of the people. To narrow the distance between ruler and ruled, officials were elected by vote of the common people (a device invented by Washington!), subject to popular recall, and expected to return to the ranks of the people after a term of service. There was no room in this system for political privilege or hereditary rank or rule. The government was kept properly simple. The "gentry" of the nation gathered in the capital named in Washington's memory. There the "virtuous scholars" of the Senate attended to domestic and foreign affairs with the aid of the men "distinguished in ability and knowledge" of the House. The head of state was a "general commander" chosen from among the "commanders" of each of the states for a four-year term. To avert clashes between central and local power, the Americans had developed a balanced and cooperative relationship between the federal government and the states. Taxes were admirably light. The government supported few soldiers (regarded by the Chinese as a scourge) and instead depended on a militia system made up of all the citizens except scholars and strikingly similar to one that had prevailed in ancient China. It had proven sufficient to keep the country safe, respected, and at peace.

By contemporary Chinese standards, Hsü's America was something of a utopia. The attractiveness of the picture was due in part to the nationalistic biases of his American sources and the desire of missionary informants to impress the Chinese by catering to their social and political prejudices. But probably even more it was due to Hsü's own desire to stir his readers by suggesting how far contemporary China itself had declined from a golden age of virtue. Chinese suffered rebellion and privation, while Americans, having succeeded in linking ruler to ruled, enjoyed order and plenty. China faced a mounting foreign threat, while the United States had enjoyed peace and security since independence. (Hsü was apparently unaware of the crises the United States had faced during the Anglo-French struggle between 1793 and 1814 or of the Mexican War, in progress even as he prepared his study.)

The vision of the United States advanced by Wei and polished by Hsü was to prove seductive to other Chinese intellectuals, who imbibed the infatuation with the United States either directly from the source or through later works which repeated Wei and Hsü uncritically. Hung Jen-k'an, called to serve as prime minister of the Taiping rebels in 1859, prepared a report depicting the United States as righteous, wealthy, and powerful and endowed with model political and social institutions. The Americans, Hung observed, though originally English, had thrown off oppressive British rule, and unlike the British "do not invade or bully neighboring countries." [12] Huang En-t'ung, an aide to governor general Ch'i-ying during the Cushing talks, made the same points in an 1865 work on "barbarian management." Both a geography of 1876 and a work prepared by the reformer Wang T'ao about that same time (but now lost) celebrated Washington's pivotal role as founding father and the economic, political, and social achievements of the country he founded. The scholar-diplomat Huang Tsun-hsien, writing in 1880, also made much of "the great founder" Washington who overthrew "the tyrannical rule of England" and whose "moral teachings" led his countrymen on the path of "propriety and righteousness." The Americans, Huang reported, conducted the most admirable of foreign policies. "They do not infringe upon other lands or people, and they do not interfere with other governments." Although possessed of greater strength and wealth than the Europeans, the United States "always helps the weak, supports universal righteousness, and thus prohibits the Europeans from doing evil." [13]

Wei's and Hsü's writing on the United States also carried an important message to policy makers looking for help in barbarian management. Both Wei and Hsü had approached their research with a pragmatic concern for solving China's crisis with Britain, and Wei at least was bold in making explicit his strategy for dealing with it. Wei contended that contradictions existed between the United States and Britain that China could turn to her advantage. But exploiting those contradictions was only one element in a larger strategy of Wei's devising. To slow the enemy's advance, China should stimulate the resistance to Britain by Russia, Nepal, Burma, Thailand, and Vietnam. Closer to home, China should make adroit use of commercial concessions to align the French and Americans against their old enemy, the English, and to secure from them assistance in building up China's naval and coastal defenses and helping ward off any renewal of British aggression. The history of Anglo-American conflict emphasized by Wei as well as Hsü lent plausibility to this strategy of enlisting "respectful and obedient" American barbarians against the overbearing English. Wei sought to support his argument by reference to the recently concluded Opium War, when the Americans and French had indeed shown their resentment of British ascendance and had even brought in their own warships. Had China then held out to those two discontented powers the lure of commercial advantage, they might have lined up in opposition to Britain. Hsü, though reticent in arguing explicitly for any particular policy, went even further than Wei in building up the United States as a new center of world power that Chinese officials might enlist to offset the Europeans.[14]

**Testing
Policy
Implications**

As Wei and Hsü intended, officials charged with devising China's response to the new foreign threat began to allot to the United States a part in "barbarian management." Pioneering studies, of which Wei's may have been the most important because of its early publication and its pointed message, may have moved some officials. But it is equally possible that many drew directly from the same body of ideas and experiences that inspired Wei and Hsü. The impulse to make diplomatic use of the United States—which asserted itself intermittently through the nineteenth century and well into the twentieth—was in turn to provoke critics to ask if Americans were really as different as they claimed to be or if they were not in fact a part of the larger foreign threat. Their doubts echo in the pidgin English observation supposedly made by a merchant on his first contact with Americans in 1784: "All men come first time China very good gentlemen, all same you [but] two three times more you come Canton, you make all same Englishman too."[15]

The temptation to test the alleged strength and independence of the Americans appeared naturally enough earliest and strongest among officials along the South China coast during the Opium War and its immediate aftermath. Lin Tse-hsü, the commissioner despatched to Canton by the Emperor to bring an end to the opium trade, was the first to take a serious, albeit minor, interest in the Americans as a makeweight in Chinese policy. In March 1839, shortly after his arrival, Lin demanded that foreign suppliers surrender their opium stocks and promise not to resume trade in that drug. Lin rewarded the compliant Americans by granting them permission to stay on in Canton and trade, while he excluded the uncooperative British. He hoped thereby (as he later explained) to keep the foreigners divided. Conceivably China might even stimulate the envy and dislike felt for the English by the Americans as well as the French and thus encourage the two countries strong enough to contend with Britain to play a more active role to China's advantage. To put the Americans in the same category as the British and deprive them both of trade would, on the other hand, be self-defeating, for it would drive the Americans to make common cause with China's enemy.[16]

The court was to cashier Lin in August 1840 after he had plunged China into a one-sided war with the British, but his vision of exploiting the divisions among the foreigners was kept alive as China suffered a series of military reversals. Late in 1840 Juan Yüan, having deepened his acquaintance with the United States, wrote from the capital urging Lin's successor to consider the possibilities of pitting the "peaceable" and powerful Americans against the "obstinate" British. "If we treat the American barbarians courteously . . . and also take the trade of the English barbarians and give it to the American barbarians, then the American barbarians are sure to be grateful for this Heavenly Favor and will energetically oppose the English barbarians. . . ." Kiangsu governor Yü-chien echoed both Juan's estimate of American strength and his emphasis on playing on American commercial interests. Within Canton itself P'an Shih-ch'eng, a wealthy merchant with official standing, suggested that China might also make practical use of the Americans in the midst of war by borrowing their technological expertise, an idea he himself acted on.[17]

Even after the war, Ch'i-ying, who came to Canton as governor general in mid-1843, continued the search for ways to manage the Americans. Anticipating Wei and Hsü, he described them as a recently established people, industrious and rich in land. "With England [the United States] is outwardly friendly but actually resentful." Initially, through 1843, he sought to win American gratitude by no more than a grant of trade "privileges" along the lines of those treaty rights already secured by Britain. Peking agreed (so long as no complications with the English resulted), and in November 1843 confirmed Ch'i-ying's promise of equal commercial opportunity with an edict meant to display the Emperor's "tranquilizing purpose." But the Americans were neither tranquilized nor grateful. Cushing soon demonstrated that Americans could be obstinate and pushy and that they esteemed national prestige no less than profitable commerce. If the British concessions were embodied in a treaty, Cushing would have the same for the United States and not be fobbed off with promises of mere "privileges." Moreover, he wanted to travel to Peking, a demand (Ch'i-ying guessed) that Cushing either had cooked up with the British or meant to use to surpass them. Carefully reining in this headstrong foreigner, Ch'i-ying made clear that he was prepared to conclude a generous commercial treaty, but only on the condition that Cushing give up his audacious dreams of a journey to Peking. As late as 1846 Ch'i-ying seems to have still been interested in ways he might use the United States to check the "proud and domineering spirit" of the British, but other officials now only occasionally referred to the Americans as "respectful and obedient," and none seriously regarded the United States as an offset to the British.[18]

The advance of the Taiping on Shanghai in 1853 reawakened interest in the United States, though now not as a makeweight against the powers but rather as a source of assistance against the rebels. The chief advocate of this revised approach was Wu Chien-chang, a Shanghai merchant who had once been a part of the old cohong system and now was active in the city government. His view of the United States, in line with Wei's and Hsü's, was of "a rich country, strong militarily, most respectful and obedient" in its dealings with China. In contrast to the English, who were not only in general "overbearing and cunning and only concerned with gain" but also of late in close touch with the Taiping rebels, Americans trading with the Taiping were few in number, and American diplomats were moderate. Attempts in 1853 to secure American assistance for the imperial cause were, however, to founder. The American commissioners, Marshall and McLane, insisted on an unacceptable quid pro quo, and in any case American warships were not available, and ranking officials in the area opposed drawing on the support of any of the barbarians, who were all "by nature deceitful."[19]

The hostile reaction Wu encountered among his official colleagues reflected widespread resistance within the Ch'ing bureaucracy to the American infatuation. The oldest and most pervasive source of opposition came from the militant view on foreign affairs, a compound of traditional assumptions about China's central place in a hierarchical world order and of ignorance about the new Western challenge to that order. Some officials turned a blind eye to treatises such as Wei's and Hsü's, regarding as dubious if not foolish the proposition that there was something to be learned from studying foreigners.[20] What educated man had not heard—indeed memorized in childhood—the lines from Mencius (and others like it): "I have heard of men using the doctrine of our great land to change barbarians, but I have never heard of any being changed by barbarians." Drawn to China by their need to trade, these men of inferior culture were supposed to be dazzled by the richness of the civilization they found. As the superior, China was obliged to show condescending kindness, display no favoritism among them, and at times indulge violent and unreasonable behavior. The phrase "nourishing men from afar" (*huai-jou yang-jen*) reflected the tendency to view foreigners as pitiable mendicants. Though increasingly overwhelmed by power realities as the nineteenth century progressed, this rhetoric affirming Chinese superiority persisted, even in adversity, in official thinking and imperial pronouncements, indeed down to the very end of the dynasty.

The militant outlook was sustained by the widespread confusion and misinformation about foreigners that resulted from chance contacts and carelessly assembled snippets of random information. The "white devils are fond of women; the red devils are fond of money; the black devils are fond of wine" was taken by the Taokuang Emperor for keen analysis. Officials on Taiwan in 1842 concluded from an interrogation of two shipwrecked sailors that America might once have been "in England," that a man might be able to "walk from London to America in a week," and that London might be as large as America. Intentionally false reports from provincial officials worsened the problem. Pandering to the known prejudices of their superiors, even those who knew better soothingly described foreigners as contemptible barbarians, minimized the threat they posed, and exaggerated China's strength. Ch'i-ying in his official communications noted how "ignorant" and uncultured Americans were, while Hsü Chi-yü's memorials to court employed derogatory terms carefully excluded from his "Brief survey."[21]

The militant outlook on foreign policy found a home at court in the 1850s. The Hsien-feng Emperor (1850–1861) reversed the policy of appeasement followed since the Opium War, cashiered the leading officials tainted by it (including Hsü Chi-yü), and redirected policy along lines of increased resistance to foreign pressure along the coast. Militants at court and in the provinces fought off demands for treaty revision and played up the dangerous symbiotic relationship that had developed between foreigners in search of profit and rebels in need of aid. Under the new dispensation the powers were at first regarded as one in cunning and covetousness; differences among them were not worth troubling over. The United States—its diplomats joining the cry for treaty revision, its missionaries spreading heterodox ideas that inspired rebellion, and its merchants supplying rebels arms and provisions—seemed in the 1850s no exception and hardly benign.[22]

The militants found their champion in the provinces in the person of Yeh Ming-ch'en, the governor general of Kwangtung-Kwangsi, who struggled to keep the British out of Canton between 1853 and 1857. Increasingly aroused by foreign incursions and aware of China's military vulnerability, Yeh promoted a strategy of popular resistance. In a protracted war on Chinese soil against a united and aroused populace, an invader would ultimately find himself overextended, outnumbered, and demoralized. Yeh saw little incentive for making (or attempting to exploit) subtle distinctions within the enemy camp since such a policy would almost certainly sow popular disaffection and confusion that would in turn undermine a policy of resistance. But in addition Yeh looked upon appeals for assistance from the deceitful foreigners as a dangerous confession of China's weakness and an intolerable blow to her prestige.[23]

The diplomatic infatuation with the United States was to encounter a second, new source of opposition in the 1850s from an unexpected quarter—officials who took the new literature on foreign affairs and the study of foreigners seriously. Their investigations had led them to doubt the diplomatic value of the United States to China. Ho Kuei-ch'ing, the governor general of Kiangsi and Kiangsu and imperial commissioner in charge of foreign affairs (1858–1860), emerged as the most consistent and forceful exponent of this view. American enmity for Britain, he argued, was much exaggerated. But even if China should manage to pit Americans against the British, the strength of the Americans (even reenforced by the Russians) was not sufficient to deter Britain. Rather than a weak China trying to manipulate stronger foreign countries, Ho preferred to accommodate all the Western powers and avoid a war dangerous to China's security. The "fuel under the pot" of Sino-foreign relations was, in his view, the dissatisfactions of foreign merchants. Meet merchant demands and they would emerge as a check on foreign diplomats with their love of prestige and their taste for disruptive wars and displays of force.[24]

Debunked by those knowledgeable in international affairs, the American infatuation was now taken up by the less well versed. The twin evils of internal rebellion and foreign aggression in the late 1850s forced the beleaguered militants at court and in the provinces—in an ironic reversal—to try their own hand at manipulating the Americans. Apparent American restraint during the renewed Anglo-Chinese conflict between 1856 and 1860 in particular shook the court's view of the Americans as simply another inconstant and unscrutable breed of barbarian and awakened interest in the diplomatic implications of American commercial greed and jealousy of Britain. Now the court watched and occasionally courted the United States (as well as Russia after France went over to the British side in 1857) not so much as a potential ally but as an intermediary to be kept neutral and used to calm the British in defeat and restrain them in victory. At times the court's policy amounted to no more than the practical recognition that American assistance was a straw worth grasping for after all else had failed.

During the first phase of the Hsien-feng Emperor's policy of resistance, Peking had watched with admiration and approval as Yeh held the British at bay outside Canton. Then in January 1858 British forces, reinforced by the French, stormed the city, seized Yeh, and carried him off to imprisonment in India. In the early stages of the confrontation, Yeh had sought to isolate the British from the Americans by opening the Canton trade only to the latter. He was in turn encouraged by the decision of the U.S. government to recall Peter Parker (a "crafty" troublemaker in Yeh's view). He interpreted Parker's replacement by William Reed in November 1857 as a welcome repudiation by the U.S. government of an aggressive policy. So when Reed arrived, Yeh agreed with uncharacteristic promptness to talk (though without making any of the concessions on treaty revision that the American wanted). With open warfare with the British looming, Yeh in a show of solicitude persuaded the American residents to evacuate Canton for their own safety and, once fighting began, made sure they understood that the British were responsible for the resulting disruption of trade. Yeh even let pass the highly partisan, indeed belligerent, conduct of the local American consul and naval commander during the conflict.[25]

Once Canton fell and the Allies made ready to go north to demand treaty revision at gunpoint, the Emperor and his advisers would now themselves try to bring the United States to serve China's cause. The groundwork had already been laid by Yeh's reports stressing the resentment felt by the Americans toward the unreasonable British. Consequently, after Yeh's loss of Canton the court had condemned his failure to take advantage of what it incorrectly regarded as an American offer of good offices, while the Emperor himself had brushed aside reports implicating Americans in the attack on Canton as "misinformation" spread by the British in order to veil their own aggression. In April 1858 British and French forces arrived off the northern coast. Peking would not yield on treaty revision, but wanting to provide them a way to back down, asked Minister Reed, who had accompanied the Allied expedition north, to serve as mediator and help avert a collision. On May 20, as the naval bombardment of the Chinese forts began, the court again appealed to Reed to set a good example by continuing his own negotiations with China and to use his good offices to secure a three-day truce and keep the lines of communication with the Allies open. On both occasions Reed demurred.[26]

All this came as no surprise to T'an T'ing-hsiang, the governor general of Chihli and the militant official to whom the court had entrusted the direct supervision of developments along the coast. He regarded the foreigners as "all in one category in their insatiable greed," and wanted to substitute for diplomatic wrangling an all-out military effort against them. But T'an's strategy for barbarian management crumbled along with his coastal defenses, and by the end of May he too wished to invoke American assistance in bringing the fighting to an end while avoiding talks with the Allies. He now reasoned that even though the Americans and Russians "are sure to make insatiable demands, still, compared to fighting, this has its advantages." The court agreed to this plan to win time; however, mounting Allied pressure forced the court in June to give in and engage in substantive talks. The two high officials, Kuei-liang and the veteran Ch'i-ying, appointed to handle negotiations twice invoked American aid in softening Allied demands, and both times came away empty-handed. Kuei-liang after the first attempt reported Reed's party insolent, querulous, and exigent, while the court dismissed as "presumptuous" their request for diplomatic residence in Peking. Under threat of force, the Chinese now bowed to British and French demands for treaty revision. To maintain the good will of the American and Russian neutrals, the Chinese agreed to revise their treaties as well. By the end of the second try—an emotional performance in which Kuei-liang invoked the clause that he himself had had inserted in the new American treaty providing for good offices in times of trouble—he was convinced that the Americans and Russians were no different from the British and French. "To get them to intercede could hardly be of any advantage."[27]

Chagrined and a bit wiser, officials in the north seem to have moved toward Ho's view that the United States was not a useful auxiliary to Chinese policy. As a halfway measure the court did agree to diplomats visiting Peking. They had, however, to reconsider when in June 1859 a second Allied expedition came north, accompanied once more by an American diplomat, and again a clash of arms resulted, due on this occasion to a dispute over what route Allied diplomats would take from the coast to Peking, where an exchange of the new treaties was supposed to take place. Now it was the Chinese turn to win, and American assistance was needed to soothe the British. But when the court and the chief Chinese negotiator Heng-fu actually turned for assistance, Ward responded that he would speak to the British only after he had completed his own projected trip to Peking. The court was not happy about receiving any foreign diplomat, particularly one so cool to Chinese appeals. Enthusiasm for Ward's visit was further dampened by reports from the victorious general Senggerinchin (Seng-ko-lin-ch'in) that the Americans, British, and French were "all fellow conspirators." The Ward visit went ahead, but not under conditions likely to consolidate ties with the United States.[28]

By mid-1860 the Chinese faced the third foreign expedition in as many years. The Chinese government, determined to fight rather than make further concessions, faced an Allied force intent on revenge for the past year's defeat. As usual, an American representative tagged along, and once again, the analysis from Shanghai suggested that while the American Ward "has no desire to do harm, he cannot exert himself on our behalf." All the major foreign powers were banded together, so Shanghai concluded. Senggerinchin, awaiting the military rematch on the northern coast, agreed. But in early August, after foreign warships had massed offshore and Allied forces had begun to move on Peking, the court snapped up an offer by Ward to act as intermediary only to see him withdraw his offer once fighting began a few days later. Now in desperation, it instructed its commanders to screen foreign prisoners for Americans who might be used as diplomatic intermediaries.[29]

Peking's fall to Anglo-French forces early in the autumn and the flight of the court brought to an end one era of Chinese resistance to foreign encroachment but not to spasmodic efforts to find a way to translate the supposedly helpful attitude of the United States into some tangible benefit to China. Indeed, hardly had the invading Allied force withdrawn than Tseng Kuo-fan, one of the major architects of the restoration of imperial authority that lay ahead, revived the clichéd view of the Americans as "pure-minded and honest" and "long recognized as respectful and compliant toward China." Why not find a way to draw them still closer to China and ensure they did not ally with the "most crafty" English and the French? Tseng wondered.[30]

The good offices provision written into the 1858 treaty by Kueiliang still encouraged Chinese policy makers to believe that the United States occupied a special position as mediator between China and the powers. Thus when the newly established foreign office, still without its own missions abroad, decided in the late 1860s to have a foreigner carry China's case for diplomatic restraint directly to the capitals of the powers, Anson Burlingame, the first American minister to reside in Peking, was its choice. He was not only able, sincere, sympathetic to China's plight, and eager to add to his reputation, but also a citizen of "the most tranquil" of the powers, so Prince Kung reported as head of the foreign office in late 1867.[31] Seven years later the benign image and the treaty pledge of good offices led Li Hung-chang, Tseng's junior associate, to appeal to one of Burlingame's successors, the "prudent and sincere" Benjamin Avery, for help in averting a conflict with Japan over Taiwan, where the Japanese had landed a punitive expedition following attacks on their Liuchiu "subjects" by aborigines there.[32]

Americans could also be useful in fighting rebels. Frederick Townsend Ward, whose force of foreign adventurers (known as the Ever-Victorious Army) battled the Taiping in the lower reaches of the Yangtze in the early 1860s, came to be valued by Li as a man of "ability, sagacity, and willingness to attach himself to the Chinese cause. . . ." Ward's death in battle in September 1862 led Li to look for another equally reliable American as a successor. The nod went to Henry Burgevine, a North Carolinian who had served under Ward.[33] Finally, the United States occupied a favored position as a place where Chinese might master the skills that made the West strong. Tseng and Li arranged to send young Chinese (between ten and fifteen years of age) for a prolonged course of study abroad. In 1872 the first group was selected and set off for the United States to settle along the Connecticut River valley. There under the supervision of Yung Wing, an 1854 graduate of Yale, their sponsors expected them to master military and naval skills, surveying, manufacturing, and mining, all desperately needed in China's own development.[34]

The outcome of this reliance on the United States once again fell far short of what the positive image had led Chinese officials to expect. Burlingame violated his formal instructions—to the annoyance of the foreign office—without making any noticeable dent in the policy of the powers. Avery, though exceedingly sympathetic to China's position on the Taiwan case, lacked the influence to effect its outcome or even to restrain an American, Charles LeGendre, who much to Li's annoyance had been assisting the Japanese as a Taiwan expert.[35] Burgevine quickly proved himself to Li a "shady character" and "very obstinate," and so after only three months Li had him replaced. For the following three years Burgevine made trouble (including a short stint serving the Taiping) until Li finally lost patience and arranged an "accidental" drowning (to the evident relief of the almost equally exasperated American diplomats).[36]

The educational mission, still another disappointment, secured none of the immediate benefits its sponsors expected because the students seem to have spent as much time on baseball and Latin as on practical subjects. The mission, moreover, proved expensive ($1,200 for each of the 120 students sent out by 1875). Yung Wing and his colleagues, who had a more traditional outlook, were like oil and water, an impossible mix. And the students themselves were becoming culturally deracinated (with some even refusing to return home). Li at last gave in to the critics within the bureaucracy, who charged that the mission "wastes money, breeds corruption, and will show little result," and to his own growing doubts and had the mission terminated in 1881. When he sought access to U.S. military academies where more mature students might derive more immediate benefit, Washington turned him down.[37] Yet despite all these disappointments and reverses, the benign image of the United States persisted, providing a basis for a renewed policy interest in the United States in the decades ahead as the imperialist powers began to close in on China's frontiers.

From Canton to California

Chinese emigration to the United States was but a small part of a greater movement of Chinese abroad that gained impetus in the 1840s and continued well into the twentieth century. The half a million or so Chinese who journeyed to the United States between 1850 and 1900 represented a fraction, certainly no more than a tenth, of all the Chinese who left south coastal China over those years to find employment abroad. Although frequently stereotyped as a people bound to home and hearth, the Chinese had not only responded at home to the lure of underpopulated frontier regions or long-settled areas suddenly decimated by natural disaster or war, but had also ventured off to Southeast Asia as early as the fourteenth century to establish permanent communities. By the mid-nineteenth century the trickle of emigration from South China became a flood as difficulties at home coincided with a high demand for labor and mercantile skills in Southeast Asia and the Americas. Records on departures from Macao alone show half a million Chinese responding to the new opportunities which Western trade, capital, and technology had created in the mines, railways, and commercialized agriculture of those regions. The flow of emigrants climbed dramatically higher from the 1870s onward.[38]

Emigrants to the United States issued almost exclusively from a 4,000 square mile region embracing Canton (the capital of the province of Kwangtung) and the immediately adjacent provincial districts to the south. This area was divided into three culturally and geographically distinct regions. One of these, the three districts known as *San-i* (or in Cantonese *Sam Yup*) surrounding Canton, lay on rich delta land and boasted the highest level of culture and prosperity. A second area made up of four districts (literally *Ssu-i* or, in Cantonese, *Sze Yup*) located just to the southwest of Canton was a hilly region where much of the land was marginal. Relative poverty, a distinct dialect, and a sense of social inferiority set the people of Sze Yup off from their Sam Yup neighbors. The third distinct zone was Hsiang-shan (Hueng-shan in Cantonese; later renamed Chung-shan) district, sandwiched between Sam Yup to the

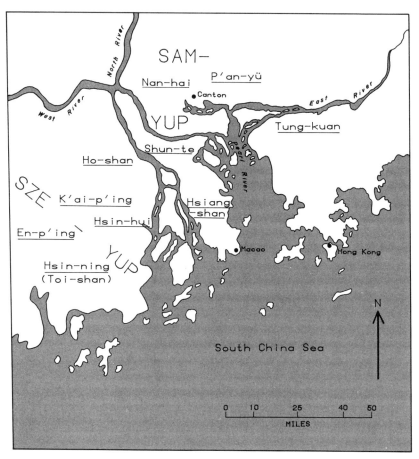

THE CANTON DELTA
(IMPORTANT IMMIGRANT DISTRICTS UNDERLINED)

north and Portuguese Macao to the south. It enjoyed a diversified economy of farming, fishing, and handicrafts as well as some claim to the cultural refinement of Sam Yup. Overlaying these territorial distinctions was a basic ethnic line that divided Chinese, who had long lived in the Delta, from the Hakkas (literally, "guest people"), who had begun to move from central China as early as the fourth century A.D. and continued southward by stages through the nineteenth century. The settlement by Hakkas on the marginal land of Sze Yup created tensions between peoples already divided by language and social customs.[39]

By the middle of the nineteenth century the impulse to emigrate had become intense. The basic impetus came from demographic pressures. In the six decades before 1850 population density in the whole of Kwangtung increased nearly 80 percent with the result that by mid-century an average of 284 people had to live off one square mile of land. In the Canton delta generally population must have pressed perilously close against the productivity of arable land, diminished by wetlands, waterways, and mountains. Sze Yup in particular, with its high proportion of marginal land, must have felt these pressures acutely as the influx of Hakkas added to the natural increase of the local population. A dismaying succession of floods and droughts accompanied by banditry, rebellion, and foreign intrusion further disrupted the delta economy, already straining to support its population. Although the great Taiping Rebellion of 1850–1864 bypassed the region, it set off local shock waves, precipitating conflicts among clans, secret societies, and ethnic groups that a weakened dynastic authority found difficult to control. Sze Yup witnessed the most intense internal conflict: a fourteen-year struggle (1854–1867) between Hakkas and the more numerous and prosperous older settlers which produced hundreds of thousands of casualties; and a major uprising by the Red Turban secret society, which ended in bloody suppression. Sam Yup bore the brunt of foreign conflict between 1839 and 1860 with Canton itself subjected for a time to British occupation. The new treaty system that the British set in place opened the way for an influx of imported manufactured goods which supplanted peasant handicrafts, sped the unsettling trend toward commercial agriculture, and tied local prosperity to developments in the international economy. The simultaneous rise of Shanghai as the major entrepôt for foreign trade further deepened the economic crisis in the Canton region.[40]

The impulse to flee a society thus rendered insecure must have been strong. The livelihood of the rural poor—peasants, craftsmen, peddlers—in particular suffered in the face of a combined threat from rebels and bandits, government troops, landlords, tax collectors, and vagaries of the market. The unfortunate grasped at emigration as a means of survival in a time of adversity. For the losers in the mid-century conflicts—the Hakkas, the hunted adherents of secret societies, and members of weak clans—the choice was probably often a precipitate one made in the face of deprivation or death at the hands of their enemies. But many, perhaps most, approached the decision to go abroad to work as a carefully calculated family affair. A son who could not find profitable employment, or a husband no longer able to care for his dependents would find employment overseas that would permit him to support himself and send regular remittances home to support his family. Any surplus was invested locally, usually in land. In some cases families would send a member from each generation abroad, a practice that in some areas went back to the seventeenth century. Those who left early would establish themselves and prepare the way for the arrival of younger relatives. In time the emigrant himself would more likely than not return. Of the roughly 4.8 million emigrants known to have left south coastal China between 1876 and 1901, some 4 million came back.[41] Migration thus served in effect as a local industry, an economic lifeline for the people of an entire region.

News of the California gold rush reached Hong Kong in the spring of 1848 and filtered out into the countryside, stimulating the first stream of emigration to the United States. Although the gold fields were to play out in a little over a decade, California's developing but labor-short economy remained a distant magnet to the Chinese. Those at home measured their opportunities through the letters sent home by the pioneers and the size of the accompanying remittances. Vivid testimony would come after a time with the first homecomings, when the successful emigrant would treat his neighbors to a sometimes resplendent celebration. As if to fix an alluring image in the prospective emigrant's mind, California came to be known as "the mountain of gold." To help convert fantasy to reality, foreign shipping companies and Chinese emigration agents circulated promotional literature testifying to opportunities abroad that surpassed anything at home and offering assistance in making the journey. Finally, the prospective emigrant might incline to California knowing that guests of the golden mountain ran small risk of falling into the ill-reputed coolie trade. The Chinese had aptly named it the "pig trade" and called "piglets" those unfortunates who sold themselves, frequently under coercion or deception, for a fixed period of labor abroad. At its worst, as it was conducted in Cuba and Peru through the mid-1870s, only one coolie in ten survived the perils of the voyage and the abuses of employment to return home. But in Hong Kong, the principal point of departure for the United States, British regulations protected the emigrant from the unscrupulous coolie brokers. And in the United States the courts would not enforce—and public opinion decried—labor contracts.[42]

The mechanism of emigration developed considerable speed and efficiency as the nineteenth century progressed.[43] In the days of sail the ocean voyage alone took two or three months, but with the advent of regular steam service in the mid-1860s a peasant could make the transition from the rice fields of his semitropical home to the gold fields of California or to the snow-covered Rockies and a railway construction gang in a matter of weeks. Brokers played a pivotal role in making the system work. An emigrant headed for the United States would deal directly with a broker based in Hong Kong who dealt exclusively with emigrants from his own home area. These brokers would rely on their local reputation, personal contacts, advertising circulars, and command of current information to draw prospective travelers to them. For the indigent they would arrange travel loans from merchants, and for those who could not stay with relatives or kin in Hong Kong while awaiting departure, they provided inexpensive accommodations in dormitories, which usually served as the brokers' headquarters and as a message center. To obtain his ticket the emigrant would have to work through a second broker, a Chinese who acted as sales agent for foreign shipping firms. These various brokers enjoyed within their particular spheres a virtual monopoly protected by their guild organizations. On each emigrant the broker made a commission, and on a large volume of departures they built substantial fortunes.

Dialect and place directed the emigrant from the moment he left home until he found employment abroad. They separated men who could not easily communicate with each other and drew together those of common dialect, customs and traditions, and perhaps even kinship. At the outset, the prospective emigrant gained his most reliable information from his neighbors with experience overseas or from brokers from his area. Once in Hong Kong he would stay either in private homes or in dormitories run and occupied by men from his own locale. In organizing his life abroad the emigrant would place no less emphasis on dialect/place loyalties. On arrival in San Francisco the emigrant would emerge from steerage to the greetings of men who shared a common surname, common background (in the case of Hakkas), or common place of origin. They would escort him through the official inspection, past the assembled white rowdies, and on to lodgings occupied by other men like him. In finding his first job and in joining social and benevolent associations, the newcomer depended on the aid and company of these fellows.

The merchant played a crucial role in emigration as financier, supplier, and trans-Pacific liaison. The dominant Sam Yup merchants were the chief source of travel funds for emigrants, the bulk and certainly the poorest of whom were from Sze Yup. Organized by guilds and linked by close personal and old business ties, these Sam Yup firms formed a network that extended from Canton to Hong Kong and on to San Francisco. Chinese merchants in the United States (perhaps 4 or 5 percent of the total Chinese population) would collect travel loans made initially on their own account or on the account of agents or associates in China. Those same merchants would retail to Chinese workers throughout California supplies brought in from China. Their shops also served as employment offices to which white employers came to hire labor gangs. The merchant maintained contact with home. He kept brokers in Hong Kong appraised of the employment situation. On the occasions such as in 1853–1854, 1876, and 1886 when economic depression and anti-Chinese agitation made jobs difficult to find and travel debts hard to pay off, merchants in San Francisco signaled a temporary halt to emigration. The merchant's shop provided for the bulk of the community a link to home. In his guise as postman and banker the merchant would receive letters from China, dispatch the reply, arrange the accompanying remittances (at a 2–10 percent commission), hold savings in safekeeping, and offer goods on credit. As a result of the multiple services he performed, the merchant built up within the emigrant community a network of clients—laborers, petty craftsmen, and small shopkeepers indebted literally and figuratively for past services and dependent for future assistance. In the absence of both officials and the scholar gentry who in China made up the local elite, the merchant patron emerged as the natural leader of the community. Not only had he helped transplant and maintain a piece of South China overseas, but his economic successes epitomized the ambitions of a community oriented essentially toward money-getting.[44]

The particularistic loyalties of the emigrant and the leading role played by the merchant fused in the variety of associations which gave social coherence and cultural continuity to the life of the overseas Chinese. These associations had their origins in Chinese society but had evolved to better meet the needs and conditions encountered abroad. The emigrant's deepest attachments were to lineage organizations. These common descent groups (or clans) flourished in Kwangtung and Fukien as nowhere else in China. Respected elders directed their affairs, fulfilling group ceremonial obligations, redressing injustices and insults, and resolving internal disputes. The more developed and affluent lineages maintained an impressive temple where homage was paid to prominent forebears (thereby consolidating group solidarity and prestige) and held income-producing property which might be used to support the indigent or educate the worthy poor from within the lineage. The lineages which existed overseas were in some cases extensions of the actual lineage to which the emigrant had belonged at home, but in other cases, where there was an insufficient number of relations to form a lineage offshoot, emigrants contrived new lineage groups based on a fictitious common ancestor. Either way, lineage members ensured mutual support and group solidarity that might find its application in job hunting, social life, and personal welfare. The lineage in the United States, however active it might be, could not help but be a pale reflection of those strong lineages which dominated the rural scene in South China through gentry leadership and the control of common lineage land and the incomes they produced. Deprived of these sources of political and economic power, the lineage in the overseas community had to make room for (though usually not ceding primary loyalty to) place associations, occupational guilds, and secret societies.[45]

The place association, whose institutional development went back over five hundred years, served in China as a lodging and gathering place for officials, examination candidates, merchants, and craftsmen of a particular region visiting in the capital or in one of the chief commercial centers. The place association, perfectly suited to the needs of emigrants, proliferated in San Francisco. The people of Sze Yup organized three different associations reflecting the chief divisions within their ranks. The people of Hsin-ning district constituted the largest of these Sze Yup associations, while several powerful Sze Yup lineages formed another. Sam Yup, with its prominent merchant element, predictably boasted the wealthiest of the place associations. To make themselves heard and to preserve their interests, the people from Hsiang-shan district as well as the Hakkas formed their own separate groups. From the time of their formation in the early 1850s, the number of associations ranged upward from five, depending in the main on the splits and consolidations that occurred within the ranks of Sze Yup. Each association aided the emigrant from his arrival in the United States up until his departure or death. Like the lineage and the other associations, place associations offered lodgings, help with employment, support through periods of disability or unemployment, assurances of burial in home soil, arbitration of disputes, and legal assistance before American authorities.[46]

Occupational guilds, yet another form of association carried over from China, brought together in each case emigrants from a particular town or district and sought to secure a monopoly over a particular occupational specialty, perhaps one to which its members were suited by previous experience, and to protect it against interlopers, particularly Chinese from other regions or other ethnic backgrounds. For example, Sam Yup controlled the butcher trade and portions of the garment industry; Sze Yup dominated in the laundries, retail stores, and restaurants; and Hsiang-shan stood preeminent in the retail fish business, flower growing, and sections of the garment industry. Other important guilds represented Chinese in the shoe and cigar factories. The occupational guild, which might count Chinese employers as well as workers among its members, also sought to put a stop to abuse by whites, settle internal disputes, and act as an employment agent. To keep wages and prices up, the guilds resorted to strikes against white employers, price-cutting and social ostracism against intruding retail or laundry firms, and violence against nonunion workers.[47]

Even secret societies in the United States were oriented toward place of origin. They had sprung from the Triads, a secret society which flourished among deracinated and insecure groups along the southern coastal provinces and the Yangtze Valley. While retaining the organization and secret oaths and rituals of the old world organizations, secret societies in the United States sloughed off their traditional anti-Manchu ideology and instead operated openly in the service of the economic and social needs of their membership, largely, if not exclusively, the men of Sze Yup, especially the poor, the itinerant, and those without strong lineage ties. By the late nineteenth century, and perhaps even earlier, 70 to 80 percent of the Chinese in the United States held membership at least nominally. In the larger Chinese communities the societies provided an alternative to, or even a refuge against, powerful lineage or place associations, while in Chinese communities too small to support lineage or place associations secret societies served as flexible, all-purpose organizations. Each society and its affiliates provided the expected services—lodging to traveling members, help in dealing with American authorities, arbitration of disputes between members, protection against wrongs by outsiders, and assurance of proper burial. Societies seem to have dominated prostitution and gambling within the larger Chinatowns, either by exercising direct control or by offering "protection" to those who did, and on occasion tried to assert control of mining rights against other Chinese.[48]

In San Francisco's China Association (Chung-Hua hui-kuan) the merchant leaders of the community's various organizations came together. Popularly known as the Six Companies, this umbrella association came into existence in 1862. It was made up initially only of the original place associations and later of their splinter groups. Because these place associations incorporated the strong lineages as well as the Hakkas, all elements in the community were represented (except for the secret societies during the Six Companies' first two decades). The heads of the place associations guided the affairs of the Six Companies and rotated the presidency among their own rank. Through the Six Companies the diverse elements of the community were able to discuss problems of common concern and to speak with one voice both to China and to American authorities. The Six Companies adjudicated disputes between constituent organizations and relied on its prestige and community pressure, or on some occasions appeal to American courts, to secure acceptance of its decisions. The Six Companies, unlike the secret societies, did not use force against the recalcitrant.[49]

To handle their relations with the host culture, each of the associations put forward an intermediary, generally known as the "interpreter." The title is something of a misnomer since usually an interpreter's duties went beyond mere translation, whereas his actual command of English, usually picked up in commerce or church schools, was more often than not rudimentary. The interpreter was primarily charged with representing and defending his group's interests. An occupational guild might employ him as foreman in a white-owned shoe factory. A secret society might send him out as the head of a railway work gang made up of society members or dispatch him to court to help defend a "highbinder" accused of murder. Or a place association might rely on the interpreter to extricate a newcomer enmeshed in customs difficulties. To assist the interpreter in his errands, the associations and especially the Six Companies would often engage the services of American lawyers and publicists.[50]

These self-imposed exiles, who labored to support distant dependents and in moments of leisure dreamed of homecoming, needed industry, a frugal lifestyle, and even luck to manage a return home in less than ten years. Prospective emigrants who could not obtain $40 to $50 for travel money from friends or relatives or by the mortgage or sale of personal property had to borrow. They would arrive in the United States with a debt of about $100. The newcomer might go to work on the railways or in industry at about $30 a month. After covering his living expenses, the emigrant would have left at the end of each month $10 more or less that he could apply against his travel debt and then later, once it was paid off, send home as remittances or retain as savings. After seven years of steady work under ideal conditions (more likely to obtain into the 1870s than after), an emigrant might have accumulated the $300 in savings considered a near minimum for a return home. With that money he would celebrate his homecoming with relatives and neighbors, and retire comfortably to live on the earnings of the land that his remittances and savings had purchased.[51]

But this story gilds the edges of the emigrant experience. The loneliness and anticipation was legendary even for those who made it back within a decade. A Hsin-ning verse couched in the voice of the emigrant's wife caught the melancholy associated with long-deferred homecoming:

> Grey hair dressed and piled high received flowers worked
>   with the feathers of kingfishers;
> Yesterday the husband who went abroad came home;
> For the ten years that I had not seen his face;
> I passed the evenings with bitter memories twisting
>   threads before the lamp.[52]

Homecoming for some was delayed beyond ten years; for others it never came because of economic setbacks, gambling losses, high living, or simply a loss of drive as the youth became an old man. Many lived out their lives in the community they had come to know best, supported by savings, odd jobs, or charity. Even those who made it back home did not always stay. They might be fleeced in Hong Kong by con men on the lookout for country boys with money in their pockets. They might invest their savings unwisely or find the demands of family and relatives exceeded both their expectations and their purse. Or they might simply go abroad again for the very reason they had left in the first place—a man's labor was better rewarded abroad than in the delta.

The community these hopeful emigrants made up was nearly all male. Emigrants did not bring wives with them, for to do so would have added to the costs of transportation while at the same time disrupting the fabric of the family that emigration was supposed to preserve in the first place. The proportion of women in the Chinese community climbed as high as 8 percent in 1870, but subsequently fell and was nowhere near that again until 1910. Among the women in the Chinese community, some came as legitimate wives or concubines, but these were too few to produce that native-born second generation that for other immigrant groups in the United States was the key to assimilation. Aside from the wives of affluent merchants, women in Chinatown were an important part of its economy. In 1870 well over half of them labored in Chinatown's most prosperous industry: prostitution. Many came from poor families unable to afford to raise a daughter, who would not perpetuate the family line and who, once married, would serve her husband's family and not her own parents. "When you raise girls, you're raising children for strangers," was one way of looking at the problem. "It is more profitable to raise geese than daughters," was another. Some women had thus been sold as children into service and others were deceived in later life, for example by "marriages" that delivered them into the hands of madams in the United States. What began in the early 1850s as a business conducted in the main by free agents soon fell under the control and protection of Chinese males who recognized the money to be made in organized prostitution. Although the prostitute's life in Chinatown was probably no harder than that in Canton, the comparison could have been of little consolation to those doomed to the ravages of venereal disease after roughly a decade of work. The more fortunate escaped or worked out a contract period of four to five years and then found a husband.[53]

In the main the overwhelmingly male work force demonstrated striking geographical and even some occupational mobility. During the period of free immigration the laborer often worked for good wages, only somewhat below that of his white counterpart, and on occasion when he thought the pay too low or working conditions unsatisfactory, he went on strike. A newcomer in the 1850s might begin in gold mining, move on to railway construction, shift to underground mining (where he could apply tunneling techniques learned on railways), or take up work in agriculture or in an urban industry. The search by laborers and merchants for remunerative employment and relief from Western hostility from the 1880s onward resulted in a steady eastward spread of the Chinese population. As the community scattered, the Chinese became regular patrons of the railways, moving from city to city, keeping in touch with friends and relatives, looking for new jobs, or attending to the business of community associations. A Chinese of this time whose life story has been documented, offers an example of this mobility. He arrived in San Francisco, followed a railroad construction crew east, and lived in Detroit before finally settling in New York. In the process he went through three occupations—servant, laundryman, and general merchandiser.[54]

The generalizations thus far advanced about the early emigrant community apply best to San Francisco, but many of the characteristics of San Francisco were reproduced by the smaller Chinese communities up and down the West coast and at points inland, and were subsequently retained as prominent features of late nineteenth-century Chinatowns across the country. As outside observers repeatedly noted, these Chinese communities—whether large or small—were merchant-led and made up primarily of bachelors whose orientation toward China was expressed culturally by their associations and economically by the flow of savings back to families in China.[55]

Although at best a flawed reproduction of life in the Canton delta, Chinatown did perpetuate in microcosm the linguistic, territorial, and ethnic fissures of home and perhaps even exacerbated those old tensions by throwing peoples of differing backgrounds into an unprecedently close relationship. The feud between Hakkas and older settlers carried over to California, as did clan antagonisms. But the most important division, high in its potential for friction and conflict, was that between the poor but numerically dominant emigrants from Sze Yup and the fewer but economically powerful representatives of Sam Yup. The Sze Yup share of the population in California, the focus of early Chinese settlement, climbed from roughly two-fifths (16,000) in the mid-1850s to four-fifths (124,000) by the mid-1870s. At the same time the Sam Yup numbers fell from a bit below one-fifth (roughly 10,000) to less than one-tenth (11,000). Sze Yup controlled three of the Six Companies in San Francisco's Chinatown, while Sam Yup controlled but one.[56]

On the whole, the men of Sze Yup arrived with empty pockets and little education. Less indoctrinated in Confucian values and made to feel socially inferior in the Chinese context by the presence of the men of Sam Yup, they were to go the farthest in picking up American ways. For example, the converts Protestant missionaries made in Chinatowns came mostly from this group. On the other hand, Sam Yup emigrants boasted greater wealth, better education, and closer contacts with home through business, flourishing lineages, and officials in the provincial capital. The Sam Yup people, constituting a large proportion of the merchant population, congregated—perhaps in the interest both of business as well as of social solidarity—in San Francisco, the largest Chinese community and the one in most intimate contact with China. Unlike Sze Yup, Sam

Yup found the smaller settlements less hospitable and American culture less alluring. They appear to have regarded the secret societies with distaste, in part perhaps because the societies in the old country were associated with unsavory types—the déclassé, desperadoes, and social bandits. More to the point, however, was fear that these Sze Yup dominated societies were a potential threat to Sam Yup interests.

The Sinophobic Reaction

Overseas Chinese communities, such as the one in San Francisco, initially maintained the values and institutions of the homeland. But once established, the community's subsequent development from decade to decade depended in large measure on the response of the host country to this foreign presence. In some cases in Southeast Asia and Latin America tolerance had led to intermarriage and the decline of a distinct Chinese community as its inhabitants either assimilated or participated in a hybrid culture neither entirely Chinese nor native. In other cases, especially in Southeast Asian states struggling for independence and identity, native nationalists have attacked the Chinese as an economic and cultural threat. The Philippines, Thailand, and Indonesia each offer an instance of an indigenous culture giving up its accommodating attitude toward resident Chinese in favor of Sinophobic nationalism in the late nineteenth or early twentieth centuries. California's mid-nineteenth century reaction to the Chinese holds parallels with these latter cases where Sinophobic nationalism has turned against Chinese settlement. And not surprisingly, to carry the comparison further, the Chinese in California did what other overseas Chinese have done when placed under external pressure. They maintained or, for those who had already begun participating in the host society, recreated a tightly knit, autonomous community where they could preserve their cultural identity and insure physical security through group solidarity. But rather than lessen tensions, the development of these compact Chinatowns living a life of their own instead served to accentuate Sinophobia by veiling these strangers in mystery.

Sinophobia in California was couched in an idiom familiar to the economic and cultural nationalists of Southeast Asia. Economically the Chinese were seen as a peril to free American labor. The Chinese, to be sure, possessed virtues esteemed by Americans. Even

Sinophobes aghast at "the swarm of Chinese" entering the United States grudgingly conceded admiration for "their wonderful manual skill, their highly developed and intelligent imitative faculties, their tireless industry, and their abnormal frugality." But the "servile labor contracts" and the inherent constitutional indifference of the Chinese to material comforts perverted these virtues and gave them the power in a racial struggle for survival to drive out the whites. The Chinese were also thought to be a constant drain on the national wealth because their earnings in the United States went back to China to support their families or to purchase the Chinese produce that they favored. Finally, American Sinophobes stressed the peril the Chinese posed to the American culture. His refusal to assimilate by becoming a good Christian and citizen signaled his rejection of national religious and democratic ideals or at least an indifference that put in doubt the transforming power of American culture and the related myth of American mission. By clinging instead to his old culture, the Chinese not only established his unworthiness to participate in the American dream but posed a palpable threat. Sinophobes conjured up a stereotype of Chinatown as a fire hazard and a sink of immorality and filth. It polluted those whom it touched and threatened the health and safety of the larger community. Here was an invitation, if not an injunction, to eliminate the Chinese presence.[57]

The Sinophobic argument developed out of a strange mixture both of popular misconceptions and of fascination and fear that long retained its vitality with serious consequences for the immigrant and many of his descendants. It assigned the Chinese mysterious qualities—the ability to work longer and harder than whites on only small quantities of the worst food, all while living in dirty, disease-ridden subterranean chambers and indulging in vices of the worst kind. The total sexual power Chinese men exercised over prostitutes was a particular and recurrent source of wonder and comment by Sinophobes, while the supposed vulnerability of white women and girls to that power was a persistent cause for alarm. The seeming lack of freedom by both men and women in the Chinese community was the fault of cunning and unscrupulous merchant leaders. They commanded an invisible and mysterious network of despotic control that enabled them to direct and profit from the lucrative and sordid vice business as well as the labor of ordinary immigrants. They had in effect created a state within a state. Chinatown, a foreign transplant that had taken root in American soil, not only threatened to subvert liberal institutions and overwhelm free labor, but might also by its mere presence scare away desirable immigrants from other parts of the world and in time claim North America as an exclusive preserve of the Chinese.[58]

The Chinese were a threat to national dreams, particularly as they were to be realized in the context of the special California frontier. There Americans, whether of old stock or new, pursued a personal quest for freedom and economic opportunity. On a larger canvas their efforts were seen as a contribution to fulfilling the nation's destiny—the last, most brilliant movement in the westward march of civilization. But reality seldom coincided with the ideal. For the European-born, who constituted slightly over a third of California's work force in 1870, the contrast between promise and performance gave rise to particularly strong discontents. Like the native-born (40 percent of that work force), they had to bear the buffeting of a boom and bust economy. They too had to get along in a raw, violence-prone society. They, however, bore the additional burden of being themselves people in cultural transition and under nativist suspicion for their own lingering foreign ways. Especially the Irish, with their papist associations and their own secret societies, provoked criticism on this count. The anti-Chinese struggle became a way for white Californians to vent frustrations, assuage disappointments, work for a better future, and secure the comforts of group solidarity. For the recent arrival from Europe the Chinese also served as a foil. By attacking the Chinese menace he could demonstrate his own impugned loyalty and enhance his sense of self-worth.[59]

Sinophobia found its home in and provided an ideological glue for California's embryonic labor movement. Labor saw as its enemies the Chinese and the "capitalist monopolies" that controlled the railways, large landholdings, and mines. Though big capital was arguably the greater threat, labor came to concentrate its fire on the Chinese, and for good reasons. An attack on the Chinese (and only incidentally on the capitalists purportedly standing behind them) founded on broad nationalist grounds brought valuable allies to the side of organized labor including independent miners, small businessmen, some farmers and politicians, while a stark appeal to divisive class interests carried with it the risks of political isolation and possible defeat. The pattern of hostility and violence began to develop in California soon after the arrival of the first Chinese and moved spasmodically toward a climax nearly two and a half decades later. Sporadic attacks against the Chinese venturing out into the gold fields occurred in the 1850s, especially between

1852 and 1854. In the following decade the conflict carried over into newly opened underground mines where both Chinese and white wage earners worked. The anti-Chinese violence also began to build in an urban context. Rowdies, transients, merrymakers, and the unemployed had earlier made city life difficult for the Chinese. The rise of a class of wage earners in the 1860s gave a new complexion to the agitation, particularly in San Francisco, California's premier city with a quarter of the state's population. In 1862 the first anticoolie club appeared there, and in 1867 and again in 1869 it launched sharp attacks against the Chinese and made their presence a central issue in state politics. In the 1870s urban agitation turned violent, most dramatically in riots in Los Angeles in 1871 and in San Francisco in 1877.[60]

Precipitants were needed to convert the hostility of Sinophobes into violence. Downswings in the local economy appear to have been one cause. The incidents of 1854 followed the end of the surface-mining boom. Those in 1867 and 1869 coincided with brief economic downturns. The 1877 riots in San Francisco came one year after California had absorbed the full impact of the national depression originating in the East. The unemployed flocked into the city. More than 4,000 men applied for relief in San Francisco over a three-month period in 1877 and an estimated 30,000, perhaps 15 percent of the city's total population, went without work that year. Major influxes of "orientals," additions to an already highly visible Chinese community, also appear to have helped touch off attacks (as in 1852, 1854, 1869 and 1877). By 1870 the Chinese constituted one-twelfth of California's population and one-quarter of her work force. Spreading word of massive new additions to the Chinese population must have induced a sense of desperation and alarm that loosened normal restraints and opened the way for mob violence.[61]

Feeling against the Chinese in the 1870s was aggravated by the failure to administer a deadly blow to them despite the considerable political progress Sinophobes had made on home ground. They had first won the California Democratic party. By 1875 the state Republicans had also assumed a sympathetic stance out of political necessity if not conviction. In 1877 the Workingmen's party under the leadership of the Irish immigrant Denis Kearney emerged as the primary vehicle for Sinophobia, momentarily depriving the state Democrats of that role. The next year the Workingmen's party, then at the peak of its power, put together a new state constitution peppered with anti-Chinese clauses.[62] These clauses together with other measures dating back to the 1850s put the Chinese under occupational disabilities, imposing special licenses, fees, and regulations on Chinese laundrymen, fishermen, miners, and peddlers and excluding Chinese labor from employment on publicly funded projects or by corporations registered in California. They deprived Chinese of the right to become naturalized citizens, to testify in court against whites, or to attend white public schools. Local courts offered little justice to the Chinese when, as was frequently the case, both judge and jury shared community prejudices against them. Municipal authorities added to the burdens of the Chinese by applying, often to them alone, special regulations concerning health, prostitution, opium smoking, and gambling. Finally, the Sinophobes sought to solve the Chinese problem by blocking the arrival of new immigrants. One approach was to require transport companies to pay taxes or post bonds set at prohibitively high levels. The simpler alternative was a flat denial of the right of "Mongolians" to land in the United States. But of this array of state and local measures, part had simply proven unworkable, and much had been struck down by the higher courts, initially as violations of the equal protection clause of the Fourteenth Amendment. The conclusion of a treaty with China in 1868 guaranteeing the right of free immigration and the passage of the civil rights measures between 1868 and 1870 gave state and federal courts new and broader grounds for nullifying these regional solutions to the Chinese problem and added to the frustrations of western Sinophobes.[63]

Finally California, with the support of other western states, took its problem to Congress with the intention of securing federal legislation, silencing the courts, and abrogating the offending 1868 treaty. The campaign in Congress, led by California's own Senator Aaron A. Sargent, began auspiciously. The Senate agreed in 1876 to an investigation, if for no other reason than to put off the West Coast delegation's call for the more drastic course of treaty revision ending immigration. In reporting the results of the joint House-Senate inquiry the next year, Sargent ignored the balance of the testimony sympathetic to the Chinese and instead stressed the "great and growing evil" of Mongolian immigration and warned that unless checked it would overwhelm "republican institutions" and "Christian civilization" on the Pacific Coast. But the best the western states could get in their first two years of effort in Washington was the passage of a mildly worded request that the President find some way to modify the objectionable provisions of the 1868 Burlingame–Seward treaty guaranteeing the right of free immigration to Chinese. Neither national political party had yet wholeheartedly embraced their cause; and the Republicans in the White House, while paying some lip service to the plaint of the anti-Chinese forces, took no concrete steps to undo the controversial 1868 treaty or to move Congress to action.[64]

In their political drive, Sinophobes had already by 1876 overcome the pro-Chinese forces in California and elsewhere in the West. The debate between them had turned on the role of Chinese labor in the local economy and the nature of the Chinese community. Farmers, manufacturers, and railway builders as well as missionaries had come to the defense of cheap Chinese labor as a major contribution to regional development and to the competitiveness of California's light industry. In general, the Chinese did not supplant white labor but rather supplemented it and increased regional prosperity, so they argued to the catcalls of the unemployed who followed Denis Kearney and found in the Chinese a scapegoat for the blows administered by unseen economic forces. To counter the charges that the Chinese were unassimilable and hence a social danger, these opponents of exclusion compared the orderly, peaceful, and industrious Chinese laborers to the riotous sandlot Irishmen and the corrupt, irresponsible politicians who went along with them. The Chinese Six Companies were not despotic but voluntary associations and Chinese labor not enslaved but free and a far sight less dangerous than the nascent labor unions with their monopolistic designs to control the terms of employment. With the passage of time and an end to discriminatory treatment, the Chinese would embrace Christian and democratic ideals, and many would return to China to spread civilized ways and thereby contribute to "the rejuvenation of a nation." Missionaries were predictably the most persistent opponents of exclusion, continuing undaunted to press the case for the Chinese in the press, pamphlets, public hearings, and lectures long after business interests in California had taken refuge in a prudent silence.[65]

Sinophobia in the West met and mastered these opponents of exclusion obviously because it enjoyed popular support in the region but also because the defenders of the Chinese seemed to place their narrow, personal interest in the Chinese as converts, customers, and workers ahead of the white workingman's welfare. However, outside the West and especially in Washington, Sinophobes found themselves not only bereft of a broad base of political support but also confronted by a new set of arguments less easily neutralized by charges of self-interest. Exclusion, so its national critics responded, would violate treaty rights, undercut the national tradition of free immigration, and damage American commercial, diplomatic, and missionary interests in China. In all these ways a narrow sectional demand jeopardized broad national interests. The exclusionists thus went into 1879 still faced with the task of moving Congress to act on the Chinese problem and of convincing the major parties of the political risks they would run by neglecting it.[66]

Part Two

Emerging Patterns of Interaction

CHINESE HAD COME to the Rock Springs area of Wyoming in the 1860s to work on the railways and in the mines, but they had not moved into the town itself until 1876, when the Union Pacific had brought in about 400 Chinese to put down a protest by white miners. Soon finding themselves outnumbered two to one, white miners organized a branch of the Knights of Labor in 1883 and voiced complaints that the Chinese stood in the way of better pay and working conditions. Then through the summer of 1885 the arrival of fresh batches of Chinese gave rise to a new complaint—that the Chinese were getting the jobs while white newcomers went without employment. New resentments added to old set the stage for the eruption of violence that occurred on September 2.[1]

An early morning dispute in the mines between Chinese and white workers led to a brawl. A foreman intervened and temporarily stopped work, but tempers among the whites, most of them European immigrants, remained hot. Through the morning they gathered in town, mixing liquor with talk of expelling the Chinese. At two in the afternoon a hundred armed men marched across the railway tracks that marked Chinatown off from the rest of the town. The Chinese, thrown off guard by their misplaced faith in protection by company authorities, began to flee in panic. The lucky made good their escape into the nearby hills; others were gunned down or caught in (or forced into) the fires set by arsonists determined to expunge every trace of the Chinese presence. The proceedings assumed a holiday spirit, with the spectators, mostly women and children, laughing and cheering through the afternoon attack and then joining in the looting that followed. The violence would continue until after midnight. By evening Rock Springs offered a macabre spectacle. The smell of burnt flesh suffused the air and the light from still burning fires flickered against the nearby barren hills where stunned Chinese hid. A week later troops arrived to escort them back into Rock Springs. The ensuing scene of the survivors attending to twenty-eight charred and mutilated bodies, caring for their own wounds, and gathering up whatever was left of their possessions, struck Huang Hsi-ch'üan, the recently arrived Chinese consul, with its pathos. "It was a sad and painful sight to see the son crying for the father, the brother for the brother, the uncle for the nephew, and friend for friend."[2]

The Union Pacific at first took a tough stand against these excesses by labor. The railroad's president, Charles Francis Adams, Jr., warned the local management against excessive timidity in handling the union agitation. Forty-five implicated in the attack were discharged and sixteen arrested (though none received jail sentences). The surviving Chinese rebuilt their lodgings and went back to work under army protection. But labor had made its point. The Union Pacific hired no more Chinese and offered to send old employees back to China at company expense. Advocates of exclusion seized on the Rock Springs explosion as yet another illustration of their old contention that the persistence of an unassimilable Chinese element in American society made social conflict inevitable. "The Caucasion race will not allow itself to be expelled from this country, or totally impoverished," Aaron A. Sargent darkly warned, "without a bloody struggle. . . ." Back came the response that the principles of free immigration applied as much to the Chinese as to their European-born antagonists, but it was a voice whose diminishing strength in the political debate boded ill for the Chinese of Rock Springs and their compatriots elsewhere in the United States.[3]

The last third of the nineteenth century was a period of unparalleled richness in the cultural interaction between Chinese and Americans. But with greater contact went a higher incidence of hostility and violence. Working from the solid base laid earlier in the century, a growing number of American businessmen, missionaries, and government agents joined in the general foreign intrusion into China and in turn bore some of the brunt of the xenophobic Chinese reaction. Threatened first by popular hostility in China and then by the prospect of runaway great power rivalry, the open door constituency began to develop the argument that their nascent enterprises collectively formed a palpable national interest deserving of a supportive China policy. Over the same decades a far larger number of Chinese reached the United States seeking economic opportunity in the developing American West only to find their way blocked by a dangerously intense nativist sentiment. They too turned to their government for support against mob violence and legal discrimination.

It fell to American and Chinese policy makers and their agents overseas to deal with the tensions that arose from this commingling of cultures on both sides of the Pacific. They found it fiendishly difficult to bring disruptive social forces under control or even insulate preferred lines of policy from those forces. An American policy, once exclusively commercial in its concerns, had to adjust itself, first to accommodate the growing support at home for exclusion of the Chinese and later to defend missionaries who had taken up exposed positions in the Chinese interior. By degrees the largely passive open door policy which had earlier prevailed gave way to a new, more active policy, which combined demands for an open door in China with its antithesis, an insistence on closing the door on Chinese immigrants.

The central problem facing Chinese policy makers was the empire's crumbling security. Just emerging from two decades of massive internal upheaval and recently compelled to accept diplomatic relations on Western terms, China in the 1870s confronted a new threat as the powers began making inroads on her tribute states. Britain was ready to advance into Burma and Tibet. France eyed Vietnam. Japan directed her attention toward the Liuchiu (Ryukyu) Islands, Korea, and Taiwan. Russia menaced the inner Asian frontier. Once those powers had established themselves on China's periphery, they would next turn to China proper, seize concessions, and stake out spheres of influence. One policy response to this developing crisis in foreign affairs was to resort to the old stratagems of barbarian management. Li Hung-chang, the foremost exponent of a balance of power solution to China's immediate foreign policy problems, sought to involve the United States on China's side in East Asian international politics. Li's attempts, however, revealed Washington's caution in responding to the dangers of great power aggression against China and its commitment to an irritatingly one-sided policy. American policy makers expected Peking to accept exclusion and to sacrifice immigrant interests without a quid pro quo. Similarly, they held the central government responsible for antimissionary violence, while the provocative mission presence grew unchecked. The first sustained attempts at Sino-American cooperation in international politics thus foundered on the shoals of social conflict that contact had unleashed and that states of unequal power and divergent visions could not resolve to their mutual satisfaction.

Chapter Three

The Politics and Diplomacy of Exclusion, 1879–1895

IN CONSIDERING THE open door, historians have tended to set Americans on the one side as advocates and the Chinese on the other as opponents. What tends to be forgotten in this context is the immigration strand in the Chinese-American relationship. Through the latter part of the nineteenth century and into the early twentieth Chinese peasants, merchants, students, and diplomats became advocates of an open door of their own. Faced with mounting exclusionist sentiment in the United States, they appealed to their own government for support, and they joined with the American interest groups that made up the open door constituency in calling for greater reciprocity in American policy. The resulting struggle over immigration not only strained relations between the United States and China but also produced cleavages within each country which made a coherent policy toward the other more difficult to formulate.

The National Solution to the Chinese Problem

The debate over the Chinese ceased to be a merely regional issue in the 1870s and became a national one as well. The argument moved from the sandlots and newspapers of San Francisco to the halls of Congress and the councils of the Democratic and Republican parties as a frustrated California appealed to the federal government. The appeal was well timed, for it coincided with the Democratic party's recovery from its Civil War debacle and the creation by the 1870s of a rough equilibrium with the Republicans. Control of Congress shifted back and forth, and presidential elections were decided by narrow margins into the 1890s. In this fine political balance California had the good fortune of being a swing state whose support both parties would have to court.¹

The Democrats in rebuilding their party had exploited the Chinese issue to build grass roots strength in the West. Regional party leaders had taken the lead in advocating federal legislation to discourage the Chinese, and subsequently in the 1870s the national party had embraced the issue to secure the political allegiance of the West. But the issue also had a subtle appeal to Democrats of other regions—it evoked those formerly explosive issues of race and states rights with which the party was traditionally associated. In its 1876 presidential convention, party leaders emphatically condemned the do-nothing Republican Congress for exposing "our brethren of the Pacific coast to the incursions of a race not sprung from the same great parent stock," and called for congressional measures to stop the influx of coolie laborers and prostitutes. The Democrats won the West by a handsome margin in that election year and lost the presidency only after a deadlocked electoral vote left the choice to Congress. Democrats repeatedly pressed California's case in Congress, finally scoring a small success in 1879 when they put through a bill limiting to fifteen the number of Chinese passengers for each ship arriving in the United States. When the Republican President used his veto power to kill this first of a long string of congressional restrictions against the Chinese, the Democratic party pledged in its 1880 presidential platform to continue the struggle.²

For Republicans, exclusionist arguments came less easily. The Republican party had since its inception advocated free immigration, and one of its recent leaders, William Seward, had on his own initiative incorporated the principle in the treaty he had concluded with China's envoy, Anson Burlingame, in 1868.³ Though constrained by their political ideals and treaty obligations of their own making, Republican leaders nevertheless began in the mid-1870s to acknowledge that the West had a genuine complaint. In 1874 President Ulysses S. Grant took a swipe at Chinese immigration, alleging it largely consisted of laborers controlled "almost absolutely" by headmen and of women "brought for shameful purposes." But he made no specific legislative recommendations, promising only to support measures that Congress might devise. He did not again refer to the issue in any important address through the rest of his term, and the 1876 Republican party platform made only a mild recommendation that Congress investigate the impact of "Mongolians on the moral and material interests of the country."⁴

The Politics and Diplomacy of Exclusion

Rutherford B. Hayes, who had only narrowly retained Republican control of the White House in the disputed 1876 presidential election, also responded sympathetically to growing western pressure for exclusion. He deplored in particular the failure of these Chinese "strangers" and "sojourners" to integrate into American life. But he too refused to make any specific legislative recommendations and vetoed the first successful congressional attempt to supply a solution in 1879 (the Fifteen Passenger Bill). He had read on both sides in the controversial literature of the time, while earnest correspondents inundated him with advice. His ultimate decision to veto the measure reflected not a preference for the Chinese or indifference to the West but rather the greater weight he gave American commercial and missionary enterprise in China and national honor when treaty obligations were at issue. But Hayes did muse aloud for public benefit that the "apparent unassimilableness of the Chinese" might yet force him to undo the 1868 Seward-Burlingame treaty. Republicans, looking ahead to another close presidential race and unwilling to concede western votes to the Democrats, soon sought to erase the memory of the recent veto by finally making their peace with Chinese exclusion. Their 1880 platform conceded that "unrestricted immigration of the Chinese" was a matter of "grave" concern requiring federally imposed limits just as long as they were "just, humane and reasonable."[5]

The Fifteen Passenger Bill of 1879 had been a crucial victory for exclusionists. Though they had failed to override the presidential veto, they had nonetheless managed to transform the terms of the debate over Chinese exclusion from "whether" to "how," raising hopes for an early consummation of their demands. At the same time the growing prospect of Congress enacting restrictions over a presidential veto and the even more worrisome possibility of a Democrat riding into the White House on the Chinese issue set the Hayes administration off on a search for some device to satisfy western complaints without affronting China or violating the 1868 treaty.

A deal with Peking, it soon became apparent, offered the quickest and most satisfactory solution. In March 1879, just as Hayes began to look for a way out, George F. Seward, then minister to China, arrived in Washington with word that the imperial government might cooperate in preventing the emigration of undesirables—the diseased, paupers, criminals, and prostitutes.[6] Hayes and his Secretary of State, William M. Evarts, agreed to give Seward a free hand in sounding out Peking while at home they would hold Congress in check by suasion, by offering assurances of Chinese cooperation, or finally if necessary by wielding the veto yet again. Seward, however, was not the man to wrest concessions from the Chinese. He regarded West Coast fears of the Chinese as exaggerated, and he worried that American tampering with the structure of treaty rights would give the Chinese an excuse to do the same. Once back in China Seward made little progress in his talks.[7]

Former President Grant at last placed in the hands of the Hayes administration the key to a diplomatic solution. During a tour of East Asia in the summer of 1879 Grant had acceded to a request by Li Hung-chang that he mediate a Sino-Japanese dispute over the Liuchiu (Ryukyu) Islands, and in return secured a Chinese promise to negotiate treaty restrictions on the immigration of laborers. By December the Hayes administration, impatient for some diplomatic progress before the upcoming presidential election, had learned of Grant's bargain. Shortly afterwards it replaced Seward with James G. Angell, the president of the University of Michigan, who was to go to Peking to consummate the deal Grant had struck. Angell was himself a moderate on the immigration issue. Though he favored exclusion, he also denigrated western demands as excessive and demagogic and hoped for a settlement that would avoid affronting China or doing violence to treaty obligations. The administration appointed as Angell's cocommissioners John F. Swift, a Californian committed to exclusion, and William H. Trescott, a South Carolinian who shared Angell's general perspective. The mission was thus constructed so that Angell's views would prevail while California would retain a voice. Angell and associates reached Peking in early August 1880, opened talks in early October, and within a month had a treaty. It allowed the United States to "regulate, limit, or suspend" but not "absolutely prohibit" the entry of laborers, thus serving in theory at least to modify rather than overturn the Seward-Burlingame treaty so offensive to those of an anti-Chinese persuasion.[8]

The Angell treaty, ratified with alacrity by the Senate in 1881, was the first step toward the ultimate abandonment in 1924 of the American role as asylum to the world's oppressed and impoverished. In terms of Chinese-American relations, the treaty was a watershed—bringing to a close the era of free immigration and opening a contentious period in which the Chinese fought against an increasingly far-reaching and rigorous American exclusion policy shaped by popular prejudice and political expediency. Peking's acceptance of the new treaty terms was an unwitting but nonetheless signal contribution to the cause of exclusion.

The Politics and Diplomacy of Exclusion

Thenceforth it became more and more difficult for a President to control, let alone halt, the call for further restrictions. Hayes and his immediate successors in the White House, James A. Garfield and Chester A. Arthur, were moderates with an ambivalent attitude toward Chinese immigration. On the one hand, their image of the Chinese in the United States was negative, like the views of exclusionists in kind although not in intensity. They saw the Chinese as socially undesirable, unassimilable, and hence a potential menace to the nation. But they weighed against this social evil the value they put on the Chinese as a source of cheap labor and the concern they had about the reaction of the Chinese government. The policy they espoused reflected this balance. They acccepted the need for some restrictions, for example to protect white labor from unreasonable competition, or to forestall any inundation of immigrants. At the same time they insisted that restrictions be of limited scope or duration, accommodate (or at least not give needless offense to) the Chinese government, and be applied with care and restraint. Hayes' handling of the immigration issue was the epitome of the moderate approach. Ironically, it was the very success of that approach in getting the Chinese to accept a new treaty which weakened rather than strengthened the moderate position. Soon the White House would fall to the exclusionists. With Congress, treaty provisions, and in time even the President working for them, exclusionists would press their case with growing success from the early 1880s down to the early 1900s.

When Garfield followed Hayes into the White House in 1881 he had already displayed the ambivalence characteristic of the moderates. In 1875 he had toured Chinatown in San Francisco to see "some of the lower life exhibited there" and came away with a vivid if somewhat contradictory impression of the Chinese as "solemn, steady-going, quiet, intelligent, stupid, cultivated barbarians." Later, in 1879, he had supported Hayes' course of negotiations with China and Hayes' veto of the Fifteen Passenger Bill. Privately, however, Garfield admitted he was "anxious to see some legislation that shall prevent the overflow of Chinese in this country." Once nominated as the Republican candidate for President in 1880, Garfield publicly pronounced against "any form of servile labor" and against any immigrant group that was nonassimilable. But he urged Congress to await the outcome of the Angell mission, which was already under way and which he hoped would preserve "commercial intercourse" and the promise of "a great increase of reciprocal trade and the enlargement of our markets." Eventually some action might prove necessary to safeguard "the peace of our communities and the freedom and dignity of labor," but Garfield cautioned that Congress even then would have to move carefully, avoiding any "violence or injustice." [9]

Garfield's assassination forced his Vice President, Chester A. Arthur, to face a new flurry of anti-Chinese legislation. Arthur's private views are not known, but once he became President he played in his relations with Congress the restraining role characteristic of a moderate. On the two occasions in 1882 when he vetoed congressional immigration measures, he acknowledged that unrestricted immigration was harmful, especially to white labor which faced "Asiatic competition." But the continued presence of the Chinese, Arthur contended, was in some cases beneficial to the nation and was linked to continued American commercial opportunity in the East, "the key to national wealth and influence." He held out to San Francisco in particular "an incalculable future if our friendly and amicable relations with Asia remain undisturbed. It needs no argument to show that the policy which we now propose to adopt must have a direct tendency to repel Oriental nations from us and to drive their trade and commerce into more friendly hands." [10]

The first of Arthur's major tests with exclusionists in Congress came when he received from the Hill immigration legislation which ostensibly fell within the terms of the Angell treaty. It prohibited the immigration of Chinese laborers (whether skilled, unskilled, or engaged in mining) for twenty years, excluded the Chinese from citizenship, and created a system of certificates to identify both Chinese eligible to enter the United States and laborers already in residence. Arthur induced Congress to accept a "shorter experiment" by reducing the period of the 1882 act to ten years from twenty and to substitute "suspend" for "prohibit" (on the grounds that the latter violated the terms of the Angell treaty). He could not, on the other hand, get Congress to strike out the congressional system of registrations and certificates, which he thought impractical and against the spirit of "liberal institutions."

As it turned out, Arthur had taken the last presidential stand against the exclusionist drive in Congress. He did nothing in 1884 when, on the eve of a presidential election, Congress broadened the definition of laborers to include peddlers, hucksters, and fishermen; applied immigration restrictions to all persons racially "Chinese" (whether subjects of China or not); and made the 1882 system of certificates more complex and the certificates themselves more difficult to obtain. Grover Cleveland's election in the fall of 1884 not only restored Democratic control of the White House for the first time since the Civil War but it also maintained the fair prospect for the enactment of further federal measures against the Chinese. During the Cleveland administration as well as that of his Republican successor, Benjamin Harrison, Congress brought forward new legislation which these Presidents supported rather than opposed.

When Cleveland entered the White House, his own views were vague even though his party's position was clear enough. In the presidential campaign and in his inaugural address Cleveland had only obliquely referred to the Chinese in his calls for protection for American labor against servile immigrants who did not assimilate and whose ways were "repugnant to our civilization." He sent his envoy off to China in 1885 with no more information on the new administration's stand on immigration than that it intended to "obey the constitution and the laws." In his report to Congress at the end of his first year, Cleveland evenhandedly condemned race prejudice while affirming the government's right "to prevent the influx of elements hostile to internal peace and security" and promising to consider any congressional measures that did not violate treaty obligations.[11]

Over the next several years Cleveland seemed to hold to the line of presidential moderates. Like Hayes before him, he looked to diplomacy and a new treaty with China to defuse electoral discontent and at the same time to preserve Chinese sensibilities. In December 1887, with less than a year to go before he would again have to face the electorate, Cleveland urged his Secretary of State, Thomas Bayard, to come up with a treaty to end the "frauds and evasions" practiced by the Chinese. After several months of talks with the Chinese minister, Bayard got the treaty the Democratic administration needed in the upcoming presidential contest. However, Cleveland's strategy was to founder on Chinese dissatisfaction with the treaty. For four months Peking delayed a decision on ratification while in the United States the election grew nearer, Congress grew restless, Republicans began to formulate their own solutions, and Cleveland and party stalwarts became ever more impatient. In the previous election of 1884 Cleveland had lost Colorado, Nevada, and Oregon by close margins. It was a loss he could ill afford again, and party leaders now urged that he move fast and decisively to keep the Democrats on the right side of the Chinese question.[12]

In early September Cleveland finally shifted ground. After his renomination he condemned the Chinese as "an element ignorant of our constitution and laws, impossible of assimilation with our people, and dangerous to our peace and welfare. . . ." With rumors in the air that the Chinese government had indeed rejected the treaty, he withdrew his own support from it and publicly endorsed a measure advanced in the House by William L. Scott of Pennsylvania, chairman of the National Democratic Campaign Committee and a friend of Cleveland's. It flatly prohibited the entry of laborers and denied reentry to laborers who had left or might in the future leave the United States (unless they left behind relatives or at least $1,000 in property). After having participated behind the scenes in writing and putting the Scott Act through Congress, Cleveland signed and returned the measure to Congress with an election year message that described Chinese immigration as "unwise, impolitic, and injurious" and laid the blame for this stringent and unilateral measure on the failure of the Chinese government to cooperate and on "the mercenary greed of parties who were trading in the labor of this class of the Chinese population."[13]

The now isolated voice of moderation in the administration was Bayard. Cleveland had repudiated Bayard's treaty and acted in Congress against Bayard's better judgment. Though he spared the President direct criticism, the Secretary of State bared his disgust with "demagogueism" on the Pacific Coast and "small politicians" in Congress. He condemned the Scott Act as a violation of "international courtesy, good faith, and self-respect," and fumed over "the consequences of subjecting our international relations to the fires of miserable partisan needs and exigencies." On the other hand, Democrats in California exulted over the passage of the Scott Act, certain that it was the key to victory. Election day was to prove them wrong. Cleveland lost not only California but all four western states and with them the election. Perhaps Cleveland had delayed too long before embracing exclusionist demands. Or perhaps the protectionist West had already decided against giving a low tariff President a second term, particularly when there was nothing to distinguish his final stand on the Chinese issue from that of his Republican opponent.[14]

Benjamin Harrison, the victor in the 1888 contest, was another moderate on immigration who had succumbed to exclusion under the weight of party pressure and personal political ambition. In the Senate in 1882 he had agreed with President Arthur that congressional legislation violated the terms of the 1880 treaty with China. Harrison nonetheless desired restrictions on Chinese labor and opposed giving the Chinese the right to naturalization. As Harrison's interest in the presidency developed, party leaders urged him to abandon the moderate stance and convince the West of his essential sympathy for their complaints. In July 1888 a Republican leader in California tried to communicate to him "the whole extent and force of the degradation and outlandish nastiness that is conveyed to our minds . . . by the term 'Chinaman'. On this coast to compare a man to a Chinaman is more insulting than to compare him to a 'Nigger' by far." By the campaign year of 1888 Harrison had converted. At the very time the Scott Act was moving through Congress, Harrison accepted his nomination and with it the "duty to defend our civilization by excluding alien races whose ultimate assimilation with our people is neither possible nor desirable. . . ." Harrison concluded that the immigration issue was now settled beyond debate and that he would execute the law and accept any additional action necessary to prevent further Chinese evasion or stop any new influx.[15]

Harrison, the first man to enter the White House a firm exclusionist, gave guarantees against any backsliding on the Chinese issue by making James G. Blaine his Secretary of State. Brought into the administration because of his party standing and service, Blaine was an old and unalloyed foe of the Chinese. Though his political base was Maine and not the West Coast, he nonetheless appreciated the depth of feeling there and urged Republicans to recognize that "the three Pacific states will be largely, if not entirely, controlled" by the question of the Chinese, who had (in his opinion) sown in San Francisco "the seeds of moral and physical disease, of destitution, and of death." Blaine had scored the Chinese as sojourners and servile laborers and warned that "Mongolian" labor, if allowed to continue, would in time displace American workers and provoke uncontrollable racial conflict. In Blaine's colorful phrase of 1879, Americans had to choose whether the Pacific Coast would be dominated by "the civilization of Christ or the civilization of Confucius." Blaine, sometimes depicted as a staunch commercial expansionist, did not allow concern for the China trade to dilute his Sinophobia. He echoed the view of West Coast exclusionists when he argued that the value of the China market was not great enough to offset the harm done by a "vicious" immigration. It would be better to lose the trade than accept the Chinese. He was unmoved by the arguments of missionaries, whose efforts at conversion had borne so little fruit. Blaine had supported restriction in its first major test in Congress in 1879 and had had the pleasure of overseeing the ratification of the Angell treaty when he had served as Garfield's Secretary of State. Now once more in that office, he could be expected to put American diplomacy in the service of exclusion.[16]

With exclusionist sentiment dominant in both the White House and the State Department, Congress worked unimpeded to tighten up existing regulations against the wiles of the resourceful Chinese. The Geary Act of 1892 (named for the Democratic congressman from California) substantially rounded out the body of anti-Chinese legislation. Aside from extending for another ten years all of Congress' anti-Chinese measures passed over the previous decade, it stripped Chinese in the United States, whether citizens or not, of substantial legal rights, and it required all Chinese in the United States to obtain from the government and carry with them at all times a certificate attesting to their right to reside in the United States. Under the law any person of Chinese descent found in the United States was subject to imprisonment at hard labor followed by deportation unless the Chinese could produce positive proof of legal residence.[17]

When Cleveland returned to the White House in 1893, he demonstrated how well he had learned in his first term the political explosiveness of the Chinese question. Although he allowed his cabinet to temper administratively some of the harsher features of the measures passed by Congress, he would not allow them to tamper with the basic structure of exclusion. He insisted that his associates handle congressional opinion with kid gloves, while he himself kept a safe distance from the problem. When the Cleveland administration finally got around to negotiating a new immigration treaty with the Chinese, it turned out to be all that Congress might reasonably have wanted; it easily cleared that body with only western diehards still calling for more.[18]

By the mid-1890s California and the rest of the West had gained much of the federal legislation they had begun calling for as early as the 1870s. One explanation for their ultimate victory is the disproportionate political influence the party balance gave to a region intensely opposed to the coming of the Chinese. An alternative view, of recent currency, is that exclusion was not simply a regional cause that fortuitously gained national approval. Rather, exclusion became national policy (so this latter view goes) because the racial hostility evident in the West was also at work in the rest of the country.[19] Beyond question, almost all articulate Americans—regardless of region or depth of interest in good relations with China—looked down on this immigrant population which had rejected the melting pot for ways that they unquestionably regarded as inferior if not base. The most flattering picture a late nineteenth-century George Gallup might have elicited from any substantial number of Americans was of the Chinese as an industrious, curious, but backward folk whose inability to participate in the American experiment posed, potentially at least, social and economic problems to the nation.

This pervasive view (arguably racist in character) did not, however, necessarily translate into support for exclusion, particularly as the West defined it. Some Americans, including notably a string of presidents, did indeed look down on the Chinese; however, moved by other, stronger considerations, they also opposed exclusion in principle or fought specific aspects of proposed exclusion measures. Was the Chinese threat so great that it necessitated immediate, drastic, and offensive solutions? Might exclusion not harm present American interests in China and diminish future cultural influence and economic opportunities? Did not treaty obligations and the principle of international reciprocity commit the United States to the free movement of Chinese? Some said yes, some no, while still others with no deeply felt opinion on the Chinese had difficulty mustering a reply. This tendency to indifference despite the widespread acceptance of an unflattering picture of the Chinese was strikingly evident in Congress, where the struggle over exclusion had focused. The Fifteen Passenger Bill passed the House in 1879 with 155 in favor, 72 opposed, and 61 abstaining. Thus on that occasion slightly more than one in five members of the House had no coherent or strong personal or constituent concern to register either pro or con. Three years later, after the Angell treaty cleared the way for action, the Senate voted in favor of immigration restrictions 32 to 15 with 29 (38 percent) of its members still staying to the sidelines. In 1888 the Scott Act divided that same body 37 to 3 leaving 36 votes (or 47 percent of that body) unaccounted for. If Sinophobia had swept the country, a substantial body of congressmen from the East Coast, the deep South, and the Midwest had shown themselves at these crucial junctures surprisingly unmoved if not firmly opposed.[20]

To suggest a direct relationship between an unfavorable image of the Chinese and the enactment of exclusion is appealing but only partially helpful. The equation fails to take into account the varying intensity of the anti-Chinese image from person to person and region to region and the extraneous considerations that in some cases held men back from supporting exclusion and in others drove them into opposition. The by no means negligible result of that variation was that exclusionists had to fight for two decades in Washington to secure a slightly watered down version of their program, while the opposition retained enough vigor to resume the debate over exclusion after the turn of the century.

Peking Responds to a Closing Door

The outcome of the exclusion debate among Americans was bound to bear principally and directly on the men of the Canton delta caught up in the myth of the mountain of gold. How heavy a blow exclusion would deal to their hopes and interests depended in turn in substantial measure on the ability and willingness of the imperial government to interpose its protection.

Peking's position toward Chinese in the United States developed within the context of attitudes and policies that applied to overseas Chinese generally. On a formal level, the Ch'ing statutes at mid-century still carried a prohibition on emigration and threatened violators with death. The flight of anti-Manchu resistance to Taiwan and Southeast Asia in the latter part of the seventeenth century had resulted in the dynasty's associating overseas Chinese with subversion and banning their return. The formal prohibition also reflected a Confucian conviction that emigration was an unfilial abandonment of family for economic advantage. Finally, the prohibition coincided with a Sinocentric indifference to men who had placed themselves outside the circle of true culture. This indifference was evident in a 1729 edict of the Yung-cheng Emperor: "Those subjects who willingly abandon fatherland for personal gain deserve no sympathy." It still prevailed among officialdom in the mid-nineteenth century. "When the emperor rules over so many millions, what does he care for the few waifs that have drifted away to a foreign land? . . . The emperor's wealth is beyond computation; why should he care for those of his subjects who have left their home, or for the sands they have scraped together?" Even as late as the late 1860s the Chinese foreign office opposed dispatching envoys abroad on the grounds that China had no overseas interest worth looking after that could not be protected from Peking.[21]

The formal prohibition, however, was never uniformly enforced and by mid-century had become a dead letter as Chinese began to emigrate abroad in large numbers and as officials in Kwangtung and Fukien came to appreciate the outward movement as an important safety valve for the growing population in their provinces. Emigration was a path of escape for the destitute who might otherwise become homeless wanderers or outlaws. Since by one contemporary Chinese calculation only 30 to 40 percent of emigrants ever returned, the relief was thought to be more than just temporary. Emigration bestowed a second blessing as a source of income for those who stayed behind (annual remittances in the mid-1880s came to an estimated $20 million) and as a stimulus to the trade of southern coastal ports such as Canton and Amoy. Emigration was, in any case, largely beyond the control of Chinese officials even if they had wished to prohibit it. Macao and Hong Kong, both major centers for the transport of Chinese abroad, were in foreign hands. The creation from 1842 on of treaty ports, each with its own foreign enclave, further impeded direct official control.[22]

Officials nevertheless took a growing interest in the movement of Chinese abroad. In the 1850s, while Peking still refused to recognize the existence of the coolie trade, provincial authorities sought to eliminate abuses associated with it, sometimes with the cooperation of British and American officials. The first set of Chinese regulations appeared in Canton in 1859; another in Amoy the next year. In the 1860s the central government, finally acknowledging the existence of emigration, joined the effort begun by provincial officials by proposing international agreements protecting contract labor. In the 1870s Peking brought the campaign against the coolie trade to a successful conclusion with the close of Macao (1873) and the signing of agreements with Spain and Peru (1874 and 1877).[23]

At the same time the imperial government also began to extend its protection to Chinese already settled abroad. The most important step in this direction was the creation of a foreign service. Some leading Chinese officials had first given serious thought to the dispatch of ministers abroad in the 1860s but without reference to the needs of overseas Chinese. Later, during the last stage of the campaign against the coolie trade, Peking took a tentative step in the direction of foreign missions when it sent officials to look into the dismal condition of Chinese laborers in Peru and Cuba. From the reports of these emissaries and other diplomats traveling abroad, Peking discovered that a large body of overseas Chinese urgently needed diplomatic representation and protection. Its response was to put together from the late 1870s on a network of consuls and ministers to cover the major concentrations of overseas Chinese in the Americas, Southeast Asia, Hawaii, and Japan.[24]

The logical culmination to Chinese policy toward emigration and the overseas Chinese came in the 1890s. In 1893 the court annulled the anachronistic prohibition on emigration, and six years later an imperial decree commending overseas Chinese for their "unshakable loyalty" made clear that the new official attitude was one of solicitude, not indifference or hostility. No longer were overseas Chinese contemptible inhabitants of "distant and solitary places" who faced prosecution or extortion if they returned home. Successful defense of the interests of overseas Chinese would, some officials argued, not only enhance China's battered international reputation but also secure a potentially valuable asset in building a China of wealth and power. Many overseas Chinese had amassed fortunes that could be invested in arsenals, mines, shipping, and other developing enterprises. They could also be counted on to help financially in times of international crisis and domestic calamity. Overseas Chinese possessed skills in such technical areas as diplomacy, law, foreign languages, industry, and finance which were in short supply within the official bureaucracy and which the government might tap to its advantage. Peking now charged provincial officials to safeguard overseas Chinese on their return home, created merchant protection bureaus in Canton and Amoy, and bestowed honors and rank on merchant leaders of the overseas communities.[25]

Chinese in the United States, who had not suffered from the abuses of the coolie transport system, followed behind other overseas Chinese communities in establishing a claim to the government's attention. The visit of the Burlingame mission to the United States in 1868 seems to have marked the beginning of a shift. The treaty negotiated in Washington that year had in itself no important immediate consequences for Chinese policy. Its provisions included a recognition by China of the right of its subjects to emigrate, but it was not the first time China had accepted in a treaty what her legal codes denied. The mission was of more practical consequence for the exposure to the problems of the Chinese in the United States gained by the two Chinese officials accompanying Burlingame. They returned home with recommendations for creating a permanent staff in the United States to look after Chinese residents and to ensure their good conduct and loyalty.

Finally in 1875 the Chinese government decided to establish a legation in the United States, essentially as a byproduct of the campaign against the coolie trade in Peru and Cuba. The prime mover was Li Hung-chang, who was disturbed by evidence of abuse of Chinese laborers in those two countries. In December he convinced the foreign office to post a single diplomatic mission abroad to deal with the coolie question and at the same time to establish a legation in the United States. Li nominated Ch'en Lan-pin as minister (1878–1880) and Yale-trained Yung Wing as his associate. It was not long after setting up his office in Washington in 1878 that Ch'en himself discovered the heavy load of work relating to the immigrants in the United States and urged Peking to establish consulates to attend to some of it. By 1883 consulates were in operation in San Francisco and New York, and a "merchant supervisor" (also under the control of the Chinese legation in the United States) selected for the Chinese in Hawaii.[26]

Once brought into contact with the Chinese community in the United States, officials began to emphasize that it was, like other overseas communities, an asset to China. Emigration to the United States not only relieved population pressure in Kwangtung but insured a steady flow back to the Canton area of remittances (probably worth a bit over a million dollars annually in the late 1880s). The value of trade received by Hong Kong and Canton on account of the American Chinese was put at that time at a minimum of several million dollars per year. Further, Chinese in the United States, demonstrating their loyalty, gave generously in times of crisis at home—for example, half a million dollars during the Sino-French War (1884–1885) and over $70,000 for emergency relief in Kiangsu, Anhui, Honan, Shantung, and Chihli in the latter half of the 1880s.[27]

The Chinese minister, the principal official charged with carrying out China's policy of protection in the United States, occupied a difficult position. While in relation to Americans he confronted an increasingly strident and powerful demand for exclusion, in relation to his own countrymen, whom he was supposed to protect, his presence raised expectations he could not possibly meet. Most ministers came from the Canton region and were as a consequence peculiarly susceptible to the complaints of their fellow provincials, both those residing in the United States and their relatives and associates at home. Whenever they felt seriously abused, threatened, or discriminated against, Chinese in the United States would appeal to the minister and, failing a satisfactory solution (frequently the case), try to force his hand by attempting to mobilize opinion back home. The natural alliance between the minister and his fellow provincials was further imperiled by the minister's hauteur in dealing with his social inferiors, the merchants and laborers who made up the American community. The ultimate result was to make the minister seem to emigrants a pretentious yet inert or impotent agent of an inattentive government, while the minister came to regard his countrymen as ungrateful troublemakers.

The establishment of the legation in the United States in the late 1870s coincided with the collapse of the policy of free emigration accepted by both China and the United States in the 1868 Seward-Burlingame treaty. Li Hung-chang had provided an opening for American exclusionists in 1879 when he had promised to accept some treaty restrictions on the emigration of laborers in exchange for Grant's involvement in the Liuchiu dispute. The ensuing negotiations in 1880 caught the Chinese foreign office at a bad time, for it was distracted by the lingering Liuchiu dispute with Japan and even more by the showdown with Russia over Ili, an inner Asian frontier area. (The Chinese government had just repudiated an agreement recently concluded by one of its representatives in St. Petersburg and was beginning preparations for war while awaiting the outcome of a new embassy to Russia.) Fears that China's two antagonists might combine against her made the capital particularly uneasy through the late summer and into the early autumn, when the talks with the Americans began. Moreover, officials in Peking were well aware of their weakness vis à vis the United States. They had already learned from the legation in Washington that the anti-Chinese movement whipped up by the "Irish party" was running too strong for the President to resist much longer, and Angell and his associates brandished the threat that Washington would unilaterally exclude Chinese laborers if China did not now cooperate in putting together a satisfactory bilateral agreement.

Nonetheless, the two senior members of the foreign office representing China in the treaty talks—a sixty-year-old Chinese, Li Hung-tsao, and a Manchu, Pao-yün—sought from the outset of the talks in October to keep concrete concessions to the Americans to a minimum.[28] Drawing on information supplied by George Seward, the outgoing American minister, the Chinese responded with a defense of immigration as a source of industrious and cheap labor for the American economy and as a means of livelihood for China's excess population. They warned that exclusion would harm a mutually beneficial economic relationship between the United States and China just as it would violate the traditional American policy of free immigration, only recently written by an American Secretary of State into the treaty he had concluded with Burlingame. But if the United States could impose restrictions on Chinese in spite of treaty commitments, why (they wondered aloud) could China not do the same to Americans? The Chinese concluded by offering proposals (discussed earlier with Seward) that would restrict emigration of certain classes of undesirables, not including laborers.[29]

The issues narrowed through the latter half of October. Angell and his associates responded that the Chinese were not negotiating in good faith and insisted that the United States would exclude Chinese labor one way or the other. Finally the Chinese capitulated. As talks proceeded, the Americans also prevailed in the way exclusion was to be defined—by enumerating those classes of Chinese who would have the right to enter the United States (only those intent on "trade, travel, study or curiosity") rather than enumerating those to be excluded, as the Chinese preferred. The foreign office saw no way around making these concessions. The Chinese in the United States "have recently become so numerous as to pose difficulties for that country. . . . If we now maintain the old treaty without change, then the Chinese will increase in number. And if some calamity is provoked, not only will Chinese laborers who leave in the future implicate those already there but moreover we fear that merchants and others will also be affected on account of the laborers." In exchange for giving way on these two crucial points, the Chinese secured what they regarded at the time as valuable counterconcessions. The Americans gave assurances that they would not flatly prohibit the immigration of laborers and that they would protect all resident Chinese. Angell, reassuringly sized up by Li Hung-tsao as "simple and honest" at least as Americans went, also promised that his government would proceed in "a friendly and judicious manner" in formulating specific regulations and restrictions to control Chinese immigration and that it would allow Peking a veto over objectionable provisions.[30]

For the Chinese this bargain looked better on paper than it later turned out to be in practice. The American promise to devise "reasonable" immigration regulations carried quite different meanings among diplomats in Peking on the one hand and, on the other hand, among congressmen in Washington, who were to make the regulations, or immigration inspectors in San Francisco, who were to interpret and enforce them. When Peking eventually sought to exercise the veto the treaty had given them, they found themselves, to their considerable dismay, ignored in Washington. The Chinese were equally disappointed by the failure of the Americans to observe a variety of lesser commitments. The Chinese and Angell had tentatively agreed to a five-year limit to this experiment in immigration control. However, Congress voted legislation running for twenty years and only settled for ten under the pressure exerted by President Arthur. The treaty definition of those Chinese subject to exclusion also gave rise to dispute. The Chinese were to discover in later years that immigration officials were to reject even Chinese of the exempt class by making a laborer out of anyone who they could establish engaged in literally any kind of manual work. Finally, the American promise of protection embodied in the treaty made Chinese in the United States no safer. The promise was not only redundant since the United States was already committed to protection by the provisions of the Seward-Burlingame treaty, not to mention general principles of international law, but it was also, as it turned out, worthless since Washington's record on defending Chinese residents was no better after the treaty than it had been before. Limited federal jurisdiction over the states tied the hands of the executive, while the political explosiveness of the Chinese question rendered statesmen, not to mention mere mortal politicians, unusually timid.

Chinese protests spanning three decades were to fail to deter American authorities from exploiting to the full their own reading of the Angell treaty. That treaty enabled the federal government to join the western states in weaving a net of administrative restrictions against Chinese immigrants and residents. The treaty also hampered the Chinese government's attempt to fight back. No longer able to claim a clear treaty right to immigration, Peking and the minister in Washington would henceforth have to duel with a resourceful bureaucracy over ambiguous treaty terms. In repeated tests of will the weaker power would usually lose. The treaty of 1880, itself a sad tale of misunderstanding and deception, introduced a no less sad period in which the continuing attack on Chinese immigration began to poison diplomatic relations.

Gathering Chinese distress over American policy turned to outrage after the Rock Springs massacre in September 1885 and the anti-Chinese outbreaks in Seattle, Tacoma, and other points up and down the coast which were its sequel. At the end forty Chinese lay dead and property losses had climbed to half a million dollars. News of these attacks set off an intense reaction in the Canton area. Foreigners were threatened and the local Chinese press demanded that the United States make immediate diplomatic redress or face popular reprisals. The situation reached crisis proportions by February of the next year as complaints reached Canton from Chinese in the United States that Washington was not making amends for old wrongs while they suffered new ones at the hands of mobs. Tales of abuse began to filter inland and in the spring contributed to a major antiforeign riot in Chungking.[31]

The seething local scene forced the governor general in Canton, Chang Chih-tung, to take an interest in resolving the crisis and preventing local violence against foreigners that might precipitate a dangerous confrontation with the powers. Chang relayed popular demands to Peking and urged the foreign office and the Washington legation to act quickly and firmly in defense of the Chinese in the United States. The demands that Chang wanted the United States to meet were sweeping—settlement of outstanding cases, payment of indemnities for losses and for suffering, punishment of wrongdoers, suppression of further agitation, and a special proclamation of regret from the President. But Chang argued that even so the demands were reasonable since they were no more than what foreigners expected of China in comparable or even less serious cases. To hasten action he played on American fears, warning that their interests in China would suffer unless their government started paying more attention to the safety and welfare of overseas Chinese. His public assurances that Peking was working for redress restored momentary calm to Canton in the spring of 1886.[32]

Diplomatic relief, however, was slow in coming. The Chinese minister in the United States at the time of the massacre was Cheng Tsao-ju (1881–1885). He immediately demanded indemnity for losses and punishment of the guilty at Rock Springs. Washington flatly refused any formal apology or acknowledgment of responsibility and allowed a year and a half to elapse before providing $148,000 in restitution for the survivors. No punishment was ever meted out to those directly responsible.[33]

Frustrated in its efforts to secure protection and redress for Chinese in the United States, the legation had even before the shocking outbreaks of 1885 and 1886 evolved an explanation of American nativism, which placed the chief blame for the difficulties of the Chinese on an "Irish party." The party was made up of the "cunning and cruel" members of the lower classes, driven to agitate against the Chinese because of economic competition. Officials and gentry on the West Coast had after some resistance bowed to popular pressure, and officials in Washington had soon done the same. Since in the Chinese political lexicon a party was by definition a self-seeking group that disrupted and debased the affairs of state, the legation's interpretation cast the American political system in a bad light. Even so Chinese diplomats expected cooperation from the government in Washington. When they failed to get it, they were perplexed and dismayed, particularly by the President's seeming inability to overturn domestic regulations in clear violation of treaties and to give the Chinese the protection guaranteed them at the very time that Americans required of China a scrupulous respect for treaty rights under strikingly similar conditions.[34]

From this situation Minister Cheng Tsao-ju drew the conclusion—accepted by his three successors—that China would herself have to undertake restriction of emigration to the United States. The conclusion was realistic because China's weakness made resistance impractical and because the violence of the anti-Chinese movement, which Cheng described as "like an inextinguishable prairie fire," might become uncontrollable. In January and February 1886, as the Rock Springs crisis reached its peak, Cheng saw disaster looming ahead unless China took some steps to remove the Chinese in the United States from danger. Repeated attacks together with ever more restrictive regulations would ravage the Chinese population. Some would be killed; others would flee home; few would be able to stay and make a living. Cheng now proposed that China restrict emigration of laborers and that the legation give closer supervision to the Chinese already in the United States. That policy, Cheng argued, would save the unwary from harm and put a stop to irritating attempts by immigrants to circumvent American regulations and hoodwink American officials. It might also take some wind from the sails of the anti-Chinese forces and perhaps even forestall dangerous new American initiatives. In properly handled negotiations, Cheng advised, the Chinese might even be able to trade self-restriction for indemnity for past losses and renewed promises of protection.[35]

In the spring of 1886 Minister Cheng presented his ideas to the Chinese foreign office. In August it approved and presented Cheng's proposals to the Americans. The detailed negotiations fell to Cheng's successor, Chang Yin-huan (1885–1889), a forty-eight-year-old native of Nan-hai district in the Canton delta.[36] Before leaving China, Chang had accepted the recommendations of his predecessor, Cheng Tsao-ju. He had consulted with his political patron, Li Hung-chang, on self-restriction and obtained his endorsement. He had then traveled south to Canton to secure Chang Chih-tung's concurrence. At the same time he had canvassed merchant groups in Canton and

Hong Kong with an interest in the exclusion question. Having sought support at home for what was bound to be a controversial initiative, Chang set out for Washington where in January 1887 he opened talks with Secretary of State Bayard. Bayard would not make the concessions Chang wanted—new American guarantees on protection as well as indemnity for losses—in exchange for self-restriction by China. It finally took the intervention by Cleveland, eager to keep the Chinese issue out of his reelection campaign, to break the deadlock. By March 1888 the minister and the Secretary of State had worked out a treaty which in large measure embodied the Cheng Tsao-ju plan of self-restriction, indemnity, and guarantees of protection. The Senate took up the treaty at once and tacked on two amendments, which Chang Yin-huan accepted with equanimity even though one spelled the exclusion of some 20,000 laborers otherwise entitled to return to the United States after a visit home.[37]

The legation's plan to save the Chinese community suffered a fatal reverse when the terms of the proposed treaty reached the Canton area. In July and August of 1888 the treaty was repeatedly attacked in public meetings and the press. A mob gathered before Chang Yin-huan's house in Canton to vilify the minister for betraying his fellow provincials. Chinese in the United States joined in the cry against the treaty. All this clamor caused Chang Chih-tung and Li Hung-chang second thoughts. After a hard look at the treaty, they now found fault with the twenty-year term the treaty was to run and the bad precedent it might set for excluding Chinese from Australia or Southeast Asia. Chang, still responsible for affairs in Kwangtung, worried about the local economic consequences once the treaty came into force, obstructing emigration and diminishing the flow of remittances. Shaken by protests as well as by Chang's and Li's doubts, the foreign office withdrew its support of the self-restriction plan and withheld its approval of the treaty, apparently on Li's advice that a Republican victory in the upcoming American election might permit China to get better terms in a renegotiated treaty.[38]

It was this Chinese dissatisfaction and delay which gave Democrats in Congress the excuse to push through, with belated assistance from President Cleveland, the restrictive Scott legislation of 1888. The latter, supplemented by the Geary Bill in 1892, placed ever higher obstacles in the way of immigrants and made ever more precarious the place of Chinese residents. Against this policy the legation, joined occasionally by the foreign office and Li Hung-chang, could only register the familiar complaint that the powers did not tolerate comparable mistreatment of their nationals in China. Why should China not also expect prompt and full redress from the United States when the rights of overseas Chinese were violated? While in private Chinese officials characterized American proceedings as shameful, they publicly continued to appeal in the name of fairness and equity for a more tolerant American policy. Continued abuse, they now began to add in exasperation, would justify retaliation against American commercial and missionary interests in China.[39]

Officials in Washington responded to Chinese protests with an ingenious set of constitutional arguments and bureaucratic stratagems which made everyone and hence no one responsible for the Chinese problem. The executive branch was the first line of defense, and there buck-passing was the rule. When approached by the Chinese legation, the State Department pointed to the Treasury as the department responsible for immigration. The Treasury Department in turn could argue that it merely took orders from the President and Congress. The claim to a separation of powers within the federal government provided a second line of defense against the Chinese. When the Chinese complained against congressional transgressions, the State Department replied that as an instrument of the executive it could not intervene, that China would have to look to the judiciary to strike down bad laws, and that in the meantime the executive was obliged to enforce them. The Supreme Court joined the game by disclaiming responsibility and pointing back to the Congress and the executive as the source of the troubles the Chinese complained against. The doctrine of states' rights provided the last line of defense. The federal government had only limited powers over the states, so this convenient argument went, and so could not intervene in essentially local questions involving the Chinese. The Chinese would have to seek relief by appealing themselves to the very hostile state and municipal authorities who generally were at the root of the problem. In this Alice in Wonderland world the Chinese understandably got nowhere.

The Chinese legation finally recognized the futility of protest and decided instead to seek a new immigration treaty to replace the Angell treaty over which Congress had ridden roughshod.[40] By putting immigration back on a treaty basis, the Chinese hoped to get the issue out of the hands of Congress and back in the hands of the President, who was somewhat less likely to treat it as a political football. The departure of the Harrison administration paved the way for treaty talks. Walter Q. Gresham, the Secretary of State in Cleveland's second administration, substituted a conciliatory manner for the studied disregard practiced by Blaine, and in an early exchange of views with the Chinese minister, Tsui Kuo-yin (1889–1893), removed some of the tension from the immigration question by himself offering pledges for the safety of the Chinese in the United States and obtaining in turn assurances from Tsui for the safety of Americans in China.

The groundwork for treaty talks was laid in mid-1893 as the legation consulted with Gresham on the enforcement of the Geary Act provisions requiring Chinese registration. The Chinese community refused to comply, thus courting mass deportation by the federal government. Gresham recoiled from the cost and ill-feeling deportation would involve, and so with the backing of Attorney General Richard Olney, he promised Tsui a delay in enforcing the law while the constitutional test of that law initiated by the Six Companies went forward. The respite, however, was only temporary. The Supreme Court decided against the Chinese, and authorities in the West began deportation proceedings. Congress, as much concerned with the potential expense of deportation as with the needs of the Chinese, extended the period of registration by six months.

Tsui's successor, Yang-ju (1893–1896), arrived at a propitious moment. Relations with the State Department were good; both sides wanted to find a way around the registration impasse; and Congress had provided time to work out a diplomatic solution. Yang had left China with instructions to replace the current body of sweeping congressional regulations with a new treaty modeled after the less restrictive, amended version of the 1888 treaty previously rejected by China. Once in Washington, Yang discovered that those offensive congressional regulations were there to stay and that he could have his treaty only if it incorporated them. But even such a treaty, Yang reasoned, might serve the interests of the immigrants by putting current restrictions on a treaty basis and thus setting a brake on any remaining impulse in Congress to extend them. A treaty was also attractive because it would help restore amicable diplomatic relations with the United States as an offset to growing difficulties with Japan. Yang proposed to Gresham making the amended 1888 treaty together with subsequent congressional regulations the basis for a new agreement, and to overcome any hesitations on Gresham's part he held out the promise of compliance with registration as a quid pro quo for the treaty. Through January and February 1894 the two worked out the details, and late in the year Gresham delivered the Senate by a 45 to 20 vote.

Yang now fretted over the reaction he expected from the

Chinese press and labor transporters to a treaty which was more stringent than the one Chang Yin-huan had seen cried down six years earlier. Restrictions were inescapable, he argued, and the registration requirements no more objectionable than China's own security (pao-chia) system. The law-abiding had nothing to fear. Although some opposition persisted, Chinese merchant leaders in the United States, having failed in their legal challenge, saw no realistic alternative to accepting the treaty and with it registration. And Peking, then on the brink of war with Japan over Korea and hopeful of American good offices, was in no mood this time to scuttle the legation's treaty.[41] Thus by 1895 the Chinese government had fully acquiesced in the U.S. policy of exclusion and rejected the bolder alternative of retaliating against American interests in China. Too distracted by the renascent foreign threat to imperial security to attend to the problems of its distant subjects, the government simply turned its back on popular opinion in the Canton delta and the cries of Chinese in the United States.

The Impact of Exclusion on the Chinese Community

Immigration controls won by exclusionists cut deeply into the growth of the Chinese population in the continental United States. After a rapid increase from a total of 34,000 in 1860 to 105,000 in 1880, there followed an extended period of decline. By 1900 only 90,000 remained, and the figure was to continue to fall. Western exclusionists could be gratified not only by declining numbers of Chinese in the United States but also by the dispersal of the Chinese to other areas of the country, a process that had begun as Chinese workers moved with railway construction teams into the western interior and as far east as Texas. But not until the exclusionist pressures began to mount in the 1880s did relocation assume major proportions. The census of 1880 showed virtually all the Chinese located in the West; by 1890 10 percent in search of safety or livelihood had moved away. By 1900 one fourth of the Chinese in the United States were to be found outside the western region, with well over half of those (16 percent of the total) in the North Atlantic states. The West had in effect found that one solution to its Chinese problem was to export it—to other parts of the United States as well as back to China.[42]

Exclusionist successes at the same time inflicted a serious wound on the remnant Chinese community in the United States. By the 1880s, as popular attacks reached a crescendo and the legal groundwork for exclusion was laid, that community entered a period of economic and social isolation and stagnation. The place of the Chinese in the economy had always been precarious, subject not only to general economic cycles but to the acquiescence of white workers. The late 1870s marked a turning point. Along with a national depression, which struck California belatedly but with full force, came intense labor agitation that drove Chinese workers from mines, farms, and light industry (especially cigar and boot making). New California laws excluded Chinese workers from state-funded enterprise and placed disabilities on Chinese businesses as diverse as laundering, fishing, and manufacture. Immigrant laborers, faced now with persistent high unemployment, were to suffer the most. But merchants too were hurt. Ever more stringent exclusion regulations inhibited their own freedom of movement between the United States and home. At the same time, local agitation diminished and scattered the Chinese population which was a major source of their business.[43] In this increasingly hostile environment Chinese began to consolidate physically into ethnic ghettos. The largest and best known of these Chinatowns, the one in San Francisco, had existed in embryo already in the 1850s, but only later, under outside pressure, did a strict pattern of residential segregation develop.[44]

These emerging Chinese enclaves were largely peopled by poor, single males. Some still found employment in the larger economy as laundrymen, peddlers, domestic servants, and migrant workers—all niches still open to them. But most were contained by Chinatown's constricted economy, especially as it developed from the 1890s on as a tourist mecca with its restaurants and curio and import stores. These enterprises coexisted with some labor intensive light industry as well as stores and services for the Chinese community. As the Western economy offered less and less, Chinese with sufficient resources returned home or moved on to such greener pastures as Mexico and Hawaii. Those left behind constituted a community that seemed doomed to extinction: it could not reproduce itself. Initially most immigrants had chosen to leave family behind. But by the 1880s the passage of exclusion laws deprived laborers of even a choice (though businessmen might still bring dependents into the country). The number of families remained small, so that the appearance of a second generation, which might have maintained and invigorated the Chinese community or begun the process of assimilation was long delayed.⁴⁵

The Chinese community, painfully and sometimes desperately aware of the debilitating impact of exclusion, sought to defend itself. But it had few of the resources of the older, well-established Chinese communities in Southeast Asia. The latter had accumulated reserves of economic power (as middlemen in foreign or domestic commerce) or social influence (built up during periods of interracial tolerance) against the day when the currents of Sinophobia began to flow. But the Chinese who had come to the United States had little time to establish themselves. They had encountered hostility from the start and had essentially no more than communal solidarity to shield them from attack. The more intense the hostility, the more closely the Chinese clung to familiar organizational supports.

The foremost defender of the Chinese community in the United States was the Six Companies in San Francisco. Its chief function was to speak for the community and to mobilize its slim financial resources to underwrite the struggle against Sinophobia. From the start it had relied heavily on whites to fight its battles. It engaged and subsidized publicists (most often missionaries) who would defend the Chinese in the interminable controversies over exclusion. It brought in special policemen to protect Chinatown from rowdies. It hired lawyers to challenge local and state ordinances in the courts, and after Congress embraced exclusion it launched two well-financed appeals to the Supreme Court.⁴⁶ With its strategy of using whites to battle exclusion doing little good, the Six Companies relied increasingly through the 1880s and into the 1890s on appeals to the legation and consulates, to the central government and influential provincial officials, and above all to opinion in the Canton delta.

As external pressures mounted, strains within the inward-turning Chinese community increased, leading to dislocation, demoralization, and finally to internal conflict over contracting economic opportunity. Conflict developed in the late 1880s along predictable regional and ethnic lines, with the chief cleavage between the men of Sze Yup and Sam Yup, and it continued into the early twentieth century. The more numerous immigrants from Sze Yup appear to have responded to diminishing economic opportunities by trying to take the offensive against Sam Yup by establishing new occupational monopolies while protecting old ones. Sam Yup, which began with the upper hand, fought back, making the most of its wealth and its close contacts with leaders in Cantonese society, with Chinese diplomats and consuls, and with American authorities, who generally accepted Sam Yup merchants as the natural spokesmen of an otherwise disreputable class of people. Under the intensity of the crisis, subsidiary conflicts broke out within the Sze Yup ranks. This multi-faceted, prolonged struggle, most sensationally conducted over control of gambling and prostitution but extending to other sectors of the Chinese economy, pitted secret societies, strong and weak lineages, and occupational guilds against each other in various combinations. "High binders," young toughs who had made extortion and warfare their livelihood, were the main instruments in recurrent contests that began with boycotts and threats of violence and sometimes ended in open warfare. As the violence spread, a small band of these toughs became essential to any organization that hoped to hold its own and keep its leaders alive. Whether this recurrent internecine struggle forced the "haves" in the Chinese community, Sam Yup immigrants and the powerful lineages, to give ground to the "have-nots" is difficult to say, but there can be no doubt that over several decades it took at least one hundred lives in San Francisco and over two hundred in New York, broke down solidarity within an already beleaguered community, and drew American authorities into Chinatown affairs. Some Chinese were tempted to turn white intervention to their advantage, while intervening whites sought to exploit differences among the Chinese to serve their own ends whether it was municipal reform, the restoration of law and order, or extortion plain and simple.[47]

By the 1890s the issue of exclusion and internal conflict in Chinatowns had created a serious breach between the legation and the Chinese community. While immigrants blamed their plight on the legation's lapses and weakness, Chinese ministers resented the immigrants' criticism and their resistance to the legation's policy of self-restriction. As exclusionist pressures exacerbated the tensions and set off spasms of violence within the Chinese community, Chinese diplomats in the United States responded with efforts at reform and control of that community that only served to widen the breach.[48]

The paternalism with which the Chinese legation regarded the immigrants began to take on a critical edge during the tenure of Cheng Tsao-ju, the first proponent of self-restriction, and of Chang Yin-huan. Although they saw the "Irish party" and economic competition as the main reasons for the campaign for exclusion, they were seriously concerned that attempted evasion of American immigration regulations by Chinese laborers kept the campaign going and drove authorities to formulate new measures. Chang had included self-restriction in the proposed 1888 treaty to deal with this problem of evasion. After the Chinese community and their allies in the Canton delta had cried the treaty down, Chang bitterly attacked the short-sighted and selfish attitude of immigrant interest groups and the Chinese treaty port press. The worst offenders in Chang's judgment were the Hong Kong and Canton-based merchants who did business with the American community (especially those involved in transporting laborers). The Six Companies was also culpable; its merchant leaders should have recognized that the lives of the Chinese in the United States depended on defusing the explosive question of exclusion, but instead they sent back to China what seemed to Chang a distorted picture of the situation and the legation's policy. Angry and frustrated, Chang began to examine the Chinese community with a jaundiced eye. And so did his successor, Yang-ju, after he encountered strong community opposition to the registration provision contained in his treaty with Gresham.

The flaws that both Chang and Yang saw in their compatriots bore an uncanny resemblance to those that Sinophobes stressed. The two Chinese envoys were most disturbed by the signs of moral degeneration among Chinese in the United States, particularly the pervasive gambling and opium smoking, which encouraged indolence, dissipated savings on which remittances and the return home depended, and led men into vicious ways.[49] Chang and Yang were also disturbed by the outbreaks of violence that became commonplace within Chinatown from the late 1880s and by the frequent evasion of the law by laborers who masqueraded as merchants, corrupted customs officials, and forged documents to gain entry to the United States. Reputedly turbulent by nature, the Cantonese had naturally fallen into lawlessness when left on their own. That natural proclivity had been strengthened by the presence of rebels who had fled to safety in the United States. No conscientious district magistrate in China would have countenanced these evils; they were doubly intolerable, in the judgment of the legation, when flaunted before the eyes of critical Americans.

The campaign to reform the Chinese community began under Chang and continued with little change during the Yang ministry. Hsü Chüeh, a junior diplomat who first served in Chang's legation and later returned to the United States to play a key role under

Yang, provided the element of continuity. The governing assumption was that the minister was indeed like a district magistrate. As a father to his people, he was obliged to look after their welfare by sponsoring hospitals and schools and to defend their interests (at least as he saw them). But he was also obliged to preserve good order. Moral exhortation was the first step to take toward restoring good conduct and the reputation of the Chinese. Consuls arranged for regular weekly lectures to the community of the sort village headmen at home were supposed to conduct, while admonitory placards served further notice on this contentious people. They were to become "conscience stricken" and aid each other in uprooting bad habits. They were to learn to submit their disputes to association officers instead of resorting to violence or, just as bad, appealing to American courts. And they were to avoid affronting American customs and laws in general and to respect immigration regulations in particular if they did not wish to bring new oppressions down upon themselves.

Exhortation and moral reform was the soft side of the legation's new policy of intervention and reform. For the incorrigible it held coercive measures in reserve. Consuls could cooperate with American authorities in securing the arrest and punishment of lawbreakers and if necessary their deportation back to China. Chang and Yang resorted to this procedure, but discovered it gave rise to abuses by involving uncomprehending and sometimes corrupt American authorities in what the legation regarded in any case as essentially affairs of the Chinese. Once the initiative passed into American hands, Chinese diplomats had difficulty influencing the proceedings. The safer though less effective procedure was for the Chinese consul or minister to work with community leaders to suppress violence. Where American law and community pressure did not work, the legation could always resort to Chinese law. Since the Ch'ing legal code sanctioned punishment of the entire family for wrongdoing by one of its members, the legation could hang over the head of particularly recalcitrant immigrants the threat of harm to relatives in China.[50]

The legation's policy of intervention did more to heighten tensions between it and the people than it did to calm social turbulence or induce upright behavior. Tsui Kuo-yin, a lackluster diplomat whose tenure fell between that of Chang and Yang, appears to have played a bit with a new approach, though not with the consistency or vigor to call it a policy. Tsui recognized that citizenship, which he hoped he might secure for Chinese in the United

States through a new treaty, would give them the right to vote and with it the political power to fight back against their foes. But Tsui did not attempt to translate this idea into policy and in any case inconsistently accepted the need for an extradition treaty that would legally orient the Chinese community toward China, not the United States.

A policy of intentional neglect—letting the Chinese in the United States fend for themselves and offering only minimal diplomatic good offices—was a possible alternative. But for most Chinese diplomats in the latter part of the nineteenth century such a course cut against increasing preoccupation with sovereignty and reciprocity in international relations and thus had little appeal. Another alternative, that of encouraging assimilation, was as unrealistic by the 1880s as neglect was unthinkable. On the American side, cultural antipathy and race prejudice had already isolated the Chinese and deprived them of even the option of naturalization, while on the Chinese side assimilation held scant appeal. For men in the legation and their superiors in China, to have actively promoted assimilation would have meant denying the superiority of Chinese culture, cutting off the remittances that helped feed the Canton delta, and abandoning subjects whose loyalty and generosity the government was just beginning to extoll. From the point of view of the immigrants themselves, assimilation held no allure. The United States, once regarded from a distance as a "mountain of gold," had for many of them proven economically barren, a place of loneliness, and altogether alien. The Canton delta, however distant, remained home.

Chapter Four

The United States in Li Hung-chang's Foreign Policy, 1879–1895

THE CHINESE TENDENCY to compensate for weakness by drawing the least threatening powers to her side persisted into the latter third of the nineteenth century. Nowhere was that tendency more evident than in the American policy pursued by Li Hung-chang, the single most important Chinese policy maker in that period. In 1870 Li became not only governor general of Chihli (the province containing the imperial capital and the growing port of Tientsin) but also commissioner of northern ports, positions he retained into the 1890s. Li's forceful personality together with his gradually accumulating experience in office thrust him to the fore in China's decentralized foreign policy apparatus. His foresight, capacity for boldness, and courage to take hold of controversial issues and brave domestic criticism were qualities at a premium in the frontier crises through which China was to pass in Li's lifetime.

Li had been born in central coastal China, in Anwei, in 1823. He had won the highest degree in the examination system at age twenty-four, an extraordinary achievement. Thereafter under the patronage of the eminent Chinese statesman, Tseng Kuo-fan (also a classmate of his father), young Li's career had moved ahead rapidly. During the suppression of the Taiping and other mid-century rebellions, Li had emerged as an able military commander, an effective and pragmatic administrator, and—no less important in those troubled times—a loyal servant of the ruling Manchu dynasty. With peace restored Li began taking a hand in foreign policy, at a time when many other officials—timid, prudent, or xenophobic in their outlook—backed away from what they regarded as the politically risky and personally demeaning contact with uncouth, grasping, yet dangerously powerful foreigners. (As late as 1891 one ranking official complained that having to attend Western diplomats at the foreign office was like "associating with dog and swine—a misfortune in a man's life."[1]) As his involvement in foreign affairs deepened, Li recruited or trained personal aides and official protégés with expertise in foreign affairs to assist him. Through an impressive network of clients and colleagues Li extended his influence into the councils of the central government, into other provinces, into the military and naval forces, and into the embryonic foreign service.

Peking, the most important of the alternative centers of power in a foreign affairs bureaucracy given to divided and overlapping responsibilities, was in the main responsive to Li's policy preferences. The Empress Dowager Tz'u-hsi played a crucial role in securing the central government's support and in shielding Li from the long knives of his political enemies. The Grand Council, the highest decision-making body in Peking, reflected the preferences of the throne. From the 1860s until the mid-1880s it was led by the redoubtable Manchu statesman, Prince Kung. The foreign office (tsung-li ya-men), headed by grand councilors, functioned as an auxiliary to the Grand Council. By the 1870s it had proven itself in practice a cautious institution devoted more to coordinating than initiating policy. Thus it left Li as well as his counterpart in the south (the commissioner of southern ports) considerable room for leadership while helping to insulate them from the often annoyingly pugnacious and impatient foreign representatives.

The high reputation Li enjoyed among those very same foreign representatives helped consolidate his position of power. While in the legation quarter the timidity and indecision of the foreign office (derided as no more than a postbox for exchange of diplomatic notes) was legend, foreign diplomats regarded Li as a commanding figure, accessible to foreigners and able to conduct business in the practical and direct manner that they missed in other leading Chinese politicians.[2] His imposing physical appearance added to his reputation. Over six feet tall and solidly built, Li carried himself with striking assurance. Among admiring foreigners, Americans included, Li was thought the Bismarck of Asia—the man to give direction to the largest state in the world and to arrange its relations with the powers.[3]

But however similar to Bismarck Li may have been, China was not Germany. Bismarck commanded the resources of a strong, expansive state; Li those of an imperiled, weak one. Thus while Bismarck could afford a more aggressive policy, Li was compelled to pursue a defensive one. In dealing with China's foreign crisis, Li followed in broadest terms a two-pronged policy. On the one hand, China had to pursue a long-term program of self-strengthening so that she could stop and in time roll back foreign penetration and restore China's influence along her frontier. To that end she would have to acquire foreign technology and skills in warfare, industry, and commerce. But Li would also have to fight for time while these efforts came to fruition. This struggle to delay foreign encroachment on China's outer line of defenses and to control foreign penetration of China proper was the other prong of Li's policy.

China entered this struggle for control at a marked disadvantage, especially in adjoining strategic regions. Her military and naval force was weak and decentralized and supported by an inadequate logistical system. Her single greatest asset in resistance—her ability to pit an aroused populace against an invader—was simply inapplicable to the immediate problem of frontier defense. This vulnerability, evident to perspicacious officials no less than to foreigners, gave rise in turn to the nightmare of Chinese policy—diplomatic encirclement. What one power might accomplish with effort against China's inadequate mechanisms for frontier defense, several acting in unison might manage with ease.

One method for getting China through its period of crisis was to draw in the assistance of a stronger state. It was a stratagem as familiar to Li from China's tradition of statecraft as it was obvious to any leader of a weak state caught between competing powers. In Li's rear guard defense of frontier areas the United States was to occupy a notable place, sometimes as putative ally, sometimes as would-be intermediary restraining the more threatening powers. Working with a changing cast of American representatives, Li was to invoke the assistance of the United States in four different crises between 1879 and 1895. This tendency to look to the United States is understandable (though ill-articulated in Li's own writings). It was partially by default—of all the major powers of the day only the United States posed no clear peril to China's territory or to the safety of her tribute states, and consistently disavowed any territorial ambitions. Moreover, the often expressed American commitment to commerce provided a handle for Li to manipulate the United States to China's advantage. From the Chinese perspective foreign trade and traders could be troublesome, but gratifying them to win their country's support was far more palatable than having to appease territorial aggression. Finally, Li believed that the support of the United States, a rapidly developing state which in its own recent civil war had put an impressive force of 2 million men in the field, might significantly strengthen China's international position.[4]

The Liuchiu Crisis and Ulysses S. Grant

The Liuchiu (or Ryukyu) Islands, a chain stretching between Taiwan and southern Japan, were the first of the imperiled tribute states that Li had to devise a defense for.[5] The Liuchius had long sent tribute to China but had also recognized Japanese preeminence. In the early 1870s the Japanese government began to consolidate its position without encountering Chinese resistance until 1877, when the King of Liuchiu appealed to Peking for protection. Chinese officials then generally agreed that these islands lacked the natural wealth or intrinsic strategic value to justify a war. But they also agreed that China could not in justice ignore the appeal from the loyal islanders. A refusal would be taken by the Western powers, so two early memorialists contended, as the beginning of open season on China's other, more important tribute states. The foreign office, after consulting with Li, decided on the path of diplomacy to protect China's interests; with imperial sanction in June 1877 it put the issue in the hands of Ho Ju-chang, the new minister to Japan. However, the belligerent tone Ho assumed once in Tokyo brought an end to discussions late in 1878. The following April Japan formally annexed the Liuchius. With China now facing the unpalatable alternatives of abandoning its claims or confronting Japan, Li Hung-chang took over the handling of the Liuchiu crisis.[6]

Li held to his earlier judgment that the Liuchius were not worth a struggle. But perhaps an appeal to international law and involving other countries in the issue might, he now reasoned, serve to check Japan. The country to which Li was to turn for support was the United States. Minister Ho had already made one appeal for American assistance in arriving at a peaceful settlement acceptable to China. Although Ho had proven unsuccessful in his bid, Li was ready to try again. Ho's thinking may have influenced him in this direction. But probably even more influential was the advice of William Pethick, an American who had gone out to China as a youth just after the Civil War and become an accomplished linguist. An almost permanent fixture in the American consulate in Tientsin and interpreter for visiting American dignitaries between 1872 and 1894, Pethick also worked for Li, initially as a family tutor and then staying on as a versatile private secretary right down to Li's death in 1901.[7]

Pethick's proposal for extricating China from its Liuchiu quandary was to utilize American good offices on the occasion of Ulysses S. Grant's visit to China during his world tour. By May 1879 Li had decided to seek Grant's mediation, begun preparations to receive him, and drawn the foreign office into the plans. On Grant's arrival in Tientsin Li treated him to a lavish reception at the same time that Pethick privately outlined Li's wish for help. Grant moved on for a courtesy call in Peking, where he received the formal presentation of China's case and request for mediation from Prince Kung. On Grant's return to Tientsin in June, Li finally involved himself directly by asking Grant to carry China's case to Japan, fortuitously the next stop in the American's travels.[8]

These first contacts in Tientsin in May seem to have confirmed for Li Pethick's favorable reports. Li saw nothing of the "pathetic and naïve" misfit that American critics of Grant the President had seen. Rather Li endowed Grant with extraordinary prestige befitting a man who had occupied the highest political position in the United States (indeed, a welcoming poster in Canton described him for the unworldly as "King of America") and who had put down a major rebellion (an arduous task that Li too knew something about). Perhaps even more important to Li than past achievements were Grant's bright political prospects. Talk of a run for a third term already circulated in the United States, and if this news did not reach Li through Pethick, we know it did through the Chinese consul in Singapore who had got wind of Grant's ambitions when the Grant party had passed through that port. Grant denied any political influence over the American government nor would he discuss with the Chinese his plans for the future, but that did not prevent them from cultivating this past and prospective head of state in hopes that he might bring his influence into play on their behalf.[9]

Grant's excursion around the world was on the surface a much deserved pleasure jaunt after a decade and a half of steady public service, climaxed by two troubled terms in the White House. But as Li shrewdly perceived, for a man aspiring to return to high office the trip might also serve political ends. Certainly Grant's friends and political backers and perhaps even Grant himself so intended it. Though he would not openly grasp for political power, there was no harm in having the public at home know of the high esteem that he was held in abroad as a world leader or being reminded of his past attainments, especially as the preliminaries for the 1880 presidential election grew near. John Russell Young, a reporter for the *New York Herald*, who drew near to Grant as confidant and aide as the trip proceeded, sent out a steady stream of reports that kept readers at home up-to-date on one triumphal reception after another. The Chinese mediation request opened up the prospect that Grant could now bring his world tour to a properly climactic end, returning home wearing the laurels of the peacemaker.[10]

Grant's first reaction to mention of the Liuchius was most likely a neutral nod, then afterwards a hurried search for a map. Little wonder. The only time those islands had emerged from the obscurity into which virtually all Americans consigned them had come twenty-five years earlier—and then only briefly—when Perry had tried to get the Pierce administration to annex them. Grant, once he located the objects of dispute, connected them not with overseas expansion, as Perry had done, but with the unrelated but for him a great deal more pressing issue of Chinese immigration to the United States. The issue had been planted in his mind when as a young army officer serving on the West Coast in the 1850s he had witnessed the arrival of the first wave of laborers from the Canton delta. Its importance in his thinking had sharpened in his second term in the White House, by which time sweeping exclusionist demands had put Grant and the Republican party in a serious bind. More recently, in his travels—at virtually every stop between Penang and Tientsin—he had had the issue again forced on his attention, this time by Chinese officials and merchants pleading for fair treatment and free immigration.[11] Li's overtures for assistance on the Liuchiu problem now placed Grant in the happy position of being able to solve the exclusion problem for the Republican party and the Republican administration then in power. In return for accepting Li's request that he act as honest broker in the Liuchiu dispute, Grant could request Chinese cooperation in controlling immigration.

The bargain was quickly struck after Grant returned to Tientsin from Peking on June 12. Li knew second-hand that Grant had shown interest in the Chinese request, while Grant for his part had already hit on the price for his services. Li, in presenting China's case on the Liuchiu dispute, followed in part the arguments developed by Prince Kung in his presentation in Peking earlier. Japan had violated international law, first by deposing the Liuchiu king, who had long recognized Chinese sovereignty, and then by refusing even to discuss the problem with China. War, both Prince Kung and Li pointedly warned, would disrupt American commerce. Li added that should Grant succeed in settling the dispute, he would win universal applause. To Li's exposition of the Liuchiu problem, Grant responded with a survey of the immigration problem and a proposed solution, a three-to-five-year suspension imposed either on

China's own initiative or through treaty revisions to be undertaken by the soon-to-arrive American Minister Seward. In an effort to be helpful Grant suggested that China redirect her excess population to wilderness areas such as Borneo, New Guinea, or the Congo region where, with "practical support" from the United States, Chinese might found "new Shanghais and Cantons."[12] The loser in the bargain now taking shape was the Chinese emigrant. Li had previously displayed concern for his welfare, knew of the rising demands in the United States for unilateral restrictions against Chinese, and even now invoked Grant's influence on behalf of better treatment for Chinese in the United States.[13] But protecting the Liuchius and avoiding war were overriding concerns. Li would now leave the foreign office to arrange the concessions to the United States on immigration, while he concentrated on resolving the dispute with Japan.

Grant the peacemaker moved on to Tokyo. There Japanese leaders denied that the Chinese had any grounds for complaint. The Liuchius were not Chinese territory but rather a Japanese protectorate. The Japanese supported their claim by reference to international law (the Chinese had earlier accepted Tokyo's demands for indemnity for injured Liuchiuans), and minimized the Liuchiu tribute missions to China as no more than a disguised form of commerce. Japanese leaders did, however, promise to give serious attention to whatever suggestions Grant might make toward resolving the dispute. Grant reacted cautiously and with indirection. On paper he advanced simple procedural terms for settlement: After China withdrew Minister Ho's offending comments, the two countries were to renew talks looking toward a peaceful settlement of the dispute. They were not to drag in the European powers, who (Grant feared) might exploit Sino-Japanese tension to their own advantage. But privately and through intermediaries Grant went further in offering terms for settlement so potentially controversial that he himself carefully avoided putting them in writing. Specifically, Grant proposed to the Japanese a partition of the Liuchius assuring China a secure channel to the Pacific. Communicating through Young, Grant informed Li of the proposal and sweetened it with the prospect of a Sino-American alliance of mutual assistance to deter any abuse of China.[14]

Grant's swing through East Asia and his abrupt engagement in its diplomacy set him to thinking about the region and its importance to the United States.[15] East Asia itself he now saw as a great zone of development increasingly imperiled by European influence and ambitions. Not so battered as Africa, where a carving up had already begun, East Asia was still in greater peril than Latin America, where the United States had succeeded in minimizing outside intrusion. Out of all of Asia, with more than two-thirds of the world's population, China and Japan alone had managed in a measure to elude European domination and still possessed the intelligence and potential strength to maintain their independence. Japan most impressed him. Modernizing rapidly and growing ever stronger, she was "more like a romance than a reality." But even in slower-moving China Grant thought he had seen the germ of "a great revolution that will land her among the nations of progress." The weak ruling dynasty might have to be removed before China could move ahead, but in the long run the virtues of the Chinese people—"enduring, patient to the last degree, industrious"—would count for much in building "a strong independent nation." Their shrewdness as merchants would enable them to absorb much of the commerce of the world as they had already managed to do in Southeast Asia, while able statesmen such as Li Hung-chang would marshal Chinese capital and talent in launching a program of reform (thus avoiding the costly strategy of reliance on Europe).

A falling out now between China and Japan would, Grant feared, create a golden opportunity for the advance of European imperialism. To prevent conflict and thus forestall this incubus settling on the development of a progressive, independent East Asia, a Sino-Japanese entente was important. Americans should stand ready to offer advice, diplomatic good offices and, if necessary, political alliance to create regional stability. Grant conceded that the region was not as strategically important to the United States as Central America and the Caribbean, and that trans-Pacific commerce would have to await the construction of an isthmian canal before it would reach the level of Latin American trade. What had captured Grant's imagination was the imminent development of modern civilization in Asia and the role the United States might play not only in planting the seed of progress but also in maintaining a propitious environment for the seed to germinate and take root. As with the countries of Latin America, so also in Asia the United States would stand against the intrusion of selfish European powers and preserve the peace. Grant was certain that Asians were as ready to accept the United States as protector and tutor as Americans were suited and obliged to claim the role. For as Asians well knew, the United States was the most friendly of the powers, and because she harbored no imperial ambitions she was also the least dangerous. "The fact is, the Chinese like Americans better, or rather, perhaps, hate them less, than any other foreigners. . . . We are the only Power that recognizes their right to control their own domestic affairs."

From Japan Grant headed directly home, his head filled with dreams of an American century in Asia.[16] Although from the moment he touched soil in San Francisco in September down to convention time the following June, Grant behaved like a presidential candidate, criss-crossing the United States on tour and paying well-publicized visits to Mexico and Cuba, political aspirations did not, even then, crowd his newly aroused interest in East Asia from his mind. He discussed with intimates his ideas on Asian policy and found time in late December to carry to Washington impressions of his trip and details of the immigration deal.[17] Grant's hopes of realizing his private East Asian vision were, however, to die along with his public bid for the presidency. Entering the Republican convention a strong candidate, Grant took the lead in early balloting only to be denied by the same coalition that had formed to deprive him of a third successive term four years earlier. (The nomination went instead to the virtually unknown James A. Garfield.) Even in defeat, Grant could look back with satisfaction on the concessions he had won from the Chinese on the immigration issue—to make the task for subsequent Republican administrations far easier—and with pride on his formulation of the American political and economic role in Asia that few contemporaries could match for boldness.

Li Hung-chang received news of Grant's defeat with disappointment, even some bewilderment. Was not a man who had twice served in a post, Li asked an American, best suited to fill it again?[18] But even stronger was his sense of disappointment that he had played his American card and it had proven worth so little. Grant's defeat consigned the idea of alliance to the category of pipe dream while his earlier efforts in Tokyo had failed to save the Liuchius.

Grant's confidential suggestion of a partition had become the subject of Sino-Japanese talks. Japan rejected a three-way split, as Grant had proposed, with the Liuchiu kingdom left in control of the central section, and suggested instead the cession to China of the southernmost portion adjoining Taiwan (Miyako and Yaejima Islands). In return, China was to give Japan the most-favored-nation status already enjoyed by the Western powers and the right to trade in the interior. Li and the foreign office considered the Japanese proposal under the shadow of a frontier crisis with Russia (over the Ili territory) and the prospect that if war resulted Japan and Russia might join together against China. Some officials thought neutralizing the Japanese threat worth the cost of the bargain, and in October 1880 the foreign office accepted the Japanese terms only to find both the division of the Liuchius and the commercial provisions of the bargain subject to sharp attack at home. Li himself dealt the final blow to the agreement when in November, in response to an imperial query, he advised delay at least until after the Russian crisis was over. He minimized the prospect and potential dangers of Russo-Japanese collusion, and judged the agreement unacceptably one-sided. It was harmful to China economically, Li contended, while conceding to Japan all but a poor and sparsely populated stretch of the Liuchius, insufficient to maintain its independence and a burden for China to support and defend. The throne accordingly rejected the October agreement. The Japanese refused further parley and proceeded to tighten their grip on the Liuchius while China looked on, resentful but unwilling to fight to preserve its dependency.[19]

In his effort to block the Japanese by drawing in the Americans Li had miscalculated on three counts. First, he had overestimated Grant's value as an intermediary. He had sacrificed immigrant interests for services rendered in Toyko which fell short of Chinese expectations. Grant's personal communications from Japan urging a compromise and the withdrawal of Ho's offensive note had exasperated Li, who suspected that a fine reception in Japan had turned his head. To Grant's partition proposal, word of which had reached Li only indirectly, he had nothing to say. Li, still hopeful of turning his relations with Grant to good account, made his second mistake by investing unrealistic hopes in Grant's political popularity at home and the influence he could wield there on China's behalf. Once it became clear that Grant had done all he would for the moment for China, Li put a good face on a poor situation, writing in a conciliatory tone to Grant and Young that China would accept their advice and resume talks with Japan. Looking to the future, Li greeted the idea of an alliance with interest, indeed proposed in his correspondence with Grant and Young making the alliance offensive as well as defensive. (Such an alliance, Li had contended, would be a logical extension of the provision in the 1858 Sino-American treaty pledging both sides to good offices in time of trouble.) Grant's failure to make it back into high office once home finished this scheme and Grant's usefulness to Li.[20]

Li finally had not counted on successive American administrations aggressively exploiting the concession that he had made to Grant on Chinese immigration policy. An exclusion policy that was to go in the 1880s and 1890s far beyond what Li had envisioned in his conversations with Grant in 1879 and Washington's attendant failure to end the abuse suffered by Chinese already in the United States came to represent more and more an insult to China as well as a breach of international law and equity. Even in 1879, the topic of Chinese in California made Li "grow angry and flushed and strike the table with his hand," and caused him to complain that the Chinese had fewer privileges than American blacks. The passage of the Scott Act in 1888 Li regarded as a deliberate affront. By the early 1890s Li had become outspoken on the problem whenever he met Americans. He would ask them to explain why China was expected to protect foreigners on its soil when the United States refused the same safety to Chinese residents and to tell him why China should not break off diplomatic relations and expel Americans in retaliation for the unending series of anti-Chinese measures to issue from Washington.[21]

Korea and Robert Shufeldt

Despite the unsatisfactory outcome of Li's first play for American diplomatic support, he was soon engaged in a second attempt. This time Korea, the most strategically placed of China's tribute states, was at risk.[22] Mounting foreign pressure against Korea's inflexible policy of seclusion dating back over two hundred years raised the spectre of armed conflict into which China was sure to be drawn. Anticipating just such a possibility, the foreign office had prudently begun in the 1860s to try to move the Koreans to abandon isolation in favor of treaty relations before the powers forced the issue as they had with China. But the Koreans refused to budge on the grounds that seclusion was the necessary concomitant of their subordination to China. China could either remain suzerain and help defend the seclusion policy or force on Korea a new foreign policy and surrender her privileged position, but China could not have it both ways. The Korean argument had a logic the foreign office could not openly challenge. At the same time the foreign office had sought to restrain the powers—though with no more success. The British were committed to extending their commerce, the French to the protection of persecuted Catholic missionaries, and the Americans to safeguarding ships and sailors driven in distress onto Korean shores.

By the 1870s the situation had grown critical. Foreign naval expeditions had probed Korea's coastal defenses; Russia's eastward march had carried her to the Korean frontier; and Japan, emerging from the domestic tumult of the Meiji restoration, took a revived interest in the peninsula. When the powers complained against Korea's self-imposed isolation, Peking defended it while disclaiming any power to modify it. The danger that the powers would take China's confession of impotence as grounds for a direct approach to Korea that altogether disregarded China's interests became acutely apparent in 1876 when Japan sent a punitive expedition and coerced Korea into her first treaty. By 1879 Japanese encroachments made urgent some new Chinese policy response.

As Li took over Korean policy in early 1879—a burden he would carry for more than two decades—he at once accepted the foreign office's contention that Korea could not remain secluded, and to it he added his own fundamental assumption that Japan was the chief threat to China's interests in Korea. Japan's 1876 Korean expedition as well as the invasion of Taiwan and the Liuchiu annexation served as evidence of Japan's dangerous expansionism. But Li was also worried by Russian territorial ambitions. To stem the advance of Japan and Russia Li would aim at creating a new status quo which the other major powers would have a stake in defending. This in turn meant that Li faced the delicate task of undercutting the seclusion policy while preserving the privileged Chinese position in Korea and oversight of Korean foreign affairs. The first step Li had to take was to maneuver the Koreans into accepting a new Chinese-designed era of treaty relations with the West and then ensuring that the first treaties safeguarded China's interests while sparing Korea some of the mistakes (particularly in relation to opium importation and missionary rights in the interior) that China had made in her first treaties. If all went well, Li's policy would create an equilibrium of foreign interests over which China as the spokesman for Korea would preside. The policy was a model of economy, necessarily so at a time when China could ill afford to pour resources on any large scale into Korea.[23]

The fortuitous arrival in 1880 of an American, Commodore Robert Shufeldt, charged with negotiating a treaty with Korea, provided Li an unexpected opportunity to advance his Korean policy, just as Grant's appearance the year before had provided the occasion for a new approach to the Liuchiu question. Shufeldt arrived in the wake of four decades of American effort to penetrate Korean isolation. The pressure had begun to mount from the mid-1860s. Following an attack on an American merchant ship in 1866 and the rebuff of an American naval mission of inquiry, the cry went up from American diplomats and merchants for a naval expedition. A blow struck "promptly and efficiently" would, promised the commander of the Asiatic Squadron, awe Japan and China and "disclose to the world who are masters of the Pacific." The opening of Korea would also send trade bounding ahead, though it would take force in the first instance to turn the trick, concluded the ministers to China and Japan meeting with the consul general in Shanghai in 1868. Secretary of State William Seward toyed with a policy of force then put it aside. During the Grant years an expedition finally set out under the direction of the minister to China, Frederick Low. It leveled a fort after the Koreans refused to talk, but failed to make any progress toward the desired commercial treaty.[24]

Aside from unanticipated good timing, Shufeldt brought to his task the skills and perspectives of a naval diplomat. After service as a youth in the navy and merchant marine, he had become a diplomatic agent on the Union side early in the Civil War and later rejoined the navy. He cruised off the Korean coast in 1867, and returned again in 1879 at the tail end of a globe-girdling diplomatic and commercial mission. Korea, not on his itinerary when he set out in December 1878, was to appear there through the efforts of California's Senator Aaron A. Sargent, who favored another try at drawing Korea out of isolation. To his Senate colleagues he emphasized the commerce, enhanced American influence, new opportunities for mission work, and good treatment for seamen and ships in need, all of which a Korean treaty would secure.[25] When the Foreign Relations committee tabled his resolution, the Hayes administration decided to act on its own, dispensing with the appropriations request and instead using Shufeldt, who was already en route, well-qualified for the mission, and supported by Senator Sargent. With the concurrence of Secretary of State Evarts, the Secretary of the Navy prepared the instructions received by Shufeldt after his arrival in Nagasaki in April 1880.

Shufeldt's first attempts at initiating talks failed. The Koreans had no interest in a treaty, and when Shufeldt presented himself on the coast they predictably sent him away. A policy of force, not "moral suasion," was in Shufeldt's opinion the most effective instrument for dealing with "Orientals" and clearing the way for commerce in barbarous lands. But from the start he had to rule out battering down Korean resistance for want of the requisite naval strength and in deference to his instructions which enjoined a "moderate" course of conduct. The most attractive alternative was to enlist the Japanese, who already had a foot in the door in Korea. Evarts and Sargent had earlier emphasized that Japan, as the most progressive nation in East Asia, was likely to be the most sympathetic to American commercial goals. Shufeldt agreed and immediately after his arrival sought the assistance of the Japanese foreign office. But its efforts proved no more successful and raised doubts about whether it really wished to help.[26]

With his mission in peril Shufeldt turned to the last and least likely source of help—the Chinese. Theretofore the Chinese foreign office had turned aside American requests for assistance in dealing with Korea or at best on occasion after 1871 sent communications on to Korea for a pro forma reply.[27] Shufeldt laid the groundwork for his Chinese gamble in June 1880 when he began cultivating Yü Ch'ien-yao, the consul in Nagasaki and a protégé of Li Hung-chang, and discreetly advertising his friendship with Ulysses S. Grant. The next month he inquired into the possibility of a visit to Tientsin to take up the Korean problem with Li. In August the desired invitation arrived, holding out the unexpected prospect that Shufeldt might after all find his road to Korea through China.[28]

Li had decided on Shufeldt as the best instrument for laying the foundation for his new policy. The United States had a strong interest in opening Korea to trade, yet seemed amenable to working with China and harbored no dangerous territorial or political ambitions. No less important was the chance Li saw to drive a wedge between the United States and Japan. By his approach to Japan earlier Shufeldt had unknowingly struck an old chord of concern in China. The foreign office had made nervous secret inquiries in 1871 when Minister Low's expedition stopped in Japan en route to Korea, and subsequently worried about possible American participation in the Japanese expedition of 1876.[29] By advancing parallel American and Chinese interests in Korea Li could also hope in a measure to isolate Japan.

Li had already unveiled his general policy to the Korean court; now he began to explain the special role the United States was to play. If Korea was to avoid the fate of Liuchiu, it would have "to play off the foreign enemies one against the other." Of those countries threatening Korean seclusion, the United States was the most eligible to use. "America, which faces directly on the Pacific, has no intention of invading the territory of others there. By concluding a satisfactory treaty you can not only block the ambition of your eastern neighbor but also set a model to be followed by the other countries who will subsequently wish to open relations. . . ."[30]

Treaty relations would thus win Korea time to strengthen herself with China's help and eventually reduce her vulnerability to foreign aggression.

With Shufeldt's arrival in Tientsin, Li was launched on a delicate diplomatic maneuver that would require nearly two years to complete.[31] On the one side, he was already engaged in an intricate and drawn out series of exchanges with a Korean court still loathe to accept any new treaties. Facing both unrest at home and a major readjustment in foreign relations, the court was badly divided. Although letters and emissaries had been moving back and forth between Seoul and Tientsin since mid-1879, it was not until late December 1881 that Li was able to extract Korea's reluctant assent to the treaty. All the while Li had to keep the impatient American envoy on the hook. In their preliminary exchange Li held out to Shufeldt the lure not only of a treaty but also of an advisership in the Chinese navy, a post the American had earlier told the Chinese consul in Nagasaki of his interest in.

With the long-sought treaty now almost in his grasp, Shufeldt set off for home in September 1880 to secure the backing of the new Garfield administration, stopping en route at Nagasaki long enough to make a second unsuccessful try to open direct talks with the Koreans. To his superiors he stressed that the Korean treaty would bring closer that day when the entire Pacific would become "the commercial domain of America." The treaty would also carry with it new political responsibilities—to protect Korea "against the aggressions of surrounding powers." Finally, but by no means incidentally from Shufeldt's point of view, the successful conclusion of the negotiations would clear the way for his own appointment to reorganize the Chinese navy. Having secured the backing of the new Secretary of State, James G. Blaine, Shufeldt returned to Tientsin in mid-1881 only to face further delays while Li continued to prod the Koreans. Though the breakthrough in the Korean-Chinese interchange finally came in December, Shufeldt's patience was taxed awhile longer—until March 1882, when Li's personal foreign affairs adviser, Ma Chien-chung, and the head of the Tientsin customs, Chou Fu, finally drew up the terms of a commercial treaty and presented it to Shufeldt as a "Korean" draft. After some revision the treaty was cast in mutually satisfactory form, and in May Shufeldt set off with Ma for Inchon on the Korean coast (where other Americans were to come ashore some seventy years later), and there the Korean representatives at last affixed their signature.

The Korean treaty, long sought by Americans and Chinese alike, proved a disappointment. It stirred no enthusiasm in the United States. The Arthur administration, formed after Garfield's assassination, dumped Blaine, the Korean treaty's most recent sponsor; treated Shufeldt with indifference; and without fanfare sent his treaty on to the Senate where it slipped quietly through in January 1883. As a contribution to the expansion of American commerce, the treaty failed to live up to the hopes of its advocates, and Washington did nothing to make good on the predictions of enhanced American political influence in the area.

For Li, the basic rationale for aiding Shufeldt—to preserve Chinese predominance in Korea while neutralizing the Japanese threat—was not sustained by events following the treaty signature. The American treaty did, as Li intended, set the pattern for a series of similar agreements with Korea concluded later in 1882 by representatives of Britain, France, and Germany working, as Shufeldt had, through Ma Chien-chung. However, instead of stemming international rivalries in the area, the treaty coincided with their intensification and the steady erosion of China's position. The Korean court was still not wholly reconciled to Li's conception of a new era of foreign relations, and its internal divisions gave ample play for foreign intrigue. To prop up the faction at court most friendly to China, Li finally had to throw off the rule of nonintervention that had long governed Sino-Korean relations and in August 1882, only a few months after the conclusion of the Shufeldt treaty, dispatched an expeditionary force to Korea. Unknowingly, Li thus embarked on the road to war with Japan.

The overlap of Chinese and American interests that had made the 1882 treaty possible was short-lived. Indeed, serious divergences had appeared even before the conclusion of the treaty. The first casualty was Shufeldt's respect for Li. Although Shufeldt had come away from the first round of talks impressed by Li's "intelligence and judgment" and gratified by "the evident respect Li entertained for my opinion," in time he began to doubt whether the Chinese was a fit diplomatic partner. The inexplicable delays repeatedly imposed by Li as well as the influence of Chester Holcombe, chargé of the U.S. legation and self-styled China expert, who was deeply distrustful of Li's motives, help account for the change. By early 1882, on the eve of the successful conclusion of his mission, Li had undergone a remarkable evolution in Shufeldt's eyes— into a ruthless and powerful yet backward-looking figure with "a clear, cold, cruel eye and an imperious manner." He was "a thorough oriental and an intense Chinaman," terms that meant for

Schufeldt "contempt for western nations and hatred of all foreigners." He was also a "repressive force," dealing sternly with dissent, and only a little ahead of his countrymen "in an appreciation of the arts political and physical which govern the modern world."[32]

Substantive issues as much as personalities got in the way of a common Chinese-American approach to Korea, as the treaty negotiations themselves revealed. Li had wanted from the United States formal recognition of Korea's dependency on China and a promise to "assist and protect" Korea against aggression. The former provision would lend support to China's claim to a special relationship with Korea, while the latter was an attempt at reviving in more restricted form Grant's broad alliance proposal and making the United States along with China Korea's patron and protector. On these key provisions Shufeldt balked. He was willing to go as far as to include a vaguely worded promise of American "good offices" in case Korea ran into international difficulties, but he flatly refused to recognize in the treaty China's claim to a special relationship with Korea (even though Li had made it a sine qua non for any treaty). It was a matter of concern only to China and Korea, Shufeldt contended, and it would certainly cause any treaty to fail in the Senate. Finally, to break the deadlock Li devised a compromise whereby Shufeldt at the time of the treaty signature would accept a letter, drawn up by the Chinese in the name of the Korean king, reaffirming Korea's dependent status. Agreed on the innocuous "good offices" clause and the supplementary letter, Li and Shufeldt had closed their bargain. But the letter affirming Chinese primacy carried no weight in Washington, which consistently regarded the tribute relationship as both a political anachronism and a legal absurdity. Li's failure to get an explicit American recognition of Korea's dependence in turn left him vulnerable to criticism at home.[33]

What Washington refused to recognize, Americans in Korea, going a step farther, intrigued to undermine. The State Department tried to prevent meddling in Korea's internal affairs, but it had only limited success with its own representatives and limited control over private Americans devoted to Korean independence. In the resulting anti-Chinese maneuvers Americans aligned with what they regarded as the "progressive" Japanese and factions in the Korean court seeking an offset to the increasingly interventionist tendency of Chinese policy.[34]

Li's chief recourse was to scold the Koreans and dispatch a personal representative to keep them in line. He simultaneously searched for a way to neutralize his local American antagonists, whom he accused of following a pro-Japanese policy. He sought to have the Seoul legation, which displayed an early tendency to look to Japanese for advice, information, and help with translation, put under the supervision of the legation in Peking and hence kept in closer touch with the Chinese point of view. He got support from the American minister, Charles Denby, who stood to enhance his own authority by the change, and from James H. Wilson, a concession-hunter eager to ingratiate himself with Li. But Secretary of State Bayard, determined that the United States should "hold aloof" from Korean rivalries, rejected the proposal.[35] Two years later, in 1887, Li watched in frustration as the Seoul legation, with State Department support, worked for a Korean diplomatic mission to Washington designed to undermine Chinese pretensions to control. Owen N. Denny, an American whose appointment as adviser to the Korean foreign office Li himself had helped arrange, was behind the scheme, and another American, Horace N. Allen, a medical missionary and physician to the Korean king, was slated to accompany the mission as adviser. Li forced an apology from the Korean court, got the American chargé recalled, and applied pressure for the removal of Denny, whom he regarded as the mastermind behind this and other efforts to defy the Chinese. But Denny, Allen, and other troublesome Americans stayed on and continued—much to Li's annoyance—to work for true Korean independence from China.[36]

A Policy Played Out

The Grant and Shufeldt episodes had understandably planted doubts in Li Hung-chang's mind about the utility of his American policy. Washington had shown itself a decidedly hesitant actor in East Asian politics, and certainly the Americans Li had dealt with had not provided any meaningful assistance to—indeed to a degree undercut—China's struggle to maintain a line of security along her border. At the same time Washington had mocked any hopes of a reciprocal relationship with China by embarking on a course of Chinese exclusion from the United States while insisting on the maintenance of a privileged status for Americans in China. Indicative of Li's changed view was the scorn with which he greeted the not-so-novel suggestion, made in the mid-1880s by one of his young diplomatic associates, that China should make an ally of the United States.[37] That was a role Li was now to reserve exclusively for the

European powers capable of a more assertive diplomatic and naval policy. Nonetheless, Li still kept a place for the United States in his calculations—as a diplomatic intermediary. American diplomats in China proved eager to occupy even this diminished position in Chinese foreign policy.

The French penetration of Vietnam, climaxing in war in 1883–1884, provided the third occasion for Chinese and American diplomats to come together over an endangered tribute state. As the French drive intensified in the early 1880s under the direction of Prime Minister Jules Ferry, agitated Chinese officials began making defensive preparations and lending covert aid to Vietnamese resistance. By July 1883 France and China had entered a twilight period of sporadic armed border clashes accompanied by abortive peace feelers. Finally, in December the Ferry government decided to settle the border issue by force.

Through this first phase of the crisis Li Hung-chang sought to avert a collision along the southern border, the better to concentrate on preserving China's position in Korea against Japan. War over Vietnam would result in the transfer south of Li's carefully nurtured military and naval forces, thus depriving him of the means to meet a sudden Japanese challenge in his own area of responsibility. In search of a diplomatic solution, Li began talks with the French minister to China in 1882 only to have Ferry return to office after a brief absence and repudiate his minister. When the next year Li sought to resume talks with a new French emissary, it was Peking's turn to interpose obstacles.

In desperation Li had in mid-1883 turned to the United States with a request for good offices that might break the deadlock and head off the impending conflict. Li directed his appeal to the United States in part by default; Britain and Germany, the only other plausible mediators from Li's perspective, were too much the rivals of France to be eligible. American initiatives also influenced Li on this occasion as they had in the past. The American minister at this time happened to be John Russell Young, Grant's associate in the Liuchiu mediation. Having acquired a taste for East Asian international politics, Young had traded on his Grant connections to secure for himself the Peking legation. From it he hoped to make his mark on what he supposed to be a plastic East Asia. The friction between France and China held out to Young the same kind of opportunity to keep peace that he had watched Grant seize earlier.

Young's strategy was to install himself as honest broker. The first step was to win Li's good will by regularly passing on to him news of international developments, using his journalistic contacts to get the Chinese side of the Vietnam issue in print in New York and Paris, and lobbying with French diplomats on China's behalf. With the permission of the State Department, Young put forward his offer of assistance. He warned Li that China had no hope against a first-rank power and might have to pay for her temerity in resisting France in Vietnam with a sizable war indemnity, the loss of Canton, and even despoilation by an opportunistic Russia and Japan. He advised Li to depend instead on world public opinion to isolate France (a mechanism with which Li archly confessed himself unfamiliar). Though now less easily impressed by American diplomats, Li finally was driven in late July 1883, following the collapse of his talks with the French, to make the appeal Young had eagerly anticipated. To Young's disappointment—but probably not to Li's surprise—the American offer to act as intermediary suffered a flat French rejection.[38]

Through the second phase of the Vietnam crisis (December 1883 down to early August 1884) the success of French military measures in Vietnam fed belligerence in the Chinese court and among influential provincial officials (including prominently Tso Tsung-t'ang and Chang Chih-tung). Outspoken advocates of war in the lower levels of the bureaucracy repudiated Li's costly policy of appeasement and pointed to the favorable outcome of the dispute with Russia over Ili as evidence that China could stop aggression. The tangible result of this rising tide of opinion was the purge from the foreign office and Grand Council of those who had supported a conciliatory policy (including—foremost—Prince Kung). The Empress Dowager Tz'u-hsi then pushed to the fore Prince Ch'un, the father of the Emperor and an exponent of greater resistance. However, by late spring even Prince Ch'un began to think a negotiated peace preferable to a losing war. Standing in his way were formidable French demands (including an indemnity and recognition of French territorial and commercial claims) and, on the other side, an uncompromising imperial edict ruling out any indemnity not to mention some greater concession.[39]

Now it was Prince Ch'un's turn to turn in desperation to the United States.[40] Taking a page from Li's book of tricks, he sought to get the United States to exert its influence against the French indemnity. In a direct approach to Young in July 1884, Prince Ch'un invoked the article of the 1858 treaty providing for American good offices in times of international crisis and asked Washington to judge the appropriateness of an indemnity. Once again Young had encouraged the Chinese overture—this time by making the baseless claim that the President inclined toward China's position. The State Department quickly revealed Young's presumption when, in response to the Chinese appeal, it refused to pass judgment on the indemnity, promising instead only to present China's case to France and vice versa. Even this watered-down American role as go-between proved unacceptable to Paris, which was still insistent on the indemnity and opposed to American meddling. Young's identification with the Chinese side and the transfer of the cream of China's merchant marine to the safekeeping of the American firm of Russell and Company in anticipation of war had already given the French good grounds for doubting the Americans could play an impartial role.[41] Undaunted, both Young and the State Department continued to search right down to the outbreak of fighting in August for some mutually satisfactory formula. In this the Americans were now encouraged by the Chinese foreign office only because it hoped diplomacy might win time to push military preparations farther along.

The third and final phase of the Vietnam crisis began in August and September 1884 with the French trying to enforce their indemnity demands by shelling Keelung on Taiwan, blockading the island, and destroying China's southern fleet in Foochow. Alarmed lest the fighting spread farther along the China coast and threaten American commerce, Young now repeatedly warned the Chinese of disasters yet to come—most fearsome of all, another expedition against North China—unless they quickly conceded France the indemnity (perhaps in the more palatable form of some economic concession, such as a railway from Tonkin into South China or possession of Keelung or the island of Hainan). Young's warnings were accompanied by long sermons on modernization, wearisomely familiar to those charged with dealing with garrulous foreigners. The foreign office was even less interested in Young's proposed concessions; for the moment it wished to appear unmoved (though it was undoubtedly shaken when in October rumors began circulating in Peking that France had joined secretly with Russia and Japan to launch a coordinated assault on China). The time for flexibility would come, the foreign office reasoned, once France realized it could not force a populous and enraged China into submission and would have to negotiate its way out of an impossibly costly war. Rebuffed in Peking, Young took his fears of another allied expedition to Li, in Tientsin. China's defenses, Li coolly responded, were better than in 1860, and backed by an aroused population. Young left with only vague assurances on the peacemaking role he might play and discomfited, as Li reported back to Peking, that he had found "nowhere to stick his beak."[42]

Late in 1884 the tide of battle along the Vietnam border turned in China's favor, and at the same time domestic opposition shook Ferry's war policy. Under circumstances now more auspicious to China, Li and Robert Hart, the head of the Chinese customs service, sallied forth on the diplomatic field to compete for the laurels of peacemaker. In this final round of diplomatic maneuvering Young retired to the sidelines, leaving the State Department to give good offices one last try. Acting on a French initiative, Washington put forward in November its own basis for peace, including a nominal indemnity of 5 million francs as a salve to French honor, a temporary French occupation of the Taiwan ports of Tan-shui (Tamsui) and Keelung, and some yet unspecified commercial privileges.[43] Peking still knew it could do better and refused. Instead, it approvingly watched Hart, exploiting his European contacts, set talks in motion in December 1884. The final settlement four months later included none of the concessions Young or the State Department had thought necessary; China simply abandoned her claim to sovereignty over a Vietnam now under French occupation.

Sino-American contacts over the tribute states entered a long hiatus, which was broken finally after nearly a decade by the complete collapse of Li Hung-chang's Korean policy. Li had tried to defuse a Korean confrontation with the powers by opening Korea to foreign influence. But in a Korea that was weak and internally divided, foreign influence had spread rapidly and fed international rivalries. Korea was too important strategically for China to abandon, and so Li had held on, hoping to maintain some control. To strengthen China's weakened position he had dispatched a Chinese diplomatic and military presence, sought support from Britain and Germany, and when that seemed insufficient, tried cultivating Russia, the least dangerous of China's two threatening neighbors. Japan, for her part, had by 1893 resolved on war if necessary to carry out a program of reform within Korea designed to undercut Chinese influence and arrest Russian penetration. Sino-Japanese negotiations subsequently failed to resolve these divergent interests, and the Chinese court resolved, in an imperial conference on July 16, 1894, to fight. The prevailing view at court, supported by the Emperor and perhaps the Empress Dowager, was that further Japanese intrusion in Korea was intolerable, coming as it did after the loss of the Liuchius and Vietnam, and likely to stir up the other powers to China's further disadvantage. With war in sight Li began to seek foreign support, but for the moment none of the eligible powers—Britain, Russia, and Germany—would leave the sidelines.

At the bottom of Li's list was the United States. In the pattern already established during the French crisis, Li would seek American assistance not as the first hope but as the last expedient. In early July, with war in the offing, he finally made his appeal to Washington through both the American consulate in Tientsin and the Peking legation. Washington's response indicated that the Cleveland administration, like its predecessors, wished to avoid political entanglements even more than to encourage peace, and so it recoiled from this and similar appeals from Britain and Korea to join in discussing ways to avert war. It instead issued a toothless cautionary warning to Japan not to transgress against Korea, and it followed with the customary offer of good offices if desired by the principals in the dispute. But above all else, Secretary of State Walter Q. Gresham assured the world, the United States would maintain impartial neutrality. Tokyo, now certain that Washington posed no risks, declined American good offices. For Li the American response was a reminder that the Americans, however well-intentioned, would not play power politics and hence could not play a role in the anti-Japanese coalition that he hoped to build.[44]

The Japanese and Chinese fleets clashed on July 25, 1894, and formal declarations of war followed on August 1. While Li focused his attention on the military front (the scene of an immediate series of Chinese reverses), the foreign office attended to diplomacy. In October, with Japanese forces moving into Manchuria, it supported a British initiative for peace on the basis of a Chinese indemnity to Japan and a guarantee of Korean independence. Gresham, invited to pass these terms on to Japan, refused and reiterated the limits of American interest by making the sweeping public announcement that the war endangered "no policy of the United States in Asia." Japan, in any case, turned the British proposal down flat, and Russia and Germany still did not stir. Early the following month the foreign office revived the British peace terms, but was no more successful in conjuring up a coalition of European powers to save China from Japan. Driven to desperation by the deteriorating military situation, the foreign office supported by the Chinese minister in the United States at last invoked the good offices provisions of the 1858 treaty, as had happened ten years earlier. Gresham agreed to sound out Japan (an initiative that Tokyo turned aside) while issuing the usual disclaimer that the United States would not be implicated in support of the policies of self-interested European powers.[45]

The prospects for energetic American mediation were further dimmed by the conduct of Charles Denby, Jr., in charge of the legation in August, the first month of the war.[46] His most controversial decision was to take into protective custody two Japanese charged by Chinese authorities with spying. The younger Denby refused to hand the accused, in his estimate mere schoolboys and "probably innocent," over to the local Chinese magistrate, who would subject them to "horrible cruelties and tortures." The Chinese legation in Washington quickly convinced Gresham that Denby's position had no basis in international law. After some foot-dragging Denby finally in early September reluctantly bowed to Washington's orders to surrender the Japanese, who were executed a month later on the orders of the central government. Denby's stand had generally compromised the legation in Chinese eyes. At home the affair had supplied ammunition to Republican critics of the Cleveland administration's foreign policy, further diminishing Gresham's own already limited willingness to take risks in behalf of peace.[47]

Despite the damage done by his son's indiscretions, Charles Denby, Sr., the American minister, was ready to tread where Washington would not. Spurred back to his post by Gresham's anxiety over Denby, Jr.'s imprudence, the elder Denby arrived in Peking ready to promote an early peace under American auspices—regardless of the wishes of the combatants and Gresham's preference for a relatively passive American role. The war had already served its purpose, Denby had concluded, by impressing on the Chinese the need for massive reform, but further drubbings by Japan, which had already taken Manchuria and now threatened Peking, were more likely to produce demoralization and disintegration than the hoped for national renaissance. To achieve a prompt peace Denby stood ready to suggest terms, prod the Chinese to action, and even draft credentials for China's peace negotiators.[48]

For four frustrating months Denby had to contend with a distinct lack of cooperation from both warring parties. The Chinese too wanted peace talks promptly begun, but they had in mind not capitulation but rather exposure of the full sweep of Japanese ambitions, which would in turn, they hoped, provoke intervention by the powers. Li, in his despair over China's military position, saw salvation only through a diplomatic coup against Japan led by Russia and Britain. Though Germany and the United States might also worry over Japanese gains, Li concluded that little was to be expected of countries so detached from East Asian power politics.[49] The Japanese, however, understood the dangers of a premature disclosure of war aims and played for time for the Japanese army to advance. At last ready to talk, Toyko refused a cease-fire and insisted that China be represented by someone of rank carrying full authority to conclude peace.

As self-appointed adviser to the Chinese foreign office, Denby helped prepare for the talks. While the foreign office tried to coax Denby into uncovering Japanese terms, he privately cursed the Chinese for their folly. They seemed not to see the utter hopelessness of their situation and the advantage of putting themselves in his experienced hands. The men he dealt with at the foreign office were "piteous and helpless," like "great school-boys looking for a teacher." After Japan's rejection of China's first peace emissary (a German adviser employed by Li Hung-chang), Denby helped get up the second peace mission led by two imperially-appointed officials and then boasted, "I have, alone, done the troublesome and difficult work of inducing China to sue for peace. . . ." He soon had to eat crow when the Japanese rejected as inadequate the emissaries' credentials, drafted (so the Chinese claimed) by Denby himself.

At last in February 1895 with Japanese troops now lodged on the North China coast and threatening Peking, an imperial edict removed all grounds for Japanese objections by imposing on a reluctant Li Hung-chang the task of traveling to Japan with full powers to make peace. Denby was eager to continue to guide China toward peace. He urged Li to accept all of Japan's demands—not only for an independent Korea and an indemnity but also for a cession of Chinese territory and treaty revision. Better to capitulate to Japan than run the risk of intervention by European powers, who would also seek Chinese territory in payment for "saving" China. In the long run China's true salvation, Denby counseled Li, lay in an alliance with Japan against Europe, a program of reform, and development of railroads, mines, and banks all under the direction of the English-speaking people.[50] Li, whose thinking both on peacemaking and the long-term usefulness of the United States diverged not a little from this vision, responded by relegating Denby's legation to the role of a simple transmission belt for messages between Tokyo and Peking pending restoration of formal diplomatic relations.

The American role in Li's peacemaking was, aside from the

legation's nominal involvement, to be limited to private Americans whose loyalties were long proven (notably William Pethick) or securely purchased. The most outstanding of this latter category was John W. Foster, an international lawyer who was briefly Secretary of State at the tag-end of the Harrison administration, and more recently adviser to foreign legations in Washington, including China's own. Foster had toured China in 1894 and paid a courtesy call on Li and the foreign office. Early the next year Foster was hired for a reported $25,000 to accompany Chang Yin-huan, one of the Chinese ministers Foster had earlier advised in Washington, in China's second peace mission. If the Chinese wanted Foster present to remind Tokyo of American interest in the talks, Gresham undercut them by publicly disassociating the administration from this partisan of the Chinese and underlining the point in private talks with the Japanese. Foster still played out his role, joining Chang in Japan in January, returning with the Li mission in March, accompanying Li's son in arranging the cession of Taiwan to Japan, and finally traveling to Peking to argue for the ratification of an unpopular peace.[51]

Li's mission failed to alter Japanese terms, but it did serve, as he had hoped, to set the stage for a joint Russian, French, and German demand that Japan restore to China the Liaotung peninsula in the southern part of Manchuria. Unfortunately for Li, the powers did not include Taiwan in their demands; they stipulated that China pay an additional indemnity in place of the Liaotung; and for their efforts secured the funding of China's war indemnity and with it a better grasp on the government's finances. Li's Korean policy had come to a disastrous conclusion—not only was all Chinese influence on the peninsula gone but much else as well. Officials high and low protested bitterly and some demanded Li's execution. Only the influence of the Empress Dowager saved him and the peace.[52]

Washington for its part was as careful after 1895 to stand clear of the struggle for Korea (now between Japan and Russia) as it had been before despite the often anguished appeals by American diplomats and missionaries, distraught witnesses to the gradual demise of Korean independence. Finally, in 1905 Theodore Roosevelt conceded preponderance to the triumphant Japanese and withdrew the American legation in Seoul. More concerned with interests in the Philippines, Roosevelt could accept Korea's new, more restricted form of dependency with the happy thought that Koreans were better off under the modern-minded Japanese than if left to themselves or the reactionary Chinese.

In August and September 1896 Li Hung-chang passed through the United States in an altogether routine but well-publicized courtesy call, most often remembered today in relation to a Russian treaty of alliance carefully hidden away in his baggage. Despite the recent failure of his military establishment and his implication in a one-sided peace, Li nonetheless still cut an impressive figure. To the warm, welcoming crowds in New York City, Li stood forth as the great mandarin, attended by an appropriately large entourage including a younger son, Chinese aides, foreign advisers, and personal servants. A band of Americans, some interested in Chinese souls, others in Chinese concessions, also waited on Li. President Cleveland, Secretary of State Olney, and five other members of the cabinet arrived to offer greetings with due pomp and formality. Li then set off on a journey that in a measure recapitulated the stages of his earlier diplomacy—a call at Grant's resting place in New York and visits with John Russell Young in Philadelphia and John W. Foster in Washington.[53]

As he proceeded on his way, Li may have reflected on the reasons why his visit was not a celebration of the special Sino-American strategic relationship he had once thought possible to build. European and Japanese imperialism had been the common enemy—a direct threat to China's security and a challenge to American pretensions to influence. Initially Li had perhaps been misled by those Americans overseas who had played down the limits of American power so he had had to learn those limits for himself. But the problem went considerably beyond Washington's cautiousness to fundamentally divergent visions of what was at stake in China's border areas. While Japan had been to Li the supreme threat, that country had appeared to Grant, Young, Shufeldt, Sargent, Denby, and Gresham as well as a variety of Americans active in Korean politics as the most hopeful local champion of a progressive East Asia free of European interference and control. While Li justified his effort to preserve influence in strategic border states in terms of China's security and prestige, the dominant American reaction was to deplore that policy as the reactionary reflex of a dying empire and as an obstacle to the spread of the principles of free trade, national independence, peace, and progress. American policy, whether shaped by Americans on the scene or by Washington, would have caused Li problems. If the Americans Li dealt with had had their way, the United States would have followed an intrusive, paternalistic path, marked chiefly by a contempt for China's seeming cultural backwardness, a growing impatience with her apparently impotent and inept leaders, and a rejection of her claim to spheres of influence under the guise of tribute relations. Washington, on the other hand, followed a cautious line of policy which neither irritated nor threatened China but also offered none of the hoped-for relief to embattled Chinese policy makers.

In the few years before his death in 1901, Li must have felt spent and downcast. He had watched the failure of his diplomacy and his army and navy, the decline of his own political influence at home, and a rising resentment against him for the power he had accumulated, the corrupting foreign influences his reforms had introduced, and the humiliating concessions and compromises he had made in foreign affairs. Over his lifetime Li had labored in an increasingly hostile international environment, in the heyday of imperialism, and he had little to show for his efforts. The border states that he had fought for—Korea, Taiwan, Vietnam, Liuchiu—were now all lost. The Russian alliance, recently purchased at the cost of concessions in Manchuria, did not stem the scramble by the other powers. And then the Boxer crisis would thrust Li forward for the last time, to accept a crushing peace in order to save the dynasty and clear Peking of foreign troops.

It is, sad to say, an open question whether Li's resilient balance of power policy had materially slowed foreign penetration or was even significantly superior to a more militant policy of resistance had it been sustained to the utmost. But there is no question that Li had carried forward the formidable task, begun by the barbarian managers earlier in the century, of identifying contradictions among the powers and trying to turn them to China's advantage. Like them, he had sought to introduce the United States into the East Asian equation of power in a way beneficial to China—and he suffered the same lack of success. Soured by his experience, Li downgraded the United States in his last years. But China's recurrent crises and the apparent attractiveness of the American position in East Asia at the turn of the century and beyond was to keep China's U.S. strategy alive. Disapprove now though Li might, newcomers to Chinese foreign policy, guided by what he now regarded as illusions, would try yet again to make that strategy work.

Chapter Five

American Policy and Private Interests, 1860–1899

IN THE LATTER part of the nineteenth century the open door constituency reached the apogee of its influence. It exercised that influence not in China, for a handful of Americans, however exaggerated their sense of self-importance, could have only an imperceptible impact on the lives of a people numbering some 400 million. It was rather in moving Washington toward a more active China policy that they made their mark. In effecting this policy transformation Americans in China were inadvertently assisted by Chinese hostile to the mission movement and by plotting foreign imperialists. The threat they posed led businessmen, missionaries, and diplomats— each in their own way—to employ a rhetoric that idealized their concrete self-serving goals. Because the activities of each group contributed (so they argued) to China's regeneration and to the realization of an American Pacific destiny, each could legitimately make a claim on Washington's support. A flurry of economic activity in the 1890s, viewed hopefully as the harbinger of more to come, strengthened the voice of commercial interests. But at the same time the articulate representatives of the expanding mission enterprise and a chain of vocal ministers in the Peking legation demanded that policy makers attend to their aspirations as well and thus by degrees diverted policy away from its earlier, essentially commercial orientation.

Economic Enterprise— Stagnation and Revival

For roughly two decades—from the 1860s through the 1880s— American economic enterprise in China stagnated, eroding the merchants' dominant position in the open door constituency. United States exports to China fell from a high of $9 million in 1864 to a low of $1 million in the mid-1870s,[1] while the number of American firms operating in China declined from 42 to 31 between 1872 and 1880, a time when the enterprises operated by other nationalities actually grew in number. The most serious casualties were the great merchant houses that had dominated American economic activity before the 1860s. Augustine Heard and Company failed in 1875, Olyphant three years later.

For a time after 1860 the merchant houses seemed vigorous as they continued the older pattern of diversifying their operations.[2] But already by the 1860s and 1870s these firms were beginning to lose their commanding middleman position in the international economy. The opening of the Suez Canal, the completion of transoceanic telegraphs and transcontinental railroads in North America, and the increasing use of fast steam vessels made up-to-date market information available and rapid action possible for traders in the United States without resort to one of the large merchant houses in China. In their place began to appear specialized export houses, nascent multinationals, investment syndicates, and Chinese firms run by one-time compradors with a growing familiarity with the workings of the international market and an unmatched grasp of the interior trade. An array of other developments—the rise of a British tea trade with the United States, the growing availability of tea from India, Ceylon, and Japan and cheap cotton from India, the disruption of American trade during the Civil War years, the failure of commerce in the Yangtze Valley to increase as expected in the 1860s, the fall in the gold-silver exchange rate through the 1860s and 1870s to the distress of exporters, and finally the disappearance of old constituent firms in the United States—all cut deeper and deeper into the profits of Russell and the others. At the same time that the China trade was becoming more precarious, the American economy continued to offer a higher yield on invested money and effort (an edge established before 1860) that convinced Russell and Company's principals to begin a process of gradual disinvestment that culminated—after nearly seven decades of business in China—in the 1891 closure of the firm. China had no special mystique to bind them when the ledger books said go.[3]

Russell and Company's retreat significantly reduced the level of American investment and dealt a particularly devastating blow to American shipping.[4] In 1870, 43 percent (3 million tons) of China's foreign and coastal trade traveled under an American flag. Russell accounted for much of this. Starting in 1861 with five ships and an investment of $1.3 million (less than a third from Russell;

over a third and perhaps close to a half from Chinese), the Russell-managed Shanghai Steam Navigation Company quickly established itself as the largest concern of its kind in China and so remained for fifteen years. Its operations along the China coast and up the Yangtze generally yielded dividends of 12 percent (one year's returns soared to 50 percent). Even after its success attracted British and Chinese competitors, the returns remained steady and substantial, thanks in part to a now-and-again division of the market. Those returns (7 percent in the last three years) could no longer, however, match what the partners in Russell and Company thought they could make on domestic investments. The sale in early 1877 of their sixteen ships and associated property to a competing Chinese government-sponsored and Chinese merchant-managed shipping line for $3.9 million, twice the start-up value, made the shift in investment strategy both possible and profitable. But the sale caused the overall share of trade handled by Americans to plummet to .6 million tons. Through the rest of the century and down to the outbreak of World War I American vessels would normally claim no more than 2 percent of the trade.[5]

With the coastal trade practically gone, the American flag was now visible only in the trans-Pacific carriage maintained by the scheduled service of the Pacific Mail Company and by sailing vessels carrying bulk goods (bullion, flour, and lumber on the westward journey; teas, silks, Chinese immigrants, and provisions for sale in Chinatown going east). However, sailing vessels gradually lost their competitive edge against steam and disappeared from the scene. At the same time, the Pacific Mail suffered a series of reverses from the 1870s on that foreshadowed the virtual end of the American trans-Pacific carriage. The entrance of Japanese and later Canadian competition precipitated a costly rate war. Congress made matters worse by withdrawing the line's subsidy in 1875; by imposing Chinese exclusion, which cut the flow of immigrant passengers and reduced the special Chinatown trade as the Chinese population dwindled; by excluding from American registry the superior British-built vessels of iron and steel and slapping duties on special steel that American shipbuilders needed to build ships of comparable quality; and by imposing costly safety and labor regulations. Finally, by the turn of the century railroads had integrated the West Coast into the national market, diverting eastward flour and lumber that had once been sold abroad for want of buyers at home. Under this rain of blows, the Pacific Mail began to retrench, first giving up its Shanghai branch line on the run between Hong Kong, Yokohama, and San Francisco, and then from the 1880s gradually cutting back ships in service. Taken over by a combination of transcontinental railroads headed by E. H. Harriman's Southern Pacific, the Pacific Mail limped along until 1925 when it finally closed down altogether.[6]

The retreat of Russell and Company and the Pacific Mail was symptomatic of the general American reluctance through the 1880s to put money into China. Aside from shipping, direct investments were scattered, small in scale, and evanescent.[7] The few efforts to start new projects in China in these years were sometimes grandiose in conception, often dependent on Chinese capital, and generally speculative in nature. Predictably they yielded nothing solid.

Attempts made by American promoters to link China and the United States by telegraph fall squarely into this pattern. Perry McDonough Collins of the Western Union Company came up with the first and most audacious of these schemes—to bring China's "great commercial cities" and the "over FOUR HUNDRED AND TWENTY MILLIONS" of her customers into a globe-girdling telegraph system embracing the northern Pacific as well as Latin America. But a variety of political and economic obstacles ultimately killed the project. While Secretary of State Seward sought to "facilitate" it, Congress withheld the subsidy which Collins wanted and on which the economic feasibility of the enterprise probably depended. The Chinese government, though advertised by Collins as "particularly favorable" to American enterprise, also failed to cooperate. It did not want foreigners controlling vital communications and had no wish for a confrontation with a public hostile for geomantic reasons to on-shore telegraph lines. One ranking official, Tseng Kuo-fan, probably came close to the truth when he concluded that the Americans could not really meet the costs and difficulties of construction. The Chinese foreign office rejected Collins' proposal and another in 1879 by Americans who obligingly invited the Chinese to provide the funds not available to them in the United States.[8]

Textile milling was a second area of investment in which Americans were hobbled by their own lack of capital and by Chinese economic nationalism. A good case in point is the effort by W. S. Wetmore, head of the import-export firm of Frazar and Company, to set up a Shanghai cotton spinning mill in the early 1880s.[9] He had gotten together the necessary funds ($410,000—much, perhaps most, of it supplied by Chinese merchants eager to put their wealth beyond the government's reach). He had secured assurances from the American consul general that his plan was in accord with treaty rights, and he had ordered his equipment. Only then did Chinese officials point out that Wetmore's enterprise violated a ten-year monopoly earlier given to a government-sponsored, merchant-managed Shanghai mill and threatened the purpose behind the monopoly, to promote an infant industry and eventually to retard foreign imports and the resulting drain on the Chinese economy. In any case, the Chinese argued, there were no treaty provisions giving foreigners the right to manufacture. While Minister John Russell Young privately accepted the validity of these arguments and expressed sympathy for "any sincere effort on the part of Chinese officials to create new branches of industry," he publicly aligned himself with his diplomatic colleagues in a contrary reading of the treaties. He challenged the monopoly as "a violation of the modern laws of practical economy" and called up gunboats to back his position. Chinese authorities in Shanghai finally settled the dispute by threatening Wetmore's comprador with imprisonment on an old charge of trading with rebels. The comprador quickly backed out of the project, and suddenly the indispensable merchant capital dried up.

The Sino-French crisis over Vietnam created a flurry of interest in political loans, the most speculative area of American investment. The widely held view that the reverses suffered during the war with France would force the Chinese to launch a costly program of reform and development attracted the attention of American investors. A series of official overtures gave the view credence. Li Hung-chang had asked Minister Young in the summer of 1883 about a loan for $50 million, ostensibly for railroads and other public works (and received Young's assurances that a "plethora of capital" was waiting to be tapped!); the governor of Taiwan had approached Russell and Company; and the foreign office had sounded out the American chargé Chester Holcombe on the availability of money, purportedly for railroad construction. Yet nothing came of these. American overtures were equally inconclusive. James H. Wilson, a railway promoter, arrived in China in October 1885 to talk to Li about a concession and over the winter toured promising routes. The loan terms proved unattractive—no long-term stake in the railroad for American investors and insufficient security for the loan—and Wilson went home. The next year, in mid-1887, a Count Mitkiewicz, representing the Philadelphia financial promoter Wharton Barker, walked into Li's Tientsin office with a plan for combining equal amounts of Chinese merchant capital and American money into a $10 million bank fund for the development of mines and railways. Final negotiations between Li and Mitkiewicz collapsed when Peking vetoed the terms of the deal.[10]

At last in the 1890s previously dim prospects gave way to China market fever. The trade statistics, already beginning to point upward in the 1880s, climbed at a rapid rate through the nineties. American exports, overwhelmingly cotton goods and kerosene, began the decade at $3 million and ended it at $14 million.[11]

Cotton goods, sent to China as far back as the 1830s, had now become a major export as American mills gained a clear edge over the long dominant British in the markets of North China and Manchuria. While the British industry with its skilled work force and accumulated experience continued to excel in finer grades of cloth, Americans took over in the cheap coarse goods, thanks essentially to lower wages (particularly in the southern United States) and more advanced machinery. The British also had to contend with the higher transport costs involved in importing American raw cotton as well as getting finished goods to China. Finally, the presence of size (a glutinous preparation added to keep up weight) in British goods turned Chinese buyers in favor of the sounder American product. By 1875 American cotton textiles had begun cutting into British sales of coarse goods, emerged dominant in the 1880s, and by the late 1890s accounted for anywhere from five- to seven-tenths of the value of China's cotton goods imports as well as a substantial portion of total U.S. cotton exports.[12]

Kerosene, the other major component in the American export drive, had begun to enter the China market as early as 1867 chiefly for use by foreign residents. Not until the late 1870s did its sales begin to boom. The 5.4 million gallons (valued at $690,000) imported in 1879 were in turn to increase to 40.4 million ($2,436,000) by 1894. China's share of American kerosene exports climbed between those years from 1.6 percent to 5.5 percent. Supplying oil for the lamps of China was becoming big business. The chief beneficiary was Standard Oil of New York, the export arm of the Rockefeller petroleum trust. Initially, it had followed a passive marketing strategy (similar to that of the cotton goods industry), selling kerosene to New York export houses and thus severing interest in its product at the water's edge. But that strategy would not tap the enormous potential of China. (Standard itself calculated that the Chinese quarter of mankind in 1887 consumed only one-sixty-second of all petroleum production.)[13]

Standard took its first step toward active involvement in marketing in the mid-1880s by sending William H. Libby, its diplomat-troubleshooter, to China to fight restrictions against kerosene's use and to argue its benefits. In 1890 Standard began handling its own shipments through a British affiliate that transported its kerosene to Shanghai and other major Far Eastern market cities for sale through local commission merchants. Other innovations soon followed, including removing from kerosene the volatile naphtha that made it hazardous in crowded urban areas, marketing a cheaper kerosene of lower grade to match the popular Russian product, transporting and storing its goods in bulk, sending its own agents to Hong Kong and Shanghai, widely advertising the virtues of the fuel, and introducing small, cheap lamps suitable to the market.

These commercial advances were accompanied by another burst of interest in investment; only this time Americans would in the end have something to show for their effort. Investments valued at $6 million in 1875 were to grow to $20 million by 1900, thus preserving for Americans approximately the same share in China relative to other nationals. American money began to go into direct investments early in the 1890s.[14] At mid-decade the Treaty of Shimonoseki (ending the Sino-Japanese War) encouraged the investment of still more American capital by giving foreigners the undisputed right to establish industry and especially textile mills in China.[15] Finally, Americans even moved into mining seriously for the first time.[16]

From 1894 China's deepening international crisis set Americans once again in pursuit of speculative, politically sensitive investments—at first with no more luck than earlier. The Merchant and Chemical Bank in New York opened talks with the Chinese legation about a £1 million loan, but withdrew when Chinese military reverses ruined investor confidence. After the war the indemnity owed Japan and a program of railroad development put China back on the international financial market on an unprecedented scale. To meet the indemnity, first Chester Holcombe, the former chargé, in association with John W. Foster, and later a syndicate representing J. P. Morgan came up with loan offers. But both were squeezed out by Russian, German, and French bankers whose governments' support against Japan's peace terms gave them a prior claim on Peking's affection. Two familiar figures, Wilson and Barker, were roused by talk of railway building. Barker, after marshaling support at home, set off for China in October 1895 with a $125 million proposal only to be rebuffed.[17] Wilson, who had established himself as a China expert by virtue of a book written after his 1885–1886 visit, hatched a plan with John J. McCook, a railway lawyer, to pry a railway concession from his old acquaintance Li Hung-chang. They planned to secure Wall Street money and the blessings of the incoming McKinley administration and of the Russians (then ascendant in North China), and then to develop the concession as part of a global transport system. But Li, whose political difficulties at home the promoters seemed oblivious to, could give them no more than the time of day during his 1896 stay in New York City. Plans to get their supporters into key positions in the new McKinley administration miscarried too, in part because Wilson's constant importuning annoyed even that paragon of patience, William McKinley. The Russians, whom Wilson had cultivated, weren't offering anything but after-dinner toasts. And financial giants on Wall Street remained noncommittal.[18]

But even speculators at last scored a triumph when the American China Development Company obtained the concession for a railway to be built between Hankow and Canton.[19] The company, organized in 1895 in anticipation of expanded postwar investment opportunities, was headed by Calvin Brice, a former Ohio senator and railway lawyer who moved in New York financial circles. It included among its shareholders the railway magnate E. H. Harriman; Jacob Schiff of the investment house of Kuhn, Loeb and Company; the presidents of the National City Bank and the Chase National Bank; the former Vice President of the United States, Levi Morton; an associate of J. P. Morgan and Company; and the Carnegie Steel Corporation. Despite the stellar support in the worlds of politics and finance, the American China Development Company seemed at first likely to suffer the same fate as its predecessors. It initially followed the Wilson-McCook line of seeking a concession for a North China section of a railroad running on through Manchuria and on to Europe, and it too failed. The Russians, whose support was thought essential to making any headway in either North China or Manchuria, were unresponsive. In a second setback, the company's agent, sent to China in the spring of 1896, nearly nailed down a major line connecting the capital with the growing port city of Hankow before the Chinese negotiator, Sheng Hsüan-huai, balked and gave the concession to the Belgians instead.

Then in April 1898 the Chinese government, for political reasons of its own, offered the Americans the southern extension of the Peking-Hankow line running down to Canton. The terms were exceptionally attractive. Americans were to arrange the £4 million loan (backed by an imperial guarantee of repayment), buy the equipment, build the line, and operate it over the fifty years the loan was to run. The company retained an option to build branch lines and to operate coal mines in land adjacent to the railroad. The Americans thus gained a firm economic position in the heavily populated and highly productive section of the central southern interior through which the railway was to run. And in case the Belgians surrendered the projected Peking-Hankow line, the American company would have first option to take it up and extend their influence northward.

A growing trade and a prize railway concession inspired the belief among Americans that they would enjoy a prominent part in the broad economic development that China was sure to undergo in the decades ahead. But roseate hopes for railway building, mining, and trade expansion contended against mounting fears that the advancing European powers and Japan might snatch away the China market just as it was at last about to realize its potential. The pattern of imperial penetration ending in formal territorial control, already worked out in Africa and parts of Asia, had begun to repeat itself in China. The powers were using political loans and their diplomatic and military muscle to stake out spheres of influence. The German seizure of Kiaochow in Shantung in November 1897 prompted the Russians in turn to take Port Arthur in Manchuria. France grabbed new concessions in the southern provinces of Yünnan and Kwangsi bordering on French Indochina. Japan laid claim to a sphere of influence in Fukien. Even Britain joined in, taking a naval base of her own on the Shantung peninsula and new lands adjoining Hong Kong, and strengthening her position in the Yangtze Valley. China's manifest weakness, growing great power rivalry, and Britain's wavering support for the principle of free trade and China's integrity suddenly made the carving up of the empire an imminent possibility.

With the future of American trade and the American role in the pending drama of Chinese economic development hanging in the balance, the commercial interests within the open door constituency appealed to Washington for support. A look at the sponsors of that appeal reveals the changing nature of American involvement in the China market. Once dominated for all practical purposes by a few merchant houses exclusively concerned with China, the business element in the open door constituency had become both broader and more diffuse, with few firms approaching the old intensity of concern or depth of commitment of a Russell or Heard. It now embraced import-export firms; export industries such as cotton textiles, petroleum, and railroad equipment; and a class of speculative investors and promoters including some of the leading figures on Wall Street. This diverse group used an array of commercial pressure groups, including the National Association of Manufacturers (organized in 1895), regional chambers of commerce, and industry and trade associations, all as instruments for political action. The spearhead of the campaign was the American Asiatic Association, organized in early 1898 with the primary support of an alliance of New York–based import-export firms (dealing chiefly in textiles) and the southern cotton goods industry. John Foord, a contributing editor of the New York *Journal of Commerce,* became the Association's secretary, its chief spokesman, and the most persistent public exponent of the China market. Articles in the *Journal of Commerce* and the *Journal of the American Asiatic Association;* numerous petitions from trade groups; and calls by select committees of businessmen on the President, the Secretary of State, sympathetic congressmen, and even the State Department's China expert W. W. Rockhill all carried one message—the need for the government to dramatically reaffirm its long standing commitment to American commercial opportunity in China and perhaps thereby save Britain from her apostasy from the religion of free trade and check the headlong rush to slice up China.

The McKinley administration was slow to respond. The aged Secretary of State John Sherman was unsympathetic, and the administration as a whole was distracted by the prospect, then the complications, and finally the immediate consequences of the war with Spain. The tangible evidence—the hard facts and figures—that business could use to prove to policy makers the immediate importance of the China market simply was not there. Indeed, the available statistics showed that China was not crucial to American prosperity and that American enterprise was far from a formidable presence in the Chinese economy. The China trade, only 2 percent of total U.S. foreign trade at the turn of the century, was growing no faster than U.S. foreign trade generally. The same held true of China investments, then only 3 percent of all U.S. foreign investments. In China itself Americans could claim only 9 percent of total foreign trade and 2.5 percent of all foreign investments.[20]

In 1899, with the China crisis none improved and business pressure unabated, the McKinley administration decided to act. Except for the rebellion in the Philippines, the crisis of the war had passed, leaving the United States in an unprecedentedly strong territorial position in the Pacific. McKinley had taken Hawaii and

Guam as well as the Philippines, envisioned as commercial and naval stepping stones to China. John Hay, appointed the previous autumn to succeed Sherman, had had time to settle in at the State Department. His response to the China problem, drawn so insistently to his attention, was measured. In February he sent, for the "serious attention" of the Peking legation, a petition from the textile manufacturers and traders complaining of the Russian threat to their market in northern China. But it was not until seven months later that Hay took the next step. In August Rockhill, and through him Alfred Hippisley, a senior English employee of the Chinese Maritime Customs Service then visiting in the United States, urged Hay to take a public stand on behalf of the commercial open door.[21] Hay accepted their advice and in September began addressing each of the major powers in turn, asking for guarantees against specific forms of trade discrimination—in regard to the treaty ports, customs duties, and railway and harbor rates. Their qualified and evasive responses he translated publicly the following March into a "final and definitive" acceptance, and the thing was done.

These first open door notes were a token nod to the future possibilities of the China market and a tribute to the influence and persistence of the China trade pressure groups. But not much more. They did not extend a mantle of protection to investments. Indeed, Hay nowhere mentioned investments in his correspondence on this occasion except to concede that wherever in China the powers had funneled their capital they had created special interests. Further, he would not free the Peking legation from the old restrictions, much criticized by promoters, strictly limiting its ability to intervene with the Chinese government on behalf of American business proposals. And McKinley in his address to Congress in December 1898 stressed that his administration's concern was limited to the protection of commerce from discrimination by the powers. It was not even clear how far the administration would go even in behalf of trade, the chief concern of Hay's notes. Neither in public nor in private did it envision forceful action against powers who might seal off their spheres of influence against American trade. It gave no thought to the persistent and no less serious obstructions thrown by Chinese economic nationalists in the way of exporters such as Standard Oil and investors such as Wetmore. Nor did it follow up the notes with a program of trade promotion. When it cautiously proposed to Congress twice in 1898 and again the next year that at least a survey be made of ways to increase China trade, Congress withheld appropriations.[22] The open door notes had not dramatically altered China policy or transformed conditions in China, but Hay had managed by his dramatic public gesture to inspire business confidence that Washington was more committed than ever before to preserving American economic opportunity.

Missionary Expansion and Chinese Nativism

Despite past disappointments, American missionaries soldiered on after 1860, adding new personnel and extending their efforts beyond the safety of the treaty ports into the interior. This policy of expansion gave rise to a shrill and often violent Chinese nativism prompting missionaries in turn to call for naval and diplomatic protection. The resulting interaction among determined missionaries, aroused nativists, and uncertain diplomats was to define the history of the mission movement through the balance of the century.

The mission crisis had its origins in an extension of missionary activities that was steadier and far more aggressive than the performance of economic enterprise and that was in global terms to make China the chief focus of American Protestant activity abroad. By 1889 the once "feeble band" of missionaries numbered nearly 500 (roughly one-third to two-fifths of the Protestant force). American missionaries had by then gathered about them some 1,400 Chinese "helpers," over 1,300 converts, 9,000 students, and close to $5 million worth of physical plant (equal at least to one-quarter of business investments). Mission boards, both the long established and the newcomers, allocated ever higher levels of funding to extend their activities, first into many of the newly opened ports (some 32 by 1900) and then into the interior, with the treaty ports initially serving as a base of operations. To sustain the expansion new non-denominational mission organizations such as the Student Volunteer Movement for Foreign Missions sprang up in the late 1880s and after.[23]

The move into the interior, which began simply as another avenue for evangelists to explore, was to become by the 1890s the mainline mission strategy. The Chinese-language version of the Sino-French treaty of 1860 was instrumental in turning the gaze of missionaries inland. Alone among the treaties of the time, it contained explicit provisions for missionaries to purchase or lease land and erect buildings anywhere they wished. Though Chinese authorities correctly pointed out that this version of the treaty was nonauthoritative, they finally succumbed in 1865 in the Berthemy convention to French demands for recognition of these broader rights. American missionaries, invoking the most-favored-nation principle, immediately used the French diplomatic success to strengthen their own claim. Protestants, who had watched with envy the successful Catholic work in the interior, now themselves could move boldly—and they thought securely—into a fertile field and put behind the dry decades of evangelism along the coast.[24]

The developing campaign in the interior was from the outset marked by a clear division of labor between American missionaries and Chinese helpers. The former had already discovered how the latter could multiply the range of their effort and the number of their converts. Once an area of work had been selected, missionaries would seek out its most populous city and there set up a compound offering something like back-home levels of hygiene and comfort, physical security from harassment, and psychological refuge. From these outposts the missionary would recruit and deploy a corps of local Chinese assistants, who could go safely, comfortably, unobtrusively where foreigners could not. By degrees, a network of congregations would spread outward from the central outpost, connected to it through a relationship—often, it seems, a loose one—with the missionary as teacher-patron-employer.

Rapid and dramatic success in winning converts confirmed the wisdom of this new departure in mission strategy. For example, two American missions based in Foochow turned from "this monster city . . . dead and barren" to the countryside and by 1880 had a total of over 3,000 new adherents served by 96 new outstations. Missionaries employed converts as new helpers as rapidly as they could be trained and tested. (Still, helpers were so scarce during periods of rapid expansion and conversion that the three Foochow mission groups found themselves competing for their services until the missionaries agreed on common terms of employment to hold down salaries and other demands.) Again, much the same story could be told of the Tientsin mission's striking success in the countryside as early as the late 1860s after a discouraging effort in the city.[25]

This successful drive into the interior created inexorably in turn an interdependence between Chinese converts and missionaries that was to entangle the latter in Chinese life, perhaps more deeply than any other group in the open door constituency was to go. Once established, the mission stake had to be nurtured, financed, and defended. Since these new congregations were too poor to support themselves, it fell to the wealthy foreigners to cover the helpers' salaries, which were the congregations' main expense (five-sixths by an 1877 estimate). The missionary also offered the poor and the illiterate "bread other than the Bread of Life" (to use Minister Young's phrase)—including medical care, famine relief, and education. To the disaffected he held out the prospect—not always realized—of support in lawsuits and controversies over land, debts, taxes, business affairs, and criminal activity as well as freedom from certain local levies associated with religious and other holidays. Where persecution occurred, the missionary had to stand ready with demands for compensation and punishment for offenders. The missionary was repaid of course by the growing flock of the faithful who could keep him informed on local affairs, identify troublemakers, and assist in circumventing local officials when they refused to register mission property under a foreign name.[26]

Minister Anson Burlingame once predicted after watching the initial mission advance into the interior that it would "plant the shining cross on every hill and in every valley." The reality as the movement unfolded during the nineteenth century was, however, closer to the sardonic observation of one old China hand less given to enthusiasm: "poor Burlingame's 'cross' shines on the hills of China, only when the population burn it."[27] The threat of mob violence, sometimes condoned or even abetted by local officials, hung almost constantly over the missionaries and their converts scattered over the interior.

Violence sprang from multiple sources of opposition, combining in a variety of ways to produce an "incident."[28] To the difficulties of cultural incomprehension between missionary and Chinese that had plagued work early in the century was added the problem of deracination as a result of the growing incidence of successful conversions. By turning their back on some of the norms and obligations of the local community, converts destroyed its unity and harmony and became in a measure traitors to Chinese ways. No less objectionable to their Chinese antagonists was the missionaries' direct challenge to the local social and political order and the prerogatives of its elite defenders. The missionary appeared on the scene in a role coveted by the elite, that of teacher—and to make matters worse one whose irrational doctrine was sharply at odds with the prevailing Confucian ideology. He was himself not subject to local authority, and he could use his patronage and influence to shield converts from informal social controls and to a lesser degree official supervision. The governor general of Fukien gave voice to a common complaint of officials and local notables when he attacked converts in 1879 for "the audacity with which they brave their constituted authorities and bid defiance to public opinion." Hostility to mission work became more intense with increasing foreign encroachment against China in general. This privileged position the missionary carved out for converts amounted, officials continually complained, to a state within a state where the good were misled and bad elements sheltered. Some saw in the missionary movement into the interior something even more alarming, one prong of a coordinated foreign plot to subdue China.

This hostility gave rise to a graphic anti-Christian literature, which incorporated popular fears and fantasies aroused by mission work. Much revolved around the theme of sexual license and debauchery—rape, sodomy, incest, adultery—all performed in the name of the foreign religion, behind the protecting walls of the mission compound, or within chapels where men and women shamelessly worshipped together. Both prose and picture depicted the missionary as a lecher and sorcerer exploiting deluded or unwary Chinese. These "foreign devils" took in orphans, the sick, and the pregnant, and they kidnapped children, it was rumored, all to maintain the supply of human organs for medicinal purposes and sacrifices. They poisoned wells and food and cast spells to win converts, seduce women, heal patients, and even confuse officials. These libelous accounts of missionary motives and scandalous tales of misconduct on the part of foreigner and convert alike served to elaborate and confirm the suspicions aroused by direct contact with the missions and to sow the seed of hostility in areas where missions had yet to penetrate.

A proximate cause or precipitant was necessary to translate these broad and pervasive cultural, social, and political tensions into a physical assault. Missionaries might out of insensitivity or ignorance of Chinese law or custom blunder into a confrontation (as might occur, for example, when a missionary was confronted by the devilish complexity of Chinese law and custom with regard to property transfer—so different from the straightforward practices of sale and leasing the missionary had known at home). Disputes within a Christian congregation—for example, a convert disgruntled because his business had lost accustomed mission patronage—might spill into the open, lead to a campaign of placarding against missionary offenses and boycotting of the mission to enforce the injured party's demands, draw in local officials, and finally result in some act of violence if the missionary did not retreat. The shocking discovery of an otherwise unexplained child's grave in a mission compound or the sight of a fetus preserved in alcohol in a mission doctor's office might confirm the worst suspicions and arouse popular anger difficult to appease or protect the missionary against. Circumstances leading to an attack might, on the other hand, be out of the control of missionaries. The periodic massing at a provincial capital of exam students, high-spirited opponents of foreign heterodoxy, often proved a prelude to an attack. The chance arrival of a bundle of inflammatory anti-Christian literature or incendiary rumors of foreign aggression might throw a tranquil local scene into turmoil. For example, the wide circulation of anti-Christian pamphlets prepared in Hunan, that hotbed of nativist sentiment, preceded the rioting of 1891; the major outbreaks in 1886 and 1895 came hard on the heels of the conflicts with France and Japan; and the climax of unparalleled foreign seizures of Chinese territory and economic concessions in the late 1890s gave impetus to the Boxer movement.

Confronted by threats to converts and mission property as well as to his own life and those of his dependents, the missionary had no difficulty choosing between a fatalistic acceptance of providential adversity (even in extremity the martyr's crown) and invoking the state's protection. Missionaries, Americans notably included, were the source of ever more persistent appeals for intervention against Chinese persecution. They began by lodging at the door of consulates and the legation the problems of securing compensation for losses, release of imprisoned converts, punishment of offenders, and widespread publicity for the rights of missionary and converts. As the Chinese opposition became more violent, the problems became less tractable and increasingly commanded the attention of the State Department.

China policy was slow in accommodating missionary demands. As blocked out in the late 1860s and early 1870s by Ministers Anson Burlingame, Frederick Low, and Benjamin Avery, mission policy eschewed a clear cut response to missionary appeals.[29] On the one hand, the legation rejected the fundamental missionary claim that the missionary presence in the interior derived from explicit treaty rights. The legation agreed with the official Chinese view that the only authoritative version of the Sino-French treaty of 1860 (the one in French) contained no reference to property rights that would support the missionary claim. The other piece of evidence adduced by missionaries, the 1865 Berthemy convention, was not a treaty and hence had, so far as the legation was concerned, no applicability to American missionaries, even on the basis of the most-favored-nation principle.

Aside from these publicly voiced legal considerations, there were sound practical reasons, generally not publicly articulated, for the legation's withholding of support for broad missionary pretensions. Missionary activity in the interior ran counter to the general preference of Burlingame and his immediate successors to strengthen rather than weaken the authority of the central government, and it created turbulence disruptive of trade. Already aggressive French support of Catholic missions in the interior was playing havoc with the status quo established by the 1860 treaties by forcing the central government into the difficult position of going against popular and official feeling and by creating controversy and ill-will that would poison Sino-foreign relations in general. To unleash American missionaries would further strain relations with China.

But there were countervailing considerations, as even the earliest ministers to consider the problem conceded. Missionaries, however imprudent in conduct and excessive in their expectations of assistance, did nonetheless carry on laudable work. Even more compelling was the fear expressed by Minister Avery that a general missionary retreat from the interior, especially under duress, "would probably be hailed by the Chinese populace as a weakening of foreign influence and power, and might react disastrously upon other interests than those of religion." Commerce, diplomacy, and missions were in that sense interdependent.[30]

Thus pulled in different directions, the legation settled into a studied ambiguity in its dealings with both missionaries and the Chinese foreign office. It would not recall missionaries in the interior. Where the Chinese had failed to challenge their presence or had acquiesced in the presence of other foreign missionaries, American missionaries acquired a presumptive right to stay on. To contain the problem, the legation wished to discourage expansion of existing missions and the creation of new ones. It did not indicate, however, how it would enforce this freeze. Nor did it indicate what the legation would do to defend missionaries, particularly those who ignored its injunction "to respect the prejudices and traditions of the Chinese people," who imitated the Catholic missionaries' claim to official status, who intervened in local politics, or who gave refuge to criminals and troublemakers, and who thereby became embroiled in controversy or subject to attack.

This purposefully vague missionary policy began to come under pressure in the 1880s. With the increase in the number of missionaries at widely separated points in the interior went an increase in local frictions, assuming on occasion alarming proportions. The Chungking riot, one of a series of outbreaks in 1885–1886, began under the leadership of exam students, got beyond official control, and spread into the countryside. Both Catholic and Protestant converts suffered heavy losses before a new governor general arrived who restored order and punished wrongdoers on both sides. Other cases dragged on quietly, deepening bad feelings and increasing the potential for greater trouble. In Tsinan in Shantung province, for example, a property dispute between Presbyterians, led by Gilbert Reid, and their determined gentry-led opposition began in 1881 and continued for a decade. At one point Reid, described by one of his coworkers as having "neither the wisdom of a serpent [n]or the harmlessness of a dove," tried to break the deadlock by singlehandedly occupying the disputed property. A crowd gathered and dragged him into the street. Rather than retreat, the lone American reentered the house only to be ejected by the same crowd, grown larger and more surly. Beaten and semiconscious, Reid was at last rescued by a friendly constable.[31]

These mounting tensions created at the local level by missionary aggressiveness and Chinese hostility forced the Peking legation to reevaluate its position. Through the first half of the 1880s the legation, under first James B. Angell, then the chargé Chester Holcombe, and finally John Russell Young, moved toward a clearer commitment of support. Angell, who had prepared himself for China by seeking out two old missionary hands, S. Wells Williams and W. A. P. Martin, was energetic in protecting missionaries in Peking against harassment and secured an imperial edict, desired by missionaries, exempting converts from levies collected for local religious festivals. Holcombe, himself once a missionary, assured his former colleagues that the government stood ready "to give such moral support and encouragement and practical protection to its missionary citizens, either at the ports or in the interior, as their valuable work may justly deserve, and a sound policy and their manifest rights under the treaties may demand." In 1885 Young, at the end of his two-and-a-half-year tenure, called for doing "all that is possible" for the protection of missionaries. However much they acted with "tact, courtesy, and forbearance," they were bound to encounter opposition because they were engaged in an intrinsically aggressive enterprise involving nothing less than the transformation of Chinese. "The convert has so much to learn, so much to forget. . . ."[32]

The legation under Charles Denby reverted through the balance of the 1880s to the policy of limited and veiled support.[33] Denby did indeed applaud the missionary as a civilizing agent preparing the way for "commerce, trade, a market for manufactured goods." And Denby also worried that appeasement on the mission question might make the Chinese reckless. But these considerations were outweighed in his judgment by the absence of any "positive treaty sanction" for mission work in the interior and the tendency of missionary-generated antagonisms to set back the cause of civilization in general and in particular to disrupt the development of the "vast" China market. The outbreaks of 1885–1886, which coincided with Denby's arrival in China, had driven him toward these conclusions—and given him a jaundiced view of missionary conduct. Missionaries were often guilty of "undue rashness" in seeking to win converts. Better to leave recalcitrant natives to their idols. Missionaries moreover insisted on the right to travel anywhere in China yet refused to "consult any official as to the propriety of their movements. . . . Opposition only seems to inflame them." Once in the interior those such as Reid "who are turbulent, troublesome, ambitious, and unable to control their temper" caused the legation "worry, bother and vexation," and threatened by their claim to "shadowy and doubtful rights" to impair diplomatic relations.

Into the early 1890s Washington continued to follow the lead of the legation without giving serious thought to the policy implications of the missionary problem. Secretary of State Frederick Frelinghuysen had given his "unqualified approval" to Holcombe's call for greater protection. The first Cleveland and the Harrison administrations, with Bayard and then Blaine conducting affairs at State, moved behind Denby back to the well-established policy of restraint. Missionaries should exercise caution, and the legation moderation, in protecting those who "effect a lodgment" in the interior. The immigration issue, then reaching its peak of asperity on both sides of the Pacific, further inclined Washington to soft-pedal mission problems. The complaints of Chinese nativists bore an embarrassing resemblance to those of their American counterparts, a point Denby noted and Blaine conceded. "Our experience with the Chinese in this country has shown us how unfortunate may be the results of provoking local antagonisms, and the experience of foreigners in China . . . amply enforces the wisdom of not seeking too suddenly to overcome obstacles created by public opinion." To support exclusion, as both administrations did, while adopting a strong missionary policy would put them in a potentially false position that Chinese diplomats would be quick to exploit. But strong action was inhibited even more by the practical desire not to further inflame Chinese feelings, already irritated over abuses of Chinese in the United States, and by an unwillingness yet to entertain the use of force, which a strong missionary policy would require.[34]

The 1890s brought a major and lasting shift in policy favorable to the mission movement. By the end of the decade Washington had dramatically broadened its definition of missionary rights and demonstrated its willingness to defend the exercise of those rights, even in the face of undiminished Chinese opposition. Missionaries, though not fully satisfied, could no longer complain that they were treated as the stepchildren of the open door constituency. China policy had come to guarantee opportunity for missions as much as for commerce.

The catalyst for this shift was a new round of anti-Christian agitation in China. It undermined the dubious policy assumption that determined missionaries and aroused nativists would somehow peacefully work out their differences and spare the legation and the State Department greater involvement. Widespread rioting occurred in the middle Yangtze Valley in the spring and summer of 1891, causing property damage to American missions at Nanking and I-ch'ang. More was to follow in 1895–1896. Rumors that missionaries had murdered Chinese children set off the two most serious outbreaks of those years—at Ch'eng-tu in Szechuan province in late May 1895 and at Ku-t'ien in Fukien in August.[35] It was a foregone conclusion that the rising level of violence and the threat to American lives would force the makers of China policy to act. Even moderates such as Avery and Denby had earlier recognized that if missionaries at their outposts in the interior were dealt a fatal blow, then the foreign community in general was bound to suffer. The missionaries were but the first in a row of dominoes that led right back to the treaty ports. Merchant, diplomat, and missionary would have to stand or fall together.

In the United States missionaries undertook a lobbying and propaganda effort, comparable to the one China market enthusiasts were to launch only a few years later, to win the support of policy makers.[36] At the core of the missionary argument was the contention that Chinese character and culture required mission uplift. To retreat in the face of violence—so contended spokesmen for the major boards, petitions from China, missionary journals, missionaries on home leave touring from congregation to congregation, and the writings of prominent missionaries whose long residence in China qualified them as experts—would be to betray decades of missionary sacrifice and forsake the obligations of a civilized nation to turn back barbarism. Older volumes such as *Social Life of the Chinese* (1865) by the Foochow missionary Justus Doolittle had laid the groundwork for the belief that China had over twenty centuries become so sunk in "senseless and useless opinions" and beset by "strange and superstitious customs practiced among all classes of society" that only conversion could save her. More recent works continued in the same contemptuous vein. Arthur H. Smith's *Chinese Characteristics*, first published in Shanghai in 1890 and then revised and reissued in New York in 1894, provided its many inquiring readers—including officials in the legation and policy makers in Washington—a convenient window on the world of a hopelessly passive, cowardly, backward-looking, morally deficient people. Smith's own estimate of this peculiar people is graphically enough illustrated by his chapter headings—"The Absence of Nerves," "Intellectual Turbidity," "The Disregard of Time," "The Absence of Altruism," and "The Disregard of Accuracy."

The negative stereotypes that appeared in Smith's and other influential missionary works provided the starting point for the argument for a program of radical change (beginning of course with religious conversion) and for the diplomatic support to see it through. The argument itself was couched in the usual military rhetoric reflecting the mission movement's original and unshaken conception of itself as an army marching against the forces of darkness to liberate those held helpless in thrall. China was "the largest single citadel still holding out against Christ" (as the 1894 convention of the Student Volunteer Movement phrased it); the Chinese people, depersonalized and even dehumanized by mission stereotyping, became mere pawns in the struggle against heathenism. The real enemies of missions were conniving local officials and gentry, who skillfully exploited popular ignorance and resentments created by aggressive Catholic missions and American exclusion laws in order to deny China to Christ.

In the great contest for China's soul already in progress, forceful government support was, so missionaries contended, essential to securing the respect of the Chinese people, intimidating elite resistance, and compelling the central government to maintain order. Immediately after the outbreaks of 1891 and 1895–1896 missionary letters and petitions chastised policy makers for failing to take earlier the firm stand that would have prevented the recent attacks, and warned that another failure to defend missionary interests would only stimulate still worse attacks in the future. For those with moral qualms over a policy of coercion, once thought a special province of grasping European powers, missionaries drew on the national sense of exceptionalism and argued Americans were best suited to carry to a conclusion this holy war without taking unfair advantage. "America is neighbor to China," observed W. A. P. Martin. "Others may wound or rob, we do neither."

The specifics of a program of protection assumed definitive form in July 1895 following the Ch'eng-tu attack. Missionaries assembled in Shanghai to give vent to their frustration over Chinese obduracy and to demand that Washington henceforth fulfill its duty to protect its citizens, prevent persecution of converts, and in general command the respect of the Chinese. To begin with, a searching, on-the-spot investigation should be made into recent incidents as well as any in the future; wrongdoers were to be identified with the help of converts and punished in accord with local judicial and administrative procedures and "without regard to rank or position." Chinese officials were to be subject to the rules of collective responsibility however at odds with the missionaries' own sense of justice or past stands they had taken against such Chinese practices. Punishment should automatically fall on an official whenever an incident occurred within his jurisdiction regardless of his own role. The central government was at once to circulate throughout the empire an imperial proclamation recognizing residence in the interior as a treaty right, and it was to clear the way for mission work, even in Hunan province, a hotbed of anti-Christian sentiment. If, despite all these deterrents, incidents were to recur, missionary losses and expenses, however remotely connected, were to be reimbursed. Such "remote" or "consequential" indemnity included not only property damages but also salaries for periods of enforced idleness, travel costs (including in some cases a trip to the United States and back), rents both for houses destroyed and those temporarily occupied, and a variety of incidental expenses. On the return of missionaries to the scene of the incident, local officials were to organize a formal ceremony of welcome and contrition in which prominent notice would be made of missionary rights and good works.[37]

Against this background of renewed attacks on American citizens and missionary appeals at home, the Peking legation took the lead in revising policy. Denby's own abandonment clearly began in 1891. In the incidents of that year Denby espied a "conspiracy to drive the foreigner out of China," which the powers would have to meet with a gunboat policy. But at the same time, he deplored the foreign community talking "loudly of reprisals and bombarding the offending cities" without regard to the rebellion, destruction of foreign trade, or indirect harm to the missionary enterprise that might result, and he still counseled prudence on the part of missionaries, for example, advising them not to accept orphans under twelve years of age to silence rumors of missionary abuse of young children.[38] The Ch'eng-tu riot of 1895 completed Denby's break with the old policy of restraint. At first carried away, he proposed that gunboats bombard the scene of antimissionary violence or destroy a substitute in cases beyond the reach of their guns. Incredulous that there could still be so much popular animosity after diligent mission work in Ch'eng-tu by Catholics and Protestants spanning three centuries, Denby could explain it only "on the theory of official connivance." Anticipating many of the demands of the Shanghai resolutions, he demanded that the Szechuan governor general be degraded and prohibited from again holding office, and that both now and in the future provincial officials who had shown "incompetency or hostile apathy, if not deliberate collusion" be called to account. Rioters were to suffer the death penalty. The missionaries should receive "remote" indemnity. Finally, an investigative commission should proceed to the scene of the outrage.[39]

Through the 1890s, Washington moved its missionary policy in the direction of intervention and coercion favored by missionaries and now warmly endorsed by the legation. After the 1891 incidents, the Harrison administration joined other powers in a show of force and demands for an imperial edict of toleration, and in 1893 it ordered work started on two gunboats designed to operate in Chinese waters. The United States was obliged, the President had earlier observed, to give "security to our citizens dwelling in those remote lands" such as China and to "challenge respect." The return of an uneasy calm to China and the appearance of Walter Gresham as Secretary of State in Cleveland's second administration momentarily turned back the clock. In 1893, immediately after taking office, Gresham received from his old friend Denby the proposal that Washington recognize missionary claims to hold land in the interior as an unconditional right. Finding no basis for such a claim, Gresham refused and handed up instead for the legation's general guidance the familiarly vague terms of the old policy—"to prevent abrupt reversal" of the tolerance previously extended by the Chinese to mission stations.[40]

In the summer of 1895, following Gresham's death in office and the contemporaneous renewal of attacks in China, Washington became responsive once more to Denby's views. The State Department went along with his recommendations (except for gunboat reprisals) after the Ch'eng-tu incident. And the new Secretary of State, Richard Olney, had good reasons of his own for backing the legation. He decided to seek in China what he had already demanded of Britain in Latin America—recognition of the United States as a major power entitled to its sphere of influence close to home and full standing alongside the other powers in more remote areas of concern. A more assertive mission policy would liberate American missionaries from their dependence on the diplomats and gunboats of Britain and France for protection, while enhancing American prestige. Olney wished to impress the Chinese as much as the powers. Confronted with unfamiliar China problems, Olney had turned for edification to two standard missionary accounts, Smith's *Chinese Characteristics* and Chester Holcombe's *The Real Chinaman,* and predictably drew from them the conclusion, underlined by bitter complaints from missionaries in China, that a forceful response now to Chinese outrages would avert future incidents.[41]

Olney endorsed Denby's demand for action against the governor general, and Peking bowed (but only after an Anglo-French display of naval power!). He insisted on the punishment of those involved in the attack. And finally he sent an independent three-man American commission to Szechuan, by the most conspicuous possible overland route and in the face of intense foreign office objections, to serve as an "impressive demonstration" to the Chinese of the depth of American commitment. After the Ku-t'ien incident the Department repeated the demands for punishment and an investigation. (This time the American "investigators" became generally involved in local judicial proceedings and administration.) After both incidents, the legation insisted on payment of "remote" indemnity. At year's end Cleveland looked back with satisfaction on these "energetic steps" calculated to ensure the future safety of Americans in China.[42]

To formalize the new terms of missionary policy Olney set the legation to work in July 1896 on securing China's full and widely publicized acceptance of what amounted to the missionary resolutions of the previous year. The heart of the project, advanced by Rockhill, was a proposal that China accept full missionary treaty rights in the interior and the obligation to punish "all individuals or officials directly or remotely involved" in any riot. Denby enthusiastically embraced this proposal and urged that China's acceptance should take the form of "an open and notorious publication in an Imperial decree." In the end, the legation got all it wanted save for the provisions for investigatory commissions, omitted without regret so that China would have to take full responsibility itself for dealing with all incidents. The most gratifying gain was the set of imperial instructions issued between June and September 1898, while the reform movement briefly held sway in Peking. Troubled by new incidents in Szechuan, Kwangsi, and Hupei, the throne put provincial officials from the highest to the lowest on notice that they were responsible for protecting foreign missionaries and their chapels and maintaining peace between converts and the rest of the local population.[43]

Denby had predicted in September 1895 that the worst of the missionary incidents had passed. Seldom was optimism so misplaced. No sooner had he offered the judgment than news arrived that antimissionary literature, accompanied by rumors that the powers had gone to war against China, had inflamed popular opinion in eastern Shantung. The incidents that followed, scattered across the face of China, mocked Denby's prediction and confirmed the new American policy of protection. In May 1896 missionaries at Kiangyin in Kiangsu province came under attack following the discovery in their compound of a dead child (planted, so the missionaries alleged, by provocateurs). The demand for punishment, successfully enforced, did not prevent still other incidents. One near Chungking in 1898 had been preceded by an official Chinese request that missionaries not settle in an area decidedly unfriendly to Christians. Catholics had already retired and an American missionary had been assaulted there the previous winter. Denby would do no more than advise "prudence and discretion." The missionary took up residence, and hardly a month after the official warning the mission was attacked and a Chinese assistant killed. Denby at once lodged demands for indemnity, punishment, and dispersal of the local militia. That same year missionaries established themselves in Hunan in the face of local hostility, and were soon driven from the scene. The legation demanded that the missionaries be given an official escort back and their safety guaranteed, that a new residence be found, and that punishment be administered to local officials who had failed in their duty.[44]

The anti-Christian movement in China from the mid-1880s through the mid-1890s, though it had resulted in the loss of not a single American life and in only one wounded, nonetheless had succeeded in altering missionary policy by undermining the essential assumption that restraint and reason might guide the encounter between missionary and Chinese in the interior. The legation and Washington had come to accept instead a new but no less ill-founded assumption and erected a more interventionist policy on it. The new assumption was that the Chinese government had the power to put a stop to violent opposition and could be induced to do so. Peking might be moved to act by appeals to self-interest, or in Denby's words to fear of loss of "character, credit, and standing before the world."[45] Failing that, intimidation would have to do, and to prepare for such a case Washington increased its naval patrol on the Yangtze from the single gunboat available in 1895.

Reliance on the cooperation of the Chinese government, however obtained, was a peculiar strategy given the widespread conviction among missionaries that the officials of that government along with the gentry were the source of the agitation and Denby's own representative view that the central government was incompetent and "reactionary" and local officials "imbeciles."[46] But leaning on Peking was also an unavoidable strategy, for the United States could not single-handedly guarantee the safety of its citizens in the treaty ports, while the far-flung and isolated missionary communities in the interior were simply beyond any guarantees. The United States could rely on the other powers to keep the Chinese in awe, but there was a danger that for their efforts they would seek, as Germany had in Shantung, territorial or economic compensation detrimental to other, nonmissionary American interests. So by default the Chinese government became the chosen instrument of American mission policy. However, the assumption that Peking would ignore the breadth and depth of opposition to mission work and play the self-destructive role of defender to a foreign presence had already been put in question by the continuing agitation of the late 1890s. The collapse of that assumption under the unprecedented fury of the Boxer movement was to force on Washington a still more interventionist policy as the protector of last resort.

The View from the Legation

The foreign service, the last element of the open door constituency to take root in China, came into its own after 1860. It gained a permanent legation in Peking, and by 1899 it had extended its consular system to a total of ten points along the coast and in the interior. It also began to consolidate its autonomy by increasingly excluding businessmen and missionaries from its ranks. Both American and Chinese officials recognized that to allow a regular merchant to serve as consul was to give him access to confidential trade information that put other American traders at a disadvantage and clothed him with a degree of immunity in carrying out illegal business activities. Missionaries fell from favor because they faced an even more acute conflict of interest with the rise of the explosive missionary question. S. Wells Williams and Chester Holcombe served as chargé on seven occasions between 1860 and 1882, but thereafter Washington appears to have acted on the view that it was "injurious" to have a missionary in the legation. (The Chinese were also opposed to having as head of mission such missionaries as Holcombe, denounced by Li Hung-chang as a cunning scoundrel with a bad reputation.) The slowing of the ministerial merry-go-round, the bane of the legation before 1860, and the emergence of a new breed of China specialists such as William Pethick, W. W. Rockhill, and Charles Denby, Jr., further enhanced the independence of the foreign service, especially by breaking the missionaries' near monopoly of expertise.[47]

Despite these particular improvements, the American foreign service in China remained largely amateur, subject to the requirements of the spoils system, and hobbled by a shortage of funds. Thus, new arrivals had no expertise; those who stayed long enough to gain it eventually fell victim to party politics or succumbed to the lure of higher pay offered by business or the customs service. Inadequate staffing, especially of able interpreters, was a recurrent and serious deficiency. Though interpreters were indispensable in nearly every official activity from communicating with accuracy, clarity, and dispatch with the Chinese bureaucracy to gathering economic and political intelligence, only three of eight consulates had them in 1863. By 1879 the situation had improved but only by dint of a variety of questionable makeshifts.[48]

S. Wells Williams, an early and insistent critic of the American failure to train its own interpreters, perceptively noted that as a consequence the American consul "feels his isolation, and consequently takes less interest in a people from whom he is thus shut out, and with whose officers he is usually in a state of chronic dissatisfaction." However, most members of the foreign service with their distaste for Chinese ways seem to have accepted that isolation as a welcome feature of the comfortable life they led in the treaty ports or legation quarter. The only Chinese they regularly saw were servants. For exercise they ventured out where the Chinese, whose presence might diminish their pleasure, were forbidden to go—to the Peking city wall or to the waterfront, park, and race track of the treaty port. In the summer they retreated to some isolated beach or cool hilltop, always away from the Chinese. Only official business forced them into brief and formal contact with Chinese officials.[49]

Within the post-1860 foreign service the legation was to emerge as the dominant institution, the chief interpreter of China to policy makers at home. The opening of Peking helped push the minister forward as the spokesman for the China service. In 1862 he gained a fixed residence, a one-acre tract in the legation quarter rented from S. Wells Williams (because Congress would not at first authorize purchase of quarters). From the legation's old and inelegant Chinese-style buildings, the minister had for the first time the advantage of ready access to the officials of the central government.[50] The tendency for ministers to stay in China longer than their predecessors—an average of over four years between 1860 and 1900 and over seven years between 1882 and 1900—further enhanced the legation's claim to authority and expertise.

Under these more favorable conditions a succession of articulate and active ministers from Anson Burlingame to E. H. Conger elaborated forcefully and with remarkable consistency an overarching vision of a special national stake in seeing China reformed and secured against foreign aggression. Already glimpsed by their predecessors, this vision of a special American role in China incorporated yet transcended specific economic and mission interests. At the heart of this view from the legation was the duty of Americans to see to China's regeneration. As reformers and guardians they would have to guide a weak China—tottering on the brink of collapse, her culture and political system rigid and unresponsive after centuries of somnolence, and increasingly threatened by aggressive powers—through turbulent times. This idea of the special Sino-American relationship rested on the status of the United States as the only advanced country whose intentions were aboveboard and pacific, whose development and destiny pointed westward, and whose geographical proximity further underwrote its claim to concern.

From their bully pulpits in the legation and consulates Americans maintained an unflagging enthusiasm for the contribution American economic enterprise—the building of railroads and telegraphs, the opening of mines, an expanding volume of trade—would make to China's awakening as well as to increased American influence. Missionary work and education would serve the same end though they generally inspired less enthusiasm. Finally, the foreign service would make its own special contribution to progress by drawing China into the family of nations. Its advocacy of a responsible foreign office, an end to Chinese pretensions of superiority (as in its claim to suzerainty over tribute states and its insistence on the kowtow as an inseparable part of an imperial audience with foreign diplomats), the extension of Chinese legations abroad, and the removal of artificial barriers to trade (such as the internal transit tax known as likin) were all steps in this direction.

The advocates of a China transformed under American guidance saw China as a wonderfully malleable land ready for reshaping. Yet this optimism was inextricably bound to fears that the very qualities that called forth the American vision in the first place—China's sluggishness and vulnerability—might also prove its undoing. On the one hand, the Chinese themselves, particularly the notoriously short-sighted, stubborn, tradition-bound gentry and officials, might not recognize the urgency of the moment and let slip the opportunity Americans held out. The hope that somewhere within the government there would emerge a powerful exponent of reform and natural ally of American efforts repeatedly asserted itself though as often undercut by experience. The vision was also threatened by the powers who might boldly seize this propitious moment of Chinese weakness and turn this vast and vulnerable land to their own selfish ends. One anxious minister after another was to watch the powers endanger China's security, engross her trade, exploit her people, and jeopardize her difficult passage to modernity. Their response was to assume the duties of a trustee of China's independence. The future of the special relationship thus depended no less on neutralizing European imperialism than on cajoling the Chinese to embrace reform the better to resist domination. Where a still inert China proved unequal to the foreign threat, ministers were to advance solutions of their own, ranging from playing the peacemaker to, in extremity, joining the powers in marking out spheres of influence.

This reform orientation of the legation, at heart deeply paternalistic, could take any of several directions. The first two American ministers to reside in Peking, the illustrious Burlingame and his less well-known successor, J. Ross Browne, define the poles. A deserving politician sent to China by the Lincoln administration in 1861, Burlingame exhibited a patiently paternal approach to China's reform. His pet scheme to establish a school of language and literature in Peking, an idea he borrowed from S. Wells Williams, captured that approach. Instruction at such a school in "sound morals as well as accurate knowledge would tend to exert a lasting and excellent influence at the seat of government in support of peace and commerce throughout all the provinces." Further, the school could familiarize Chinese with Christianity, and at the same time it would offer the practical advantage of training that corps of interpreters so badly needed by the foreign service, merchants, and missions. Characteristically, Burlingame's proposal called for Chinese funds to serve as the school's endowment while excluding Chinese from oversight of the school's operations (a task better left to enlightened representatives of the open door constituency with a vested interest in its success—four diplomats, three missionaries, and the heads of the business houses of Russell, Heard, and Olyphant).[51]

That same paternalism was expressed in Burlingame's belief that the Chinese government ought to be saved so that it could play a major role itself as an instrument of change. Toward that end, Burlingame embraced in the early 1860s and later in the decade sought to sustain a policy of cooperation with China initiated by Britain and France. By pledging not to take Chinese territory or interfere with the government's control over its own subjects, the powers were ensuring China a respite from an impatient and grasping foreign community without renouncing a fair reading of the treaties on which foreign life and property depended. During his ministry Burlingame saw himself and his British colleague thus acting as arbiters between the foreigners, who implanted the seeds of change, and the imperial government, which was struggling to put down rebellion and set its house in order.[52] In 1867 Burlingame took the unusual step of enlisting as China's own minister charged with carrying to the Western capitals the argument for restraint. In the United States he proclaimed China's steady advance "along the path of progress" under the American aegis. Decrying coercion and the chaos it would generate, he argued for "a generous and Christian construction" of the treaties as the best way to serve American interests and aid the "enlightened" government in Peking. But even in this, his last service to the new China (to end with his sudden death in February 1870), this disinterested "umpire" of China's foreign relations continued to act out of a deep-seated sense of paternalism. Though associated with two Chinese of a rank equal to his own and enjoined to remain in close touch with Peking and to make no commitments without prior approval, Burlingame established the mission as his own once the China coast disappeared from sight. He pushed aside his associates, concluded a treaty in Washington without a word to Peking, and in general kept his nominal superiors in the dark about his intentions.[53]

Browne's proceedings during his brief residence in Peking offer a glimpse of how a commitment to reform and a sense of paternalism could evolve into a policy of coercion once hope in the Chinese government was gone. This California travel writer and adventurer sent out by the Johnson administration believed no less than Burlingame that Americans in China embodied the "progressive spirit of the age." But influenced on his arrival in September 1868 by the accumulation of missionary and merchant complaints, Browne decided that China was in a "state of hopeless decay" and that its government was plotting "to establish arsenals, build gunboats . . . and, in the end, . . . make a final attempt to drive every foreigner out of the country." The United States, he argued, could not permit such backsliding by "an ignorant pagan nation." Indeed, the United States, guided by its clear duty "to elevate the Chinese to our standard," must join with the other powers in drastic and immediate action to batter down the walls of "isolation, ignorance, and superstition."[54]

The legation between Browne's departure and the arrival of John Russell Young in 1882 hewed to the Burlingame line on reform, though the rapid turnover in personnel (eleven ministers or chargés over those fifteen years) temporarily weakened the legation's position as an advocate of any policy.[55] Young's three-year tenure began a period of restored stability and influence. Faced with an aggressive European policy in the East, he emerged as a forceful exponent of the special American role as China's protector, thus building on earlier efforts by Burlingame and Grant and adding to the varieties of the reform strategy. Already in 1880, following his world tour with Grant, Young had in making a bid for the legation in Peking or Tokyo emphasized his concern over "the fierce game of Russian and English diplomacy" in East Asia and his belief that "the American government can interfere very effectively, and it is the only one which can."⁵⁶ Once in China Young put his anti-imperialist views in practice by trying to mediate an end to the Sino-French crisis and by urging friendship between China and Japan. While the individual ambitions of the powers could be harmful to China, collectively foreigners played a positive role, bringing the peace and foreign trade necessary "to bring China abreast of western civilization," and also in extremity (in the face of "direct and palpable danger") using force to ensure China's good behavior. Without foreign encouragement, direction, and supervision, Young somberly warned, China would fall "prey to internal dissensions and external schemes of aggrandizement."⁵⁷

Charles Denby, Sr., whose fourteen years (1885–1898) in the legation under no fewer than seven Secretaries of State made him its longest tenured occupant, was a transitional figure in the evolving tactics of reform.⁵⁸ He began by essentially espousing the more temperate position—vis à vis China—of Burlingame and Young. Like them he believed it was possible to work with China's leaders.⁵⁹ He also followed their methods when in late 1894 and early 1895 he tried to mediate an end to the war with Japan in order to minimize the damage done to China and the temptation by the powers to exploit her.

However, by 1895 the complete incapacity of China's government to deal effectively with the problems festering at the heart of Sino-foreign relations at last drew Denby into the coercive mold favored by Browne. Concerned particularly with the threat that the antimissionary agitation posed to "a handful of Americans surrounded by four hundred million Asiatics," he concluded that "the fear of interfering with international rights or offending China should not for a moment be allowed to stand in the way of ordering immediate, and armed protection to . . . all foreigners in China." Without the backing of force, he reminded Washington in 1891, "both mission work and commerce would languish."⁶⁰

The inability of Peking after 1895 to check the growing aggressiveness of the powers completed Denby's conversion to a coercive policy. His mediation effort in 1894 had shown the Chinese as children blundering in foreign affairs from crisis to crisis, unable to recognize their true interests or true friends. The government seemed to Denby a hopeless patchwork of ignorance, reaction, corruption, drift, and incompetence, and in 1897 even Li Hung-chang fell from grace when revealed as the architect of the alliance between Peking and St. Petersburg.[61] The powers themselves failed to meet Denby's hopes that they would come to some accord on how best to promote China's development. In April 1895 Japan, earlier regarded by Denby as a "champion of civilization," made territorial demands which convinced him that she had under false pretenses "pursued her own aggrandizement" at the expense of the Western powers. Two years later Russia—whose railway plans Denby had expected would be a force for progress in Manchuria—also showed her true colors, refusing to cooperate with American railway interests, deserting her Chinese ally in the face of German demands in Shantung province, and responding with demands of her own for new concessions in Manchuria.[62]

In the face of the powers' unchecked "colonial ambitions" and "plans of national aggrandizement," Denby at last felt constrained to recommend that the United States imitate them. Though he was to back away from advocating steps as radical as making alliances or joining in China's partition, he became convinced that the United States would have to pursue a more aggressive financial diplomacy. Denby had long believed the legation should have wide latitude "to push American material interests . . . and thereby to extend American influence and trade." The arrival at mid-decade of European concessionaries who enjoyed strong official backing intensified Denby's demand for greater freedom of action to advance the cause of American concession hunters in "this limitless field of financial and industrial operations." Visits to the foreign office became for Denby, a self-styled "old lawyer," occasions for arm twisting as he reiterated the American claim to a favored role in China's economic development.[63]

The checks repeatedly suffered by American promoters and the success of the powers in defining ever clearer spheres of influence drove Denby during his last several years in the legation to embrace bolder measures. He urged that Washington express "disapproval of acts of brazen wrong, and spoilation, perpetrated by other nations towards China." Overcoming his recurrent doubts about Britain's true intentions, Denby was now even ready to see the United States join with her in setting China's house in order, imposing an enlightened policy of development, seeing to the details of its execution, and excluding powers that would exploit China to their own selfish advantage. Denby was sure the people of China would not object to his "great schemes of improvement" since they understood the United States stood for neither "territorial absorption nor governmental interference, while both these results are possible, or even probable, in dealing with European powers." If the Chinese used the period of Anglo-American control to overthrow the Manchus and restore native rule, so much the better.[64]

E. H. Conger was to begin where his predecessor Denby had left off in arguing for a coercive and assertive policy to protect China and preserve American interests. In background and outlook Conger was much like Denby—a Midwestern lawyer and deserving politico.[65] Though he epitomized the China problem in the more cosmic terms of a struggle between "orientalism" and "occidentalism," Conger nonetheless agreed with Denby on the importance of forcing China to observe the letter of the treaties, on which China's material progress and Americans in China both depended. Conger also agreed on the need to check the schemes of the powers against China, and in this regard he too looked to financial diplomacy. He wanted to advance the cause of American investment by acting as adviser, protector, and intermediary with Chinese officialdom. By seizing the present "extremely rare opportunities for profitable investment" American capital could capture for the United States "permanent and potent channels of trade possibilities and political influences," thereby offsetting the territorial advance of the other powers. The United States might also, as Denby had proposed, take an independent moral stand condemning aggressive acts or even encourage the commercial powers against such wrongdoers, especially Russia. (He saw no need to consult China, now powerless to resist.) Conger welcomed Hay's first open door notes as a step precisely in line with his own recommendations for saving Manchuria and North China from the final Russian embrace.[66]

The apparent insufficiency of these measures to save China from partition emboldened Conger to propose what Denby had only gingerly toyed with—joining in the partition. Hardly a month after his arrival, Conger had urged Washington to at least keep Manila and its hinterland as a step toward "securing and holding our share of influence" in East Asia. Such a base would facilitate "the commercial conquest, which Americans ought to accomplish in China" and support the naval presence needed "if the Chinese government is to respond to our reasonable demands, without fear of interested opposition or official interference from European powers." But even this lodgment struck Conger as an ineffective response to the likely consequences of partition, so in early 1899 he urged the seizure of a coastal pied-à-terre, which would mark off an American sphere of influence and "from which we can potently assert our rights and effectively wield our interests." Conger audaciously fixed his attention on the northern province of Chihli encompassing both Tientsin and Peking. From that point the United States could protect the North China market for its cotton goods and guide China's political destiny.[67]

China Policy and the Open Door Ideology

The reiterated invocation of the open door ideology by commercial groups, missionaries, and diplomats had by the end of the 1890s transformed a once passive, narrowly commercial China policy. This shift in thinking on China policy is evident in the message of a group of publicists and popularizers who stepped forward at this crucial juncture. Figures as diverse as Richard Olney, Henry Cabot Lodge, W. W. Rockhill, Charles Denby, John Barrett, Josiah Strong, D. Z. Sheffield, Charles Conant, Brooks Adams, and Alfred Thayer Mahan now made an important contribution to the apotheosis of the open door ideology by carrying the message of the open door constituency to a wider audience.[68] Their China was the scene of two related struggles—between the regenerating forces of the West (best represented by the United States) and a stagnant China and between selfish and exclusive imperialism and American aspirations for long-term, benevolent involvement in Chinese affairs. Only a China reformed could stand off imperialism unaided. Only an immediate assertion of American interests, especially against Russia in North China and Manchuria, would preserve the national dream of China's eventual reformation and of access to this "gigantic" market with its "almost boundless possibilities." In global terms they saw China as the main or even, in Adams' estimate, the decisive scene in the struggle for control of world civilization. The open door ideology, embodying long-established ideas of westward expansion and moral and material uplift and pointing to a solution for the problem of excess production raised by the depression of the 1890s, made it easy for these popularizers to hold out in common the dramatic possibilities of China—ancient, vast, potentially rich—as a stage for national action.

But translating that ideology into a coherent, workable policy to combat the threat of imperialism precipitated the commentators into disagreement and perplexity. In framing a response to this problem they had, by the end of 1898, to entertain the implications of the Spanish-American War. The new territories acquired as a result of the war—Hawaii, Guam and, most important, the Philippines—strengthened pretensions to mastery in the Pacific. Thus emboldened, Lodge proposed joining with Britain in keeping China's ports open. "All Europe is seizing on China and if we do not establish ourselves in the East that vast trade, from which we must draw our future prosperity, and the great region in which alone we can hope to find the new markets so essential to us, will be practically closed to us forever." In his desire to resist partition and protect American trade against discrimination, Olney, the elder statesman, endorsed Hay's open door notes, called for a stronger navy, and even proposed a temporary alliance with Britain. Denby wanted to develop the Philippines as a link to the "splendid" China market. Conant, in a survey of possible responses to partition, touched all the bases—acquisition of formal or informal control of territory, cooperation with Britain, general assertion of diplomatic influence, employment of force from advance naval bases, and the creation of financial and commercial instruments of national policy. Which one the United States would follow was for him "a matter of detail."

In addressing this "detail," involving nothing less than formulating a new, more active policy consonant with the broader definition of the American stake in China, successive administrations in Washington moved reluctantly and in piecemeal fashion. A method of dealing with the Chinese was more easily and definitively settled than the method of neutralizing the powers. At first, in the 1860s and 1870s, policy makers had resolutely supported Burlingame's mild cooperative approach over the more costly and risky line preferred by Browne. Secretary of State Seward saw too few concrete interests in China to justify a coercive policy and no naval force, especially in the Civil War years, available to implement it. "A policy of justice, moderation, and friendship, is the only one we have a choice to pursue, and it has been as wise as it has been unavoidable." Browne's eagerness to force the pace of progress predictably evoked from Seward a warning not to "endanger the stability of the present government or the internal peace and tranquility of China." It also made President Grant, a friend of Burlingame's, indignant, and caused Fish to recall Browne after less than a year in China and refuse his request for reappointment. Fish insisted that thereafter the legation itself show consideration in dealing with China.[69]

The determined missionary advance, however, began to overcome Washington's qualms about a coercive policy during the Harrison and Cleveland years and by 1895 had secured policy makers' acceptance of a simple syllogism that was to underwrite the new policy: Americans engaged in uplifting China and advancing national influence deserved protection; American missionaries were undeniably and deeply devoted to those causes; ergo, American missionaries deserved protection. By 1895 policy makers were as resolute in keeping China in line as they had been earlier in keeping hands off.

Since Washington would not adopt a simple policy of force in meeting the aggression of the powers, the legation had laid out before policy makers a variety of alternatives. Acting as an honest umpire between a vulnerable China and a rapacious West had appealed to Burlingame, Young, and Denby, and Washington had been willing to go along as long as no commitments to China or the powers were involved. Seward approved Burlingame's mediatory efforts, first as an envoy of the United States and then of China. The revival in the 1870s of foreign designs on Asian territory strengthened the impulse to become involved in the international politics of East Asia. While Grant with his strong views on the trustee role of the United States in Asia pursued his exercise in good offices during the Liuchiu dispute, Secretary of State Evarts denounced the British tendency to bully "Oriental nations" and President Garfield could point to his earlier warning that the United States, by right and duty "chief in the councils of international powers" in the western Pacific as well as South America, would have to be on guard against "other maritime nations . . . ready to snatch this [Asian] prize from our hands. . . ." Frelinghuysen joined Young in the effort to bring an end to the Sino-French crisis over Vietnam, and his successor, Bayard, argued against the wisdom of associating with the other powers whose policies toward China "have been aggressively harsh, and ours the direct opposite."[70]

By the 1890s China's peril at the hands of intriguing European powers had become as much a staple of thinking in Washington as it was in the Peking legation. But the very Chinese incompetence that invited aggression had by 1895 rendered bankrupt the strategy of mediation and active diplomacy intended to block that aggression. Gresham's warning to Denby during the Sino-Japanese peace negotiations not to "go too far in aiding China" derived in part from his fear of entanglement per se and the political criticism that would follow but also from his view of the hopeless condition of the Chinese in contrast to Japan.[71] After 1895, policy makers faced a redefined task—that of preserving the special American role as China's guardian and reformer in spite of and sometimes even against the Chinese.

In this task Washington was aided by a steady stream of suggestions from the legation, beginning with the proposal that the United States adopt a vigorous policy of economic diplomacy. However, policy makers were constrained by their faith in the religion of free trade. Its high priests argued that international trade directed by the dictates of the market materially benefited all parties involved while contributing to peace by insulating economic questions from political quarrels, and bringing far-flung parts of the globe into harmonious contact. So superior were the products of American farm and factory that so long as free trade prevailed, Americans would be assured of not just a profit in China as in any other market but a secure long-term influence without Washington having to assert itself. The government need not regulate and promote trade or protect it beyond seeing that all parties played by the same minimum rules of political noninterference.[72]

These laissez-faire biases caused Washington to restrain the legation so that it might not play favorites among competing American schemes and in the process compromise Washington's resistance to European mercantilist practices in China. The State Department kept Denby on a short leash despite his repeated calls for more latitude. Bayard enjoined him to get Washington's approval before promoting or aiding any individual scheme, while Gresham later gently suggested that he not allow his "generous and obliging nature" to compromise him. In the spring of 1895 Washington issued a reminder that the legation was to limit itself to making formal introductions between promoters and the responsible Chinese official. Later that same year Olney left blanket instructions to Denby to "carefully abstain from using your diplomatic position to promote financial or business enterprises," and only at the end of 1896 did the Secretary give way a bit and grant Denby permission to use "your own judgment and experience" in employing "all proper methods." The latter did not include, Olney warned, favoring one American firm over others or devoting substantial staff time to helping business agents with communications with the Chinese.[73]

Washington was also constrained, incongruously enough, by a contradictory impulse to protect American manufacturers from

Chinese competition. Thus, through the 1880s down to the late 1890s policy makers regarded American investment in Chinese textile mills, one of the most promising of available economic opportunities, as a "cheap labor dodge" likely to cost American producers sales and American workers jobs. "The transference to China of American capital for the employment there of Chinese labor," President Arthur predicted, "would in effect inaugurate a competition for the control of markets now supplied by our home industries." When W. S. Wetmore sought support in setting up his textile mill in Shanghai over Chinese objections, Arthur and Secretary Frelinghuysen refused, a position maintained by the Cleveland and Harrison administrations and emphatically affirmed even after the treaty of Shimonoseki legitimized foreign investment in the treaty ports. Expressing a view shared by Denby, Secretary Olney contended that sound policy called for the government "to keep foreign markets open for our manufactures" against goods made on "the cheap-labor basis."[74]

Even reservations about the bona fides of investors, particularly as the press of railway promoters thickened in China in the latter half of the 1890s, served as a check on the State Department. An organization with backing as reputable as the American China Development Company might, as Secretary Sherman presciently guessed, promise more than it could financially deliver and thereby hurt the chances for other American firms and generally damage the reputation of American enterprise.[75] In the end, Washington's response to the legation's call for a sort of dollar diplomacy amounted to no more than tinkering with the guidelines provided the legation and issuing the open door notes of 1899, themselves a simple call for conditions in China conducive to free trade.

Washington was equally reluctant to play the game of alliances and spheres of influence that the legation moved steadily toward after 1895. In March 1898, for example, Washington turned aside overtures from London intended to get the United States to join in opposing any commercial restrictions that the other powers might institute in their spheres of influence and concessions in China. McKinley, then on the brink of war over Cuba, replied to the inquiring British ambassador that he saw no threat yet sufficiently serious to justify his breaking with the traditional policy of nonentanglement. John Hay accepted the wisdom of that position even after the war with Spain was out of the way. In his view, Anglophobia at work in domestic politics ruled out alliance, though he personally thought there was much to be said for a common Anglo-American approach to China and had earlier as ambassador in London helped stir up the British overtures.[76]

The remaining alternative was to strike out on an independent course, strengthening the American position in East Asia the better to withstand the coming scramble for territory. McKinley had already anticipated such a line of action by taking the Philippines. Control of territory adjacent to China improved the American claim to a voice in China's ultimate disposition, while a Philippines naval base put Washington in a better position to back that claim. McKinley's predecessors had put in his hands a developing navy, successfully tested in the war with Spain, to turn those bases to good account.[77]

The next logical step in an independent policy might have been to stake out a slice of Chinese territory. But McKinley publicly proclaimed in December 1898 that that was a step the United States did not have to take. Though coastal China might fall under foreign control, he confidently predicted that somehow the "vast commerce" and "large interests" of the United States would be preserved without departing from traditional policy and without the United States becoming "an actor in the scene." That disclaimer did not, however, prevent Hay and McKinley from returning the following year repeatedly, though with characteristic caution and discretion, to the question of a land grab. In March Hay, while conceding that the public would frown on the United States joining "the great game of spoilation now going on," nonetheless added that the government, with "great commercial interests" to safeguard, did "not consider [its] hands tied for future eventualities." Again, after the Japanese in March 1899 demanded a concession at the port of Amoy, Hay watched developments with interest—and with the expectation that, whatever grants China made to Japan, she would be prepared to make similar ones to the United States. Even the preparation of the open door notes in September of that year did not remove the temptation. When Hay wondered aloud whether the United States could in truth deny in the notes any interest in Chinese territory, McKinley responded, "I don't know about that. May we not want a slice, if it is to be divided?" In November rumors stirred up by the dispatch of the notes brought the Chinese minister calling. Under examination, Hay freely admitted that the United States reserved its right to claim "conveniences or accommodations on the coast of China," though for the moment such a demand was not in the cards.[78]

The McKinley administration's ultimate recourse to note writing after toying with the possibility of a more aggressive policy was for the moment an adequate response to the China problem. Hay's qualified statement of antiimperialism cost little. It satisfied the domestic demand for action without stirring up controversy. It was offensive to no one abroad. Rather than singling out Russia and Germany, the most egregious offenders against the open door to date, Hay had blandly asked assent from all the powers. Yet Hay's stand with its implied promise of support served as a gentle prod to the other commercial powers. Whether it would prove sufficient to preserve China and—more to the point—the broad interests that the open door constituency claimed to represent remained to be seen.

Part Three

The Patterns Hold

ON May 14, 1900 Horace T. Pitkin, Yale College class of 1892, sat down at his desk in the American Board mission compound on the outskirts of Pao-ting. He began a long contentious letter to the board headquarters in Boston complaining against his coworkers' mismanagement and blunders and against the stinginess of the board itself in its housing and travel allowances. This independently wealthy lineal descendant of Elihu Yale had—after schooling at Phillips Academy, Yale, and Union Seminary—offered himself for service on the condition that he be sent to North China. However, after his arrival there in 1897 he found conditions less and less satisfactory. His wife, another wellborn newcomer to mission work, had suffered a nervous breakdown only a few months earlier and had returned home for rest. Now in a sour mood, Pitkin concluded his letter with a veiled threat of resignation unless the board could see its way to treating him with greater generosity.[1]

That same day in that same compound Annie Gould, the daughter of a bank cashier in Portland, Maine, wrote to the same mission board. She had been in China since 1894, the year after graduation from Mt. Holyoke. Following the nervous collapse of one of her coworkers, she had taken over direction of the mission's girls' school. Her letter to the board alluded to the possibility of having "to pass through the fires of persecution." Since December the presence of Boxers had made long tours in the countryside unsafe. By April only Pao-ting was still secure. The Tientsin consul had then advised the Pao-ting missionaries to evacuate to the coast for safety, and Minister Conger had concurred, but the missionaries had rejected flight. Two weeks after writing the Board, Gould described the deteriorating situation to her parents. Armed Boxers, then numbering a thousand, had gathered in the city and threatened to overwhelm the small guard posted outside the mission. They had also damaged the railway, making any attempt at escape—now necessarily on foot—as dangerous as staying on. In the privacy of her letter Gould reflected on the coming crisis with calm resignation.

"I can't tell you exactly what I fear, not death, nor even violence at the hands of a mob, for the physical suffering would be over soon and God can give strength for that. I think it is that I am conscious my daily life is too selfish. . . ."[2]

A month later Boxers finally attacked the compound. Pitkin's ordeal was brief. He emptied a revolver on the assailants and was then himself felled and beheaded on the spot. Ten Chinese Christians died there with him. Gould was captured and, after a day of public calumny, executed in the evening. A devastated mission movement was quickly to raise up these and other martyrs. The sufferings of Gould, whose passive figure of outraged womanhood set swirling fantasies of terrible torture and sexual mutilation, cried to be redeemed, while the courageous Pitkin, shorn of his discontent, stood forth a chivalrous figure whose sacrifice was held up to other college men and served to inspire the mission-minded at Yale to establish a school in Ch'ang-sha, Hunan, at the very heart of the anti-Christian movement. "The sufferings so patiently and bravely borne may be looked upon as the birth-pangs of a new era in China," was the hopeful conclusion that mission publicists drew from these terrible events.[3]

The Boxer attack on Pao-ting, Tientsin, T'ai-yüan, and a score of smaller places were sideshows to a world looking on aghast at the main drama going on in Peking. A state of siege had descended on the legation quarter in mid-June, trapping within a motley, status-riven, and anxious community. Foreigners, both regular troops and young, single, able-bodied volunteers, manned the outer defenses and suffered heavily in dead and wounded. Missionaries, with the Americans in the lead, threw themselves into directing work on the fortifications and organizing the life of the besieged. The dignitaries and their wives sat leisurely on the sidelines in the safety and comfort of the British legation where were the inner defenses. Two thousand Chinese (more than twice the number of foreigners) also waited. Some were Christians brought by the missionaries to safety, others laborers and servants who had fallen by chance into the trap and could not escape. All were regarded with suspicion and all, except for those who worked under missionary supervision or who had a foreign patron, were kept on short rations. Though the troops shot birds and carrion dogs for them, distended bellies were everywhere to be seen in the carefully segregated Chinese section. Within the first month Chinese children were already dying of starvation.

That first month of the siege was the hardest. By mid-July the court began to waver uncertainly as foreign forces gathered off the coast, overcame stout resistance at Tientsin, and began a forced inland march against disorganized Boxer and imperial units. Government troops took over the lines about the legation quarter from the Boxers, and rather than unlimbering their modern artillery, instead fraternized with the enemy in a truce that was to last down to the eve of relief. Finally on August 14 the international army fought its way into the capital. The summer of anarchy in Peking, introduced by the arrival of the Boxers in the spring, was now to be prolonged only a while longer—by the court's flight to safety in the western interior and by the misconduct of foreigners and converts, now in control in the city.[4]

The empathetic Robert Hart, head of the Chinese customs, had anticipated the desperate fury of 1900. "One of these days," he had mused in 1894, "despair may find expression in the wildest rage, and . . . we foreigners will one and all be wiped out in Peking— 'If it had not been for these cursed *fan-kwei* [foreign devils]', everyone will say, 'this would never have come upon us: let us teach them what destruction is, before we have to meet it ourselves!' " Having survived the siege, Hart emerged to explain with remarkable prescience and dispassion the significance and consequence of China's convulsion. The Boxers had tried "to free China from the . . . corroding influence of a foreign cult and [to] free China from foreign troops, contamination, and humiliation." Though the Boxers had failed, they nonetheless stood as "today's hint to the future."[5] Chinese nationalists would seek to redeem the humiliations of the nineteenth century and end foreign domination.

China was to change under the influence of nationalism, but the American commitment to the open door ideology and its paternalistic pretensions to guide and guard China was to persist. Already in the two decades following the Boxer Uprising American aspirations and Chinese nationalism were becoming fixed in dangerous contradiction. The new "civilized antiforeignism" subsumed and intensified much of the old concern with limiting foreign penetration and securing respect for overseas Chinese. Inevitably then the privileged position of the open door constituency and the American policy of exclusion became a target of attack. As Americans stood by the unequal treaties and an unyielding exclusion policy, Chinese nationalists began to question American benevolence and wonder if, behind the rhetoric of friendship and uplift, there did not lurk a hostility—all the more dangerous for being masked—to China's independence and national self-esteem. Even attempts at coordination and cooperation in strategic policy, where American and Chinese interests were not contradictory but compatible, proved dismayingly difficult and in the end added only one more irritant to an already troubled relationship.

Chapter Six

China's Defense and the Open Door, 1898–1914

FROM THE LATE 1890s onward China's strategic position continued to suffer blows at the hands of the imperialist powers. The response of officials in Peking and the provinces was once again to reach to the United States as a makeweight in China's policy. The successors of Wei Yüan, Hsü Chi-yü, and Li Hung-chang took as their special task the testing of the American open door policy as an antiimperialist doctrine conducive to cooperation in defense of China's independence. Chinese policy makers reasoned that heightened American concern for China's integrity together with the inherent lures of the China market, sharpened by the offer of special economic concessions, would irresistibly draw the United States into the desired cooperative relationship.

The legacy of the recent past all but ruled out a favorable American response. Most obviously American policy makers carried into the twentieth century a vision of the Sino-American relationship as an unequal one between patron and client, in which coercion was thought to play a necessary part. Americans also brought with them an ambiguous open door doctrine, a compound of concern for commerce, missions, and national influence in China that McKinley and his successors were each to evaluate in a different way, with the result that the open door translated not into one but into a shifting set of policies under Roosevelt, Taft, and Wilson. This inconsistent policy, mixed with pervasive American condescension toward China that sometimes lapsed into contempt, proved the despair of a line of Chinese officials who had desperately hoped for better.

The Specter of Partition

Between 1898 and 1903 Chang Chih-tung did the most to carry forward interest in the United States as one of the commercial powers that might be counted on to oppose seizure of Chinese territory. Chang, sixty-one years old at the beginning of this period, was a seasoned and influential official, well entrenched as the governor general of the provinces of Hunan and Hupei in central China. He had begun his career in 1863 after winning the highest degree in the state examination system, and later made his reputation as a zealous advocate of resistance to foreign aggression and as a critic of Li Hung-chang's and Prince Kung's policy of appeasement. He had been ready to fight Russia over the Ili dispute in early 1880 when his articulate and adamant stance first won him wide notice. He took an equally bellicose stand against French inroads in Vietnam later in the decade and against a negotiated peace with Japan in 1895. Through these years Chang had put little stock in barbarian management.[1]

Yet the fact remained that until China unaided could preserve her own interests, by war if necessary, policy makers such as Chang would have to employ as the only practical short-term alternative the balance of power techniques of the barbarian managers. But deciding on which power or set of powers was the principal threat and which the most secure and likely source of support was difficult, as Chang's own experience suggests. When the throne insisted on peace with Japan in 1895, Chang had endorsed Li's policy of alliance with Russia and incidentally denigrated the United States as a potential ally. By 1898, however, Chang had repudiated the Russian alliance as a small benefit purchased at a high cost in concessions and loans. But rather than abandon barbarian management, Chang sought thereafter to apply it more successfully. In Japan's place as the chief threat to China, he now put Russia along with her French diplomatic partner and shifted his hopes to Britain, Japan, and the United States, all commercial powers with a concrete incentive to block seizures of Chinese territory or even exclusive spheres of influence.[2]

That Chang now included the United States among this set of relatively safe powers obviously owed much to the exigencies of the time. Still, the persistence of the old image of the United States as a strong yet benevolent power facilitated the shift in Chang's interest. The failure of Li's recent policy experiments and the corrosive missionary and immigration questions had done surprisingly little to tarnish the American image. Comments on the United States in late nineteenth-century works, replete with references to George Washington's exemplary leadership, model institutions, and rapid industrial and technological development, betrayed the lingering influence of Wei Yüan and Hsü Chi-yü. Certainly China could learn from the United States in railroad building and industrial development, so a new generation of Chinese observers suggested, though they made no claims of even a modest sort for the United States as a source of diplomatic support.[3]

Such claims did issue, however, from a trio of foreign affairs experts, all former protégés of Li Hung-chang. Each carried a favorable view of the United States and likely played a pivotal role in interesting Chang in the United States. Hsü Chüeh was the first of the three to see the possibilities of American diplomatic ties. His view had taken shape by 1885 and was subsequently confirmed during his service in the Washington legation between 1886 and 1888 and again during the Sino-Japanese War. He saw the United States as a sympathetic power, wealthy, "sincere," and less at fault in the immigration question than the Chinese themselves. Americans were in general well-inclined toward China, and by early 1898 he had concluded that China might do well to draw on American wealth as well as Russian strength in holding back Britain and Japan, whom he looked upon as the most aggressive of the powers. At the end of the year Hsü elaborated these views in an imperial audience in response to queries by the Empress Dowager.[4]

Wu T'ing-fang, another of the foreign affairs experts, shared with Hsü's outlook little in common but a hopeful interest in the United States. Born in British Singapore in 1842, educated in British Hong Kong as a boy, schooled in London as a lawyer, and employed for a time thereafter in Hong Kong colonial affairs, Wu was if anything an Anglophile. Nevertheless, in 1895 Wu identified the United States along with Britain and Germany as the powers from whom China might expect help in checking Russian, French, and Japanese aggressiveness. In February 1898, now as minister in Washington, Wu gave the United States even greater prominence, putting the Americans, who were skilled in naval and military affairs, in place of the Germans, the despoilers of Shantung. At the moment, Wu reported, the United States was preoccupied with the questions of Cuba and Hawaii. Still, the inroads of the powers had created alarm that made Americans receptive to Chinese overtures. By holding out commercial opportunities to the United States as well as Britain, China could, he contended in a memorial to the throne at that time, create an advantageous balance of power and at the same time enhance China's own prosperity. Wu couched his recommendation in the antique vocabulary of barbarian management, describing the United States as "most respectful and obedient" and referring to possible use of the United States as comparable to the classic strategy of using distant states against proximate dangers. But in realistic terms, Wu was making an argument for a Chinese open door policy well over a year in advance of the formal American pronunciamento on the subject.[5]

Finally, Chang may also have owed something of his American strategy to Yung Wing (Jung Hung), a Cantonese who had studied and finally settled in the United States but remained involved with the work of the Chinese legation and interested in China's self-strengthening. Yung had acted as Chang's financial agent in the United States during the Sino-Japanese War and afterwards had returned to China to serve for a time in Chang's secretariat. In 1897 Yung had proposed the use of American capital in railway construction. Because relations with the United States were uniquely uncomplicated, China could borrow American capital without political risk. "We will borrow their merchants' financial strength and we will control the authority."[6]

Chang's interest in railway construction in his area of influence in central China provided him his first opening to the United States. Though Chang preferred in principle to rely exclusively on Chinese funds to build railways, he soon discovered that only abroad would he find the necessary capital. So in consultation with Sheng Hsüanhuai, director of the state railway administration, Chang decided to turn necessity into a virtue by borrowing from a country committed to the status quo. Sheng accordingly opened talks with a representative from the American China Development Company, but finally when the Americans inflexibly stood by a draft contract "filled with great flaws," Sheng concluded an agreement with the Belgians instead. American demands for an unusually high rate of interest and maintainance of control of the railroad once built revealed, Chang himself concluded, a lack of good faith. Nonetheless, the next year Chang again took up the thread of his American interest. He ordered Minister Wu to sound out investors in the United States, and in April 1898 entrusted to the American China Development Company the railroad to run from Hankow south to Canton. Though he had had to accept terms similar to those rejected out of hand the year before, Chang had drawn in the only country with capital yet "no intention of taking advantage of our territory." He hoped this move would counteract the spheres of influence the powers had only recently staked out, secure rapid and skilled construction of the strategic railroad, and give Washington more reason to defend the status quo against the dangers of partition.[7]

Washington failed to respond according to Chang's calculations. While the open door notes of 1899 did affirm American trade interests, they shied away from attacking the spheres of influence, a cancer threatening China's survival as well as American trade. Moreover, the United States acted unilaterally, lending credence to rumors picked up by Chinese officials that the United States might in desperation join in the land grabs by the other powers. This concern, perhaps stirred by loose talk by Minister Conger or by good intelligence from Wu Ting-fang in Washington, was well founded. The seizure of some strip of Chinese territory either as an adjunct to the Philippines stronghold or as a marker for a sphere of influence in North China was a line of action actively canvassed both in the Peking legation and in the McKinley administration in 1899. When at last late in the year Minister Wu went to inquire about the rumors, Hay offered a highly qualified denial, not calculated to give any encouragement to Peking's resistance to pending Italian territorial demands.[8]

At the same time Chang also confronted the first and no less disturbing evidence of bad faith on the part of the American railway concessionaires. The sorry tale, to unfold over the next seven years, began with the American China Development Company demanding a contract revision that would raise construction costs to $40 million, secure for the concessionaires additional economic advantages, and extend the deadline for completion of the work from three to five years. With the court's concurrence Chang agreed, both out of a sense of urgency to have completed this line crucial to defense and development and out of a still unshaken faith in the United States. Thus in July 1900 Wu, once again acting as Chang's representative, concluded a supplementary agreement. But suspicious that Belgian money was displacing American in the company, Wu added a provision forbidding transfer of control of the American concession to investors of any other nationality. His intention was to prevent the railway passing to the hands of a power that was already in control of the Hankow-Peking concession directly to the North and that was moreover known as the cat's paw of France and Russia.[9]

Meanwhile, officials within the capital had become increasingly strident in their attacks on the abortive strategy of purchasing foreign support through the grant of special concessions. They now called for a more militant policy based on an alliance with the people. By channeling the considerable popular discontent in the North created by rumors of foreign incursions, the possibility of partition, and a series of natural disasters, Peking might be able to check the run of foreign demands. Success in withstanding Italian territorial demands in November 1899 suggested that a more adamant and self-reliant foreign policy with popular support might indeed work even against still more powerful aggressors. In line with this reorientation of policy, the court took steps to assure the loyalty of the Boxers, and it sent out instructions for military preparations to meet possible foreign invasion. A series of developments in the first half of June 1900—the reinforcement of the legation guard, the dispatch of more foreign troops toward Peking, and intensified Boxer activity in the capital—finally threw the court into all-out resistance. While offering supposedly misguided converts a last chance to repent, "to escape from the net," the court sanctioned the expulsion of foreign missionaries and diplomats.[10]

As the court's dramatic and costly experiment in populist foreign policy turned into a disaster of a magnitude unprecedented in Ch'ing foreign relations, Chang sought to salvage what he could. His main goal was to insulate central and southern China from the gathering conflict and to minimize its intensity and consequences in the North. He directed his efforts along two different lines. On the one hand, from mid-June Chang joined other leading provincial officials in pleading with the court for caution and for conciliation with the powers. The first step, these eminent officials advised, was to suppress the Boxers and ensure the safety of the foreign diplomats in the capital; the second was for Li Hung-chang to set talks in motion. By July this advice together with the gathering foreign invasion force began to move the Empress Dowager away from the new militant policy line. While executing those officials within her grasp who openly called for moderation and had been tainted by their earlier involvement in barbarian affairs, she also accepted piecemeal the advice of Chang and others. Appeals to restore good relations went out to the powers; the siege of the legations was turned into an armed truce; and Li was restored to his old post as governor general of Chihli at the center of the crisis. On August 7, as foreign troops marched toward Peking, the Empress Dowager empowered him to arrange a truce. Finally on August 27 the court, now in flight from the capital, seconded Li's call for an end to all resistance.[11]

China's Defense and the Open Door

Over this same period Chang's efforts to contain the fighting and bring it to an early end led him to enlist the cooperation of the United States as well as the other commercial powers. The Americans had no interest in Chinese territory, he once more noted, and more than ever they feared renewed foreign aggression that might endanger China and prolong the conflict. Encouraged by the American refusal to participate in the initial foreign attack on Chinese coastal defenses (at Taku on June 21), Chang suggested that Peking seek United States mediation as a way out of the current difficulties. Finding the court at first unresponsive to that proposal, Chang tried a new tack. Through Minister Wu, he sought to convey his peaceful intentions to Washington and to convince American policy makers that China and the world were not at war, that the rebellion was the work of ignorant people driven to desperation, and that the court was having to move carefully against the rebellion to avoid yet greater popular violence. After the fall of the capital and the court's capitulation, Chang still turned to the United States for help in bringing an end to punitive expeditions launched by foreign forces from Peking into adjoining areas, including Shansi province where the court had taken refuge.[12]

The McKinley administration had at first responded uncertainly to this crisis in China. Though Hay had not pressed Peking to put a stop to the worsening attacks on converts in Shantung province in November and December 1898, he had at the same time readily acceded—over strong protests from the Chinese foreign office—to Minister Conger's request for a reinforced legation guard. But the troops had proven unnecessary, and so after this false alarm Hay, even in the face of growing turmoil in Shantung and adjoining Chihli in late 1899 and through the first half of 1900, was hesitant to send them in again. Moreover, up until June both the Secretary of State and his adviser on Chinese affairs, W. W. Rockhill, doubted the alarms sounded by jumpy missionaries, and found in Conger's reports, alternately disturbing and reassuring, no clear guidance. Finally, Hay feared that the dispatch of troops might only serve to create unwanted complications among the powers in China. Even consultation with the other powers on the use of force, Hay warned Conger, should be handled so as not to give umbrage to the administration's Anglophobe critics, who would be quick to pounce on anything that looked like a foreign alliance. Not until June 15 did Conger finally rouse Hay to action—but too late. By June 20 the legations were under full siege.[13]

Once news of the attack on the legations arrived, McKinley took from Hay's hands responsibility for keeping the commitment to the open door constituency and for ensuring the United States a role in the international force gathering off the China coast. But he moved with a caution and restraint as characteristic of the President as it was welcome to Chang Chih-tung. McKinley and his aides were predisposed to accept Chang's warnings against the complications that would result from foreign armies camping on Chinese soil and Chang's promise that some of the more level-headed Chinese officials might yet restore stability and the status quo. McKinley was further restrained by the army's already heavy occupation duties in newly acquired territory and his own unwillingness to give more ammunition to his antiexpansionist critics in the presidential campaign already in progress. Accordingly, he ordered the commander of the American relief force, Major General Adna Chaffee, to act independently of the powers where possible and jointly only where necessary, to treat the Chinese with firmness and justice necessary to keep their friendship, and to behave generally as if the United States and the other powers were not at war with China but instead only engaged in cooperation with friendly officials, foremost Li Hung-chang, in saving foreign lives and restoring order. Though McKinley rejected Chinese appeals for mediation as impossible until the legation was safe, he nonetheless offered assurances that American troops were in China only for purposes of relief and rescue, and he carefully limited their use to the North.[14]

Following the capture of Peking in late August, McKinley—true to his word—at once began cutting back on the ten-thousand-man force assigned to China. Units en route were recalled, and two-thirds of the 6,300 already in China returned to suppressing Filipino "rebels." For a time McKinley had even leaned toward Secretary of War Elihu Root's call for an immediate and complete withdrawal (in response to the Russian evacuation of all their units back to Manchuria) before reluctantly acceding to the arguments of Hay, Rockhill, Chaffee, and Conger that a token military presence was needed to keep the pressure on the Chinese and to guarantee an American voice among the powers. Still concerned over excessive use of force, McKinley forbade the American troops wintering over in North China to engage in the punitive expeditions conducted by the other powers, brought an early end to aggressive patrolling immediately outside the city, and at last in the spring called back the remaining troops, this time overruling Rockhill and Conger. The result was, as McKinley wished, an exemplary record for the American occupation force that none but the Japanese could match and that Chinese officials singled out for praise.[15]

While McKinley kept the American military response to the Boxers in check, Hay brought note-writing, exercised so successfully in 1899, back into play to restrain the powers. To forestall a new round of great power competition that might end with China's dismemberment, Hay sent out on July 3, 1900 his second major statement on the open door, calling this time for China's preservation, both administrative and territorial. This proposal invoked Chang's fiction that the powers warred against neither the Chinese people nor the imperial government and hence could have only limited objectives. To avoid contradiction or evasion, Hay did not ask for a response from the powers since, he calculated, a unilateral assertion of the integrity principle was as likely to build an international consensus as a prolonged and involved diplomatic exchange. In early August Hay at last retired from the scene for two months, exhausted by his labors in the Washington heat and tormented by imagined "scenes of tragic horror" to which, as well he knew, his own delays in the spring may have contributed.[16]

The Boxer crisis had sharpened for both Americans and Chinese leaders the specter of partition. But that it was averted was due less to the efforts of either than to the growing recognition in European capitals of the difficulties—the high cost, perhaps even the physical impossibility—of directly controlling China. Hay and McKinley had, to be sure, made some contribution, as Chang had hoped, to restraining the powers. However, alongside this positive aspect of United States policy, Chang would have had to balance other, distinctly less favorable features of that policy.

To begin with, American officials retained their contemptuous attitude toward the Chinese, regarding them as a negligible quantity in determining their own future. Hay had no more bothered to consult China on the second series of open door notes than he had on the first. Had he done so, he would have seen that the notes would have met Chinese hopes only halfway. For the integrity principle the notes propounded posed no challenge to the unequal treaties (which the powers were then vindicating with main force) or the existing foreign concessions and spheres of influence (which Chinese regarded as grievous violations of sovereignty.)

Moreover, the Boxer summer enhanced the role of coercion in the tactical armory of the open door. American policy makers had had confirmed a belief of growing importance since the late 1880s—that the Chinese respected only force. Accordingly, Washington retained occupation forces in Peking until a settlement was near conclusion, kept at the legation a guard of 150 men (increased to 300 in 1905), earmarked for emergency service in China an army force of 2,000 (that could be expanded to 5,000) in the Philippines, and in 1911, in the wake of the revolution, would post an infantry battalion to Tientsin (augmented with a second battalion in 1914) to guard the communications route to Peking. The Asiatic Squadron underwent a parallel buildup. By 1902 it would number some forty-eight ships, including gunboats to patrol Chinese waters as well as one battleship and two armored cruisers that might be quickly deployed to the China coast. Together this force stood ready to protect "the interests of civilization and trade" and deal out "severe and lasting punishment" to any Chinese who threatened Americans.[17]

Finally, Chang could have known little if anything of the continuing American interest in territorial concessions. For a time a strong point on the China coast seemed essential for the future defense of Americans in a xenophobic China. It was also important to keep up with the apparent ambitions of the other powers. (Britain had landed troops in Shanghai in the summer, while Japan extended its claim to an Amoy concession. Russia held Manchuria under military occupation, and in November 1900 demanded a concession in Tientsin.) That Hay—despite his recent stand on the integrity platform—could entertain once again, as he had the previous year, acquiring bases or at least a concession that would demarcate an American sphere of influence suggests how little fixed were the tactics of the open door policy and how deep were the concerns for preserving American influence.[18]

The period of diplomatic maneuver in the aftermath of the Boxer outbreak saw a steady diminution of Chang's interest in the United States. Though he still looked to Washington for help in negotiating terms favorable to China, his hopes had by now been tempered by experience. The official American position between late 1901 and 1903 would further lower his expectations. Washington entered this period still contemptuous of even the most friendly Chinese leaders. Hay doubted Chang's bona fides and regarded Li as "an unmitigated scoundrel" and "thoroughly corrupt and treacherous." Washington was, moreover, united with the other powers against China on the need for some form of retribution for past misbehavior and for guarantees for the future. Further, Washington's growing interest in reform inclined it to seize this opportune moment to push through changes in China's commerce, her system of foreign relations, and the status of missionaries—even though the Chinese might, as McKinley himself conceded, see these changes as "an alien invasion." Finally, it was becoming ever clearer that the McKinley administration approached China in an opportunistic and cautious frame of mind. Though Hay declared bravely in the fall that the United States would "hold on like grim death to the open door" in order to preserve American opportunity, above all commercial opportunity, forceful action in league with either China or the powers was out of the question. Domestic politics ruled out alliance or even the suggestion of entanglement with Britain, though Hay continued to talk of an Anglo-American identity of interests against Germany and Russia. Moreover, the administration had no army or navy it could risk in an international struggle for China. Hay in a rueful but perceptive mood observed, "The talk of the papers about 'our preeminent moral position giving us the authority to dictate to the world' is mere flap-doodle." That left two complementary lines of action: cultivating China's friendship in order to secure preferential treatment; while discreetly following the other powers in wresting concessions from China.[19]

Though Chang and Li might be drawing closer together in their estimate of the United States, they were still at loggerheads over general policy. Chang had not entirely given up on the United States, but he looked more and more to Britain and Japan for help. Li, enfeebled by age and suffering partial paralysis, regarded Russia and not the commercial powers as the key to quickly discharging the responsibility the court had thrust on him for making peace and securing foreign military evacuation. A prompt diplomatic settlement seemed to him imperative since it would clear the way for the withdrawal of foreign troops. As long as those troops remained, the court could not return to the capital, the country would remain restive, and the costs of the occupation army, which China had to bear, would increase. He distrusted the motives of the British, Japanese, and Germans, while the Americans, he warned, drawing on his memory, followed a weak and erratic policy and often assumed a friendly pose to ingratiate themselves with the Chinese and win without effort what the other powers would take by force. In Li's view the abuse of Chinese in the United States and the aggressive and disruptive part taken by missionaries in China certainly proved that these displays of amity were false. The seizure of the Philippines seemed to him an unusually ominous sign that American policy had begun to move in the direction of European and Japanese imperialism.[20]

Peking, where the marathon talks among the powers began in the fall of 1900, was one of the two arenas in which Chang and Li were to test their divergent strategies. During those talks, Chang, observing from the sidelines, routinely invoked Hay's support. But when at last Li brought the talks to a close in August 1901, Chang had little to show for Hay's assistance. The United States was in those trying and complex negotiations only one voice of the eleven there. On several points Hay had succeeded in restraining the powers, to the advantage of both China and the United States.[21] Where he could not, Hay refused to put the United States in an "exceptional position," so that his country would share in whatever advantages were won by others.

Hay's most sustained support for China's position came in his efforts—ultimately unsuccessful—to reduce the total indemnity demanded by the powers. An excessive sum (by American estimates anything over $150–200 million) would impoverish China, probably cripple her foreign trade, and possibly result in further foreign encroachments in case she failed to keep up payments. With American encouragement Chang pressed Li for better terms, while Rockhill and Hay also applied pressure on the other powers to scale down their claims. The powers refused to budge or even arbitrate the issue and finally decisively voted the United States down in May 1901 and set China's total indemnity at $333 million. Hay, who had padded American claims, could have accepted an across-the-board reduction of the total indemnity by one-half and still fully covered all legitimate American claims. Now with the total indemnity undiminished, Hay was guaranteed a surplus, later used to promote American educational influence in China.[22] When a dispute subsequently developed over whether payment of this indemnity was to be made in gold or silver, Hay again tried to lighten the Chinese financial burden. And again he failed to move the other powers. When China at last in 1905 asked the United States alone to accept payment in silver, Hay invoked his "no exceptional position" formula and insisted that the United States too be paid in gold.[23]

The talks to secure Russian evacuation of Manchuria was the second arena in which Chang and Li clashed. Here too the American position proved disappointing to Chang, especially compared to the strong stand taken by Japan and Britain. Russia had used the Boxer crisis to take military control of the Northeast. With this strategic territory now in jeopardy, Chang put up a determined defense against any concessions that Li might make in his haste to get the Russians out. Chang was particularly concerned that anything less than full restoration of Chinese political and military control would serve as an invitation to other powers to encroach elsewhere in

China. "Those who in the slightest are loyal to their sovereign and love their country are pained," Chang observed in a jibe which had Li as its mark. Chang urged the court to line up the commercial powers against Russian demands by holding out the promise of opening Manchuria to profitable trade, mining, and railway ventures. An "open door," a term which Chang now frequently used, would attract American, British, and Japanese merchants and lead their governments to demand Russian withdrawal. Predictably Li denigrated Chang's faith in the commercial powers. They would not (he contended) stand up against Russia and "can only wag their tongues at us."[24]

Events proved Li right at least as far as the United States was concerned. Hay responded to Russia's "shameless" post-Boxer Manchurian policy by taking the familiar path of note-writing and of coercing the Chinese to do for themselves what they would not do on their own. While this last phase of Hay's note-writing cast in yet broader terms the formal open door commitments made in 1899 and 1900 to equality of commercial opportunity and the defense of China's integrity, its chief effect at the time was to do no more than lend tacit support to the anti-Russian cause. In February 1901, Hay reiterated his concern for China's integrity in a strongly worded warning to the Chinese foreign office against accepting Russian demands without first consulting the other powers. After a second similar warning in early December 1901 evoked a troublingly vague Chinese reply, Hay launched an attack in February 1902 on any "exclusive right to open mines, construct railways or other industrial privilege" as a violation of the open door and American treaty rights. Twice more before his death Hay would reaffirm his concern for China's integrity and the open door (once in February 1904 as the Russo-Japanese War began and again in January 1905). But despite his bold words in public, Hay himself privately conceded the exceptional position that Russia occupied in Manchuria and by extension that of other powers in other parts of China. Only with the passage of time and the arrival of a new team of policy makers would Hay's bolder public doctrine and especially his 1902 attack on the special investment rights associated with the spheres of influence begin to be taken seriously in Washington.[25]

Meanwhile, the Manchurian talks dragged on inconclusively, with Russia continuing in military occupation and with the Chinese court refusing to recognize Russia's claim to a broad range of special rights. The conclusion of the Anglo-Japanese alliance, publicly announced in February 1902, broke the deadlock. By guaranteeing

British support, it enabled Japan to dig in her heels against Russian influence in Manchuria and Korea. It also vindicated Chang and greatly strengthened the hand of Prince Ch'ing, who had picked up the thread of the Manchurian talks following Li's death in November 1901. By April he had a Russian promise of a gradual evacuation (to be completed within two years) on terms more favorable to China than any offered before.

The building tensions between Russia and Japan suggested that time was now on China's side. A war might end in Russian eviction from Manchuria or at least a weakening of her influence. So Chang decided he could for the moment safely stand aside and wait. But Hay was not to allow him the luxury of passivity. When in April 1903 Russia reneged on the promise to evacuate made to China a year earlier, Hay responded by asking Peking to open new treaty ports in Manchuria to American diplomats and merchants (under the terms of a Sino-American commercial treaty then under negotiation). Such a step would, he thought, serve as "a good test of the sincerity of Russia's declarations as to the integrity of the Chinese Empire, the acquisition of territory, the open door, etc." Chang opposed complicating relations with Russia by acceding to American demands. Let the Americans help end the Russian occupation, and then China with her control restored would open new ports, Chang advised Peking.[26]

The Contest for Manchuria

Japan's war with Russia (1904–1905), rather than solving China's Manchurian problem, complicated and accentuated it. Russia, though weakened, still held on in the North. Japan, having finally obtained a lodgment in the South at great cost in life and treasure, began to consolidate her influence there. This Japanese defection to the ranks of China's despoilers neutralized the value of the Anglo-Japanese alliance to China and thrust the United States forward to a central place in the calculations of Chinese policy makers. From 1906 to 1910 they were to press with unprecedented vigor for American financial and diplomatic support for their quest for effective control in the Northeast.

The effort was launched by Yüan Shih-k'ai, yet another—and ultimately the most influential—of Li Hung-chang's protégés. Yüan had represented Li in Korea in the years before the Sino-Japanese

War and thereafter had helped organize Li's modern army in North China. He had established his favor in the eyes of the Empress Dowager in 1898 when he played a crucial role in her coup against the reform-minded Emperor. He had counseled restraint during the Boxer crisis, husbanded the forces under his control, and then rallied to the support of the Empress Dowager following her flight from Peking in the fall of 1901. An appointment as governor general of Chihli (a post he held from 1901 to 1907) and growing political influence soon followed.

By 1903–1904 Yüan was seriously engaged in the Manchurian question. He argued forcefully alongside Chang Chih-tung against conceding the American treaty port demands and then, on the eve of the Russo-Japanese War, stood forth as the foremost advocate of neutrality, a position the court embraced in February 1904. After the war, as Russian and Japanese armies evacuated, Yüan's troops—and officials closely identified with him—moved in, while Yüan himself orchestrated overall policy, first from his provincial headquarters in Tientsin and then in 1907, following his transfer to Peking, from the Grand Council and the foreign office.[27]

Yüan picked up the thread of the American policy that Li had experimented with and dropped and that Chang more recently had tried to manipulate with only limited success. Already during the post-Boxer crisis Yüan had come to accept Chang's estimate of the United States as one of the commercial powers with an interest in protecting China's independence. But while Chang's enthusiasm waned, Yüan's seemed to wax, so that by 1905 Yüan had identified the United States as the principal potential guarantor of China's weakened position in Manchuria. He accordingly wanted to minimize peripheral conflicts of interest with the United States (especially the immigration controversy and the boycott it gave rise to) in hopes that postwar tensions between the United States and Japan, engendered by another immigration controversy and by reported trade discrimination in Manchuria, would lead the Americans to see the coincidence of their interests with China's in that endangered region.[28]

In their efforts to draw the United States into Manchuria, Yüan and his associates in charge there were to revitalize and draw together the two formerly parallel approaches to the region's problem, each of which had reached a dead end. One approach, the policy of reliance on the commercial powers earlier advocated by Chang, had lost some of its promise. Japan was now more of a threat to than a support for the open door in Manchuria, and Britain was immobilized by her alliance with Japan. Moreover, Britain and the United States had failed to extend trade interests into Manchuria sufficiently to provide even a minimum offset to Russian and Japanese influence. By 1907 foreigners could reside and do business in no less than fifteen cities in the region and in addition had at hand a remarkably well-developed railway network to carry their trade. But the long-awaited Anglo-American commercial invasion never materialized. American commerce even fell off. The other approach, a development policy to make Manchuria more defensible and strengthen Chinese control, had led nowhere for want of funds in the treasuries of Peking and regional authorities, especially after they were burdened by the Boxer indemnities. After the Russian invasion in 1900, provincial officials had begun calling for an elaborate program that included an increase in local military power, the extension of political administration, the encouragement of colonization, and the promotion of commerce and industry. Heading the list was the construction of Chinese-controlled railways, which would hasten the region's economic development, facilitate defense, and weaken the influence of Russian and Japanese-run railways. To secure both the necessary funds for this ambitious program and the diplomatic support for an attack on the spheres of influence, it would be necessary to win the help of one of the powers. The United States was to be that power.[29]

The details of designing a financial open door policy and seeing if avarice would draw the United States to it fell to Yüan's men in Manchuria, Hsü Shih-ch'ang and T'ang Shao-i. Hsü was a classically trained official sent by the central government in 1906 to tour Manchuria and then kept on there as governor general to carry out the reforms he had recommended. T'ang, the governor of Fengtien (the southernmost of Manchuria's three provinces and its most populous), had been a student with the Yung Wing educational mission and had studied at Columbia and at New York University. On his return to China in 1881 he had helped implement Li Hung-chang's Korean policy. After the Boxer crisis he had assumed more and more prominence in China's diplomacy and railway affairs. T'ang took to Manchuria views on the United States that coincided with Yüan's. Only the United States, which "continues to have in the Far East a national policy independent of all foreign alliances," was in a position to join with China in defense of Manchuria.

Hsü and T'ang devised two plans. Initially they offered E. H. Harriman the opportunity to build a railway, first in southern Manchuria, later to be extended northward to the Russian border. The

1907 financial crisis in the United States forced the American railway magnate to draw back. Later, a substitute plan for an all-British financing and construction of the railway also failed due to lack of official support in London. So the Manchurian administration went back to the United States, this time with a proposal for an American-financed development bank, a plan earlier conceived by Yüan.[30] In late 1907 and early 1908 both Hsü and T'ang traveled to Peking to win the support of the central government and to sound out W. W. Rockhill, now the American minister. They also enlisted the backing of the American consul general in Mukden, the irrepressible adventurer Willard Straight, and through him laid their ideas before Secretary of War William Howard Taft (during the Manchurian leg of his 1907 East Asian tour) and a number of Wall Street financiers. Harriman in particular they wanted to keep on the string. To him they held out the tantalizing prospect of "directing the railways of a nation" through participation in the development bank.

Simultaneously, Yüan in Peking lent his assistance by securing the backing of the court and cultivating good relations with the United States. He set in motion negotiations for a Sino-American treaty of arbitration, a device calculated to appeal to Secretary of State Elihu Root's legalistic biases. He ensured a warm reception for the visit of elements of the Great White Fleet brought to anchor in 1908 in Amoy harbor, right in the midst of the Fukien sphere of influence that Japan had begun staking out. And finally he and T'ang launched a public campaign in behalf of a Sino-American alliance, which would (in the words of one contemporary article) seal "the bonds of friendship." More concretely, such an alliance would "guarantee China against the ambitions of Japan, and at the same time safeguard American commercial interests." In the United States Chinese officials laid out for American businessmen, thought to be burdened down by spare cash and excess production, the opportunities awaiting them in "the greatest world market in the near future." "In the thousands of square miles of virgin soil in Manchuria, there is grand opportunity to employ your surplus capital, the ability of captains of the great farms in your Middle West and your good farming machines." With the stage properly set T'ang set out for the United States in the fall of 1908 to commit Washington and Wall Street to Chinese plans.[31]

Had the open door policy, as Hay's notes had come to define it by 1902, prevailed in Washington, T'ang might have evoked a working response to his invitation to join against Japan in defense of China's integrity. However, T'ang had to deal with a President who was not prepared to stand by the advanced position of 1902. The problem was that Theodore Roosevelt, into whose hands the making of East Asian policy had passed following Hay's illness and death in 1905, embraced only that part of the open door ideology that stood for directing a backward China toward reform (and even here Roosevelt made no claim for a preponderant American role as a civilizer of China). With a belief in reform went, predictably enough, a conviction that force would occasionally have to be used against a people who were both backward and cowardly. The Boxer Uprising was for Roosevelt but the most recent reminder that the United States and other advanced countries had to stand ready "to put down savagery and barbarism" and exact full performance from the Chinese government of its treaty obligations.[32]

Whatever influence the open door ideology had in shaping Roosevelt's thinking on China was more than offset at this time by strategic calculations. He had long since concluded that the United States lacked both the national interest to justify and the power to make good on a serious commitment to preserving China's independence or territory. Thus he had earlier accepted Russia's "exceptional position" in Manchuria (a far cry from Hay's broad 1902 formulation of the integrity principle), and in 1903 he had rejected taking a tough stand against the Russian occupation. He was no more willing now to challenge Japan in Manchuria. Indeed, he was glad to have Japan engrossed in Northeast Asia rather than eyeing the strategically vulnerable Philippines. His mediatory efforts during the Russo-Japanese War had contributed to that end by preserving a Russian presence in northern Manchuria that kept Japan tied down.[33]

The apparent contradiction between Roosevelt's long-term hope for China's cultural betterment and his short-term indifference to China's fate as a political entity he reconciled by his faith in foreign influence in, even control over, China as the prerequisite for change. Roosevelt had thus welcomed the occupation of Manchuria in 1900 as a boost for progress. Even though the Russian occupiers were, in Roosevelt's opinion, the least advanced of the civilized peoples, they still exercised a beneficial influence over a people as deficient as the Chinese, and Russian proximity and strategic interest made that influence all the more potent. Later he looked to Japan as an advanced country and regional power uniquely qualified to instruct backward China. Roosevelt as well as Root saw a natural parallel—ominous indeed for the success of T'ang's mission—between Japan's "paramount interest in what surrounds the Yellow Sea" and

American hegemony in Latin America. Roosevelt expected only that Japan carry out its duties as policeman and tutor with restraint—"no more desire for conquest of the weak than we had shown ourselves in Cuba"—and minimal respect for American treaty rights.[34]

T'ang Shao-i reached the United States intending, as far as the public knew, only to offer thanks for the remission of the Boxer indemnity, recently authorized by Congress. That part of his mission was accomplished easily enough. On the other hand, his confidential goal of finding money for the Manchurian bank was foredoomed by the old contempt for Chinese incapacity and the new reluctance to play China's protector at the risk of collision with Japan. The State Department's China experts feared that the Chinese would fritter any money away. Roosevelt joined the coalition of skeptics. "The Chinese are so helpless to carry out any fixed policy, whether home or foreign, that it is difficult to have any but the most cautious dealings with them." Root had difficulty finding time in his busy schedule for even a chat with his visitor, and then sprang a surprise on T'ang in the form of his yet unpublicized agreement with the Japanese envoy Takahira. By reaffirming Roosevelt's committment to the status quo in the Pacific and to amicable relations with Japan, the agreement ruled out any Chinese-prompted American challenge in Manchuria and openly advertised China's continuing isolation there. Yüan reacted to what Root described with unintended irony as this "good news" for China with unfeigned irritation and disappointment, while one of his newspapers decried American paternalism: "It is truly as if our country were a guest whose affairs were to be managed by these nations which make arrangements together."[35]

The death of the Empress Dowager in mid-November 1908, followed almost at once by the demise of the Emperor, set off political tremors in Peking that unsettled Yüan and wrote finis to his plans for Manchuria. Yüan had made enemies not simply by amassing political and military power but also by the stands he had taken against first the Emperor's reform movement in 1898, then a possible alliance with Japan against Russia in 1903–1904, and finally the anti-American boycott in 1905. In the winter of 1906–1907 he and those associated with him—both junior foreign affairs experts such as T'ang and senior officials such as Prince Ch'ing—had survived one sharp political attack thanks to the backing of the Empress Dowager herself. But now that support was gone and Yüan's enemies were free to let fly again with charges of corruption, disloyalty to the throne, and lack of patriotism. Encouraging the attack was the new presiding presence, Prince Ch'un, the regent for the boy-Emperor P'u-i and not incidentally the brother of the deceased Emperor, whom Yüan had betrayed in 1898. He is supposed to have wanted Yüan executed, but when Prince Ch'ing and Chang Chih-tung interposed their objections, Yüan was instead allowed to "retire" from official life. T'ang prudently did the same.[36]

The Regent completed the purge of Yüan's Manchurian associates by moving Hsü Shih-ch'ang to the capital and putting in his place as the Manchurian governor general Hsi-liang, a Mongol bannerman in his fifties with a deserved reputation for personal austerity and moral rectitude. He came to his new post after long and loyal service in the provinces, where he had developed a recognized expertise in frontier defense and shown himself a determined foe of foreign penetration. Hsi-liang quickly grasped the problem posed by the two powers flanking Manchuria, and resurrected old plans to defend the region by pursuing internal development on a broad front—from building railroads to opening new ports and promoting Chinese settlement.[37]

Once more hopes for the necessary financial and diplomatic support came to rest on the United States. Influenced by Willard Straight, now an American financial agent, perhaps by an earlier reading of the works of Wei Yüan and Hsü Chi-yü, or even by the calculations of his predecessors, Hsi-liang concluded in short order that the United States was an ideal ally. Americans had used their mastery of modern technology, especially in railway construction, to develop their own frontiers. "America's rise to eminence thus took no more than one hundred years," Hsi-liang observed, "and now it has no equal in wealth. It can serve as an example." Americans coveted the wealth of Manchuria and would grab at any proffered opportunity to get a share of it. American businessmen, who had begun to take an interest in East Asian trade and investment, had been complaining of Japanese discrimination, and the new Taft administration, much disturbed, might be ready to embark on an anti-Japanese policy. This outcome would be much to China's advantage, Hsi-liang contended, since the United States harbored no "hidden intentions" dangerous to China's independence.

Support in the central government for a strong Manchuria policy was now missing, however. To Hsi-liang's proposal for a foreign railway loan, the Regent seemed agreeable at first but by the fall of 1909 he had backed away under the advice of the head of the foreign office, Na-t'ung, and the president of the finance ministry, Tsai-tse. Na-t'ung feared provoking a crisis with Japan and preferred, instead of creating new problems, to negotiate a settlement of some of the old ones that already strained Sino-Japanese relations. Tsai-tse for his part centered his fire on the foreign loan, which (he claimed) the government could simply not afford. Whether out of agreement with these objections or simply out of deference to their authors, the Regent withheld his approval.[38]

Hsi-liang persisted. He could not in principle accept a policy of accommodation, and he began to worry, moreover, that concessions to Japan would create popular discontent in Manchuria and perhaps even antidynastic agitation. He now boldly decided to force Peking's hand. Without consulting the central government, he summoned Willard Straight to Mukden and immediately concluded a preliminary railway agreement combining Harriman money with British construction. In early October he presented this fait accompli to officials in the capital and tried to convince them of the importance of the proposed railway in creating a balance of power as well as in strengthening Manchuria's defenses and hastening its development. "Although we call this a commercial railway, it is in fact part of a diplomatic and political policy." Hsi-liang, however, could not overcome the objections of Na-t'ung and Tsai-tse, who recommended abrogating the railway agreement. With the Regent still temporizing, Hsi-liang continued to maneuver, keeping the railway talks open while renewing his proposal for a foreign loan to establish the development bank.

Into the deadlock between Peking and Mukden, the administration of William Howard Taft unwittingly stepped late in 1909. Taft was Roosevelt's hand-picked successor, and it was generally assumed that his loyalty to Roosevelt and his apprenticeship as Roosevelt's proconsul in the Philippines and emissary to East Asia in 1905 and once again in 1907 was a guarantee of continuity in China policy. But Taft and his Secretary of State, Philander C. Knox, moved off instead in the direction Hay had been pointing the open door between 1900 and 1902—toward a challenge of the spheres of influence.[39]

Taft and Knox came up with their own shared version of the open door ideology. In contrast to Roosevelt, who felt little of the lure of the China market, they accepted without question its "almost boundless commercial possibilities." They made the chief task of Washington and the foreign service the realization of those possibilities. "Today diplomacy works for trade," Knox declared, "and the Foreign Offices of the world are powerful engines for the promotion of the commerce of each country." Also unlike Roosevelt, who studiously avoided a conflict with Japan over China, Taft and Knox soon became convinced that national interest required the United States to play China's guardian against Japan. By imputing to Japan selfish motives dangerous to the future of American commerce, Taft and Knox restored to the open door ideology that sharp contradiction—reasoned away by Roosevelt—between foreign encroachment on the one side and, on the other, China's progress and independence and the future of American interests. Taft's doubts about Japan went back to his 1905 tour, which had taught him that "a Jap is first of all a Jap and would be glad to aggrandize himself at the expense of anybody." A second tour two years later had convinced him that Japan sought the domination of China and might in time threaten the Philippines. With the Chinese almost childlike in their inability to protect themselves, it was left to the United States "as the only country that is really unselfish in the matter of obtaining territory and monopolies" (as Taft had privately observed in 1907) to do so. Now, as President, Taft publicly promised that he would if necessary rely on more than "mere verbal protest and diplomatic note[s]" in defense of the open door and American interests.[40]

Knox, who immediately claimed a large measure of independence within the administration, accentuated this shift in China policy by elevating to prominence a group of young foreign service officers earlier critical of Roosevelt's approach to East Asia. Huntington Wilson, Knox's Assistant Secretary of State, and Willard Straight, now in the Division of Far Eastern Affairs, soon established Japan's tightening grip on Manchuria as the favored explanation for the collapse of American exports. These State Department activists conceded that American exporters, generally deficient in energy and enterprise, were also at fault. But if businessmen would not be provident and Japan respectful of the open door, then the government would have to act in the name of national prosperity "to gain a foothold in what must be our future markets."[41]

Knox, a corporation lawyer and former Attorney General fresh to foreign affairs, brought to China policy a new style as well as a new direction. (As a British diplomat serving in Washington acutely observed, "To him a treaty is a contract, diplomacy is litigation, and the countries interested parties to a suit.")[42] In the Manchurian trial that Knox decided to initiate in his first few months in office— just as Hsi-liang was taking over in Mukden—Japan was without question the chief defendant, and Russia an accessory, in the despoilment of China and American trade. As in a law office, Knox based his decision to go to court on the briefs prepared by Wilson and the other junior partners. They were guided by their reading of treaty rights and broad principles of international law, and seldom did they allow power calculations to intrude. As the senior partner, Knox confined his attention to the major decisons and happily left the details to his subordinates.

To win his case in behalf of the open door, Knox resorted to a tactical innovation in China policy, the creation of a chosen financial instrument. Hay had already pointed the way by emphasizing in 1902 the importance of investments in establishing foreign political influence in China and stimulating foreign commerce. The organization in June 1909 of the American Group, made up of some of the foremost firms on Wall Street— J. P. Morgan and Company, Kuhn, Loeb and Company, the National City Bank, and the First National Bank—gave Knox the power to do precisely what Hay had indicated had to be done—direct the flow of capital into China and thereby guarantee future economic opportunities. Knox expected railway investments, that old favorite of China market enthusiasts, in particular to "be attended by enormous internal development." In time these and other kinds of investments would open "countless commercial opportunities to American manufacturers and capitalists." But Knox also expected to obtain increased diplomatic leverage as a result of his investment campaign. The powers had already demonstrated the efficacy of capital in carving out spheres of influence; Knox would now use the same tool to penetrate those spheres and increase American political influence.[43]

Hsi-liang's offer of a Manchurian railway gave Knox his chance. But rather than concentrate on the line that Hsi-liang wanted built, Knox subordinated it to a far more ambitious project, nothing less than the neutralization of the Japanese and Russian-controlled railways of Manchuria. In November and December 1909 Knox launched this audacious direct attack on their respective spheres as blandly as he could. His proposal, he assured them, amounted to no more than a plan for "an economic and scientific and impartial administration" of the region's railroads. According to the scheme, the participating powers would purchase the lines from Japan and Russia and return them at least nominally to Chinese ownership. All the pieces were in place—or so at least Knox calculated. Russia would sell out as a step toward liquidating her northern Manchurian sphere; Britain would support neutralization; and China (though in fact worried over the prospect of foreign supervision over the railway that Knox envisioned) was expected to agree to the plan.

Knox held in reserve Hsi-liang's proposed new line, which he understood to have Peking's formal sanction as well as the backing of the British foreign office, as an alternative just in case Japan and Russia refused neutralization.

Knox could not have been more wrong in his calculations. Russia and Japan joined in public opposition to the neutralization proposal in January 1910; a month later they warned against proceeding on the alternative railway project without prior consultation. Their British and French allies and Germany as well deferred to them, and Peking, faced by a nearly united front, began to show alarm over the American success at uniting Japan and Russia against China. Knox now belatedly discovered that his backup railway plan had in fact not received final approval, and despite both Knox's and Hsi-liang's pressure, Na-t'ung and Tsai-tse continued to obstruct it.[44] The final blow fell on July 4, 1910 with the announcement of a formal Russo-Japanese accord on Manchuria. It committed both powers to cooperate in maintaining the status quo against unnamed interlopers, and in a slap at the United States omitted all reference to China's integrity.

Mutterings at home against Knox's blundering rapidly spread from the press to the American Group and even into the State Department.[45] Even Roosevelt, who had silently watched his own policy be overthrown, now made a point of reminding Taft that current strategic and political interests, not broad doctrines lacking in teeth such as the open door or imagined future interests, were the touchstones of a sound policy. American interests in Manchuria were "really unimportant and the American people unprepared" to run the slightest risk of collision over them, whereas Japan had a "vital interest" in Manchuria as well as Korea. The prudent course, Roosevelt counseled, was for the United States to avoid even the appearance of a challenge in that area and to "preserve the good will of Japan" and her cooperation on immigration control (the latter of "vital interest" to Americans).

To Roosevelt's criticisms Knox prepared a long and spirited reply in January 1911, denying "any essential connection" between the Manchurian and immigration questions, defending his course in Manchuria, and refusing to rule out an eventual appeal to arms. Treaty rights and the principles of the open door laid down by Hay and sustained by Roosevelt himself had to be translated from "theory to practice," so Knox contended, "if we are to hold our own in China." He had given Japan no grounds for offense. And although Japan might still object, "it would be much better for us to stand consistently by our principles even though we fail in getting them generally adopted." Privately Knox shifted the blame for the policy setbacks from himself to the British foreign office, which had failed to see that its true interests lay with the United States and not Japan. He denigrated press criticism as ignorant, and lobbied on Wall Street to keep the restless American Group in line.

The Manchurian setback nonetheless forced Knox to reconsider his tactics for securing American interests in China and to scale down his objectives. As early as August 1910 he had warned Taft against encouraging "any roseate hopes" on China policy, and in the months that followed he moved toward an out-of-court settlement of his case against Japan and toward working more closely with the powers. In reshaping his policy Knox was partially guided by a bright young foreign service officer in the Peking legation. Lewis Einstein, whose views reached Washington under the signature of the minister, urged joining with Britain, France, and Germany in an international financial consortium in order to moderate the aggressive policies of the British and French allies, Japan and Russia. While serving in Istanbul, Einstein had seen the advantages of great power cooperation in the face of the breakup of the Ottoman Empire and concluded that the formula was perfectly suited to China. With the Manchus appearing feeble and demoralized and Knox's Manchurian policy in shambles, an accommodation with the European consortium seemed to Einstein not only sensible but perhaps the only remaining hope for safeguarding the open door at a reasonable price.[46]

This advice in effect sent Knox back to a line of policy that he had begun to develop in the spring of 1909, just after coming to office, but which he had set aside in favor of his Manchurian initiatives. In May and June of that year Knox had laid claim to a share in the Hukuang railway. An Anglo-French-German consortium had just nailed down this consolidated project running from Peking to Canton and branching westward into Szechuan.[47] Threatened with the loss of the Boxer indemnity surplus unless it made room for American capital, Peking in late June 1909 had handed over to the United States a quarter share of the £6 million loan. At the time the State Department had justified its course in language that by mid-1910 seemed eminently sensible. "Full and frank co-operation [is] . . . best calculated to maintain the open door and the integrity of China and . . . the formation of a powerful American, British, French and German financial group would further that end." With Knox's blessing the American Group met with their European counterparts in July 1910 to begin a ten-month-long contest over the precise terms of their international financial marriage of convenience. Though cooperation might entangle the United States in the interests of the other powers and limit the flexibility of American China policy to an unprecedented degree, it also carried few risks, a feature Knox could better appreciate with his Manchurian experience behind him. He now warned Taft against saying anything that would suggest "a pro-China policy in opposition to the aims and acts of other powers," and at the same time pointedly informed Peking that it was on its own in Manchuria and could expect nothing more from the United States than "hearty moral support."[48]

Knox's shift to a cooperative policy occurred in the summer and fall of 1910 just as Peking was stirring itself to action.[49] Hsi-liang had remained an outspoken advocate of relying "on the United States and Britain and on the policies of the open door and of the balance of power as a device to save ourselves from oblivion." The Russo-Japanese agreement on Manchuria in July, followed by Japan's annexation of Korea in August, had added to his sense of urgency. "What has happened to Korea," he warned, "can happen to Manchuria." In early September, after a year of obstruction, Na-t'ung and Tsai-tse finally accepted Hsi-liang's argument for some vigorous counteraction. They chose, however, to put a development loan in place of his controversial railway. While also involving the Americans and underwriting the region's long-term defense, the loan was thought less likely to provoke Japan.

Having at last embraced Hsi-liang's policy, Peking also took the lead in its execution. After only a month the Regent took the development loan out of Hsi-liang's hands, tied it to a loan for national currency reform (a cause known to be of keen interest to the United States), and assigned Tsai-tse to arrange terms with the American Group. By the end of October Tsai-tse had an agreement for a $50 million combined loan and the Regent's formal approval as well. At the same time the Regent directed Liang Tun-yen, a Yung Wing mission alumnus recently retired from the foreign office, to set off for Washington and Berlin to secure a reaffirmation of the open door and China's integrity. A public statement to that effect might offset the recent Russo-Japanese accord. To render Washington more receptive to this diplomatic initiative, the Regent also sent his brother at the head of a naval commission to the United States to purchase $15 million worth of warships. In China, officials gave lavish welcomes to visiting Americans—Taft's Secretary of War and a delegation of businessmen—while even the press got into the act, highlighting (as it had in 1908) the advantages of a Sino-American alliance.

Peking now found Knox in full retreat. Consistent with his new policy, he refused to acknowledge that the attractive currency-development loan carried with it any diplomatic obligations to China, especially in Manchuria of all places. The Secretary of War's visit to China was kept routine and his remarks entirely noncommittal. The State Department carefully shied away from exploring the political implications of the naval commission's visit in September. When Liang reached Washington with his proposal for a diplomatic accord, Knox parried it with soothing reassurances that the United States was still guided by a desire to see foreign investments contribute to China's peace, development, prosperity, and political integrity. The Chinese could help him, Knox suggested, by accepting the cooperative policy as "the most effective means to ensure [their] protection."

The clearest signal of Knox's intentions came in November 1910 when he acceded to the claims of his Hukuang partners to a portion of the new combined currency-development loan and formed with them an international financial consortium guaranteeing a share for each in all future loans that China might offer. When his new European allies refused to do business in Manchuria except on terms acceptable to Tokyo and St. Petersburg, Knox simply dropped the Manchurian portion of the $50 million loan. Finally, in June 1912 Knox agreed to admitting Japan and Russia to the consortium to facilitate great power cooperation.

As it became clear that this most recent effort to protect Manchuria by forging closer ties to the United States had gone awry, Peking tried to limit the damage. It resisted Knox's decision to bring the Europeans into the combined loan and his decision to seek with them a foreign monopoly of China's finances (thus ending for all practical purposes competition among the powers and financial groups for major loans). Peking also resisted Knox's demand that an American financial adviser, clothed with broad powers, be appointed to oversee China's use of the loan. On this latter issue alone did the Chinese win out, aided by the jealousy of the other powers. In April and May 1911 the Regent finally bowed to the consortium terms, first on the currency reform loan and then on the Hukuang loan.

Hsi-liang's hopes for Manchuria lay shattered. When he made one last plea for his balance of power policy, Na-t'ung—his doubts about the open door policy now fortified by the rebuff Knox had just dealt Peking—sharply replied, "What advantage is there to opening up Manchuria since as before the only ones who will take advantage of the opportunities will be our neighbors and not the Americans and Europeans?" Hsi-liang—in poor health, deeply discouraged by the court's policy, and buffeted by local criticism of the Regent's government—resigned his post in May. From the sidelines he would watch China's position in Manchuria disintegrate under the weight of the politically disruptive revolution of that year and steady Japanese penetration.

In Defense of the New Republic

By early 1912 Yüan Shih-k'ai had returned to power—and to wrestling with the problems endemic to a weak and threatened country. The Manchus had recalled him to service in November 1911 to combat the revolution, but they had soon capitulated, leaving Yüan the dominant figure on the political scene and the consensus choice to head the new republican government. Operating from Peking, Yüan made T'ang Shao-i his first prime minister and with his assistance turned at once to the consortium powers for support essential to consolidating political control and settling the accounts of predecessor regimes, both Manchu and revolutionary. Those powers, however, would not make a loan except on their own onerous terms, nor would they legitimate Yüan's government by bestowing de jure recognition.[50]

Through all this Knox clung to his policy of great power cooperation on diplomatic and financial questions, still hopeful that it would check aggressive moves by any single power, particularly by Japan. This more passive and modest approach to China was, however, to stir up even more domestic criticism than had the activist Manchurian policy Knox had recently abandoned. The financiers in the American Group complained ever more loudly about the difficulties of working with Washington, and by early 1913 were readying an approach to the incoming President with hopes of a graceful release from their political burden. More vociferous were the complaints of financiers excluded from the favored Wall Street band. The open door constituency, the press, and Congress joined the attack on Knox's policy in increasing numbers through 1912 and early 1913. Knox, they charged, had implicated the United States in a cynical attempt to secure a strangle hold on the finances of a new and progressive China. By early 1913 even the Peking legation was casting doubt on the wisdom of implicating the United States in "combinations of big Powers with common interests to accomplish their own selfish political aims." Hoping to appease his critics, at least on the issue of recognition, Knox made a pro forma effort in July 1912 to reverse consortium policy. Unsuccessful, he stayed his course and suffered the brickbats at home. Even after Taft's defeat at the polls in November at the hands of an opponent of "money trusts," Knox held on.[51]

That opponent of "money trusts," to whom at last in March 1913 Knox had to surrender control, was Woodrow Wilson. In his first months in office Wilson redirected China policy in a fashion as dramatic as Knox and Taft's performance four years earlier. In repudiating his predecessor's policy of financial cooperation through the consortium, Wilson was guided naturally enough by the reluctance of a new Democratic administration to be tied to Republican policies which had generally failed of support. But weighing far more heavily in the scales against dollar diplomacy was Wilson's own clear conception of the obligation of the United States to promote the modern trinity—democracy, the rule of law, and Christianity. This task, to be shared by the Anglo-Saxon peoples, involved civilizing "undeveloped peoples, still in the childhood of their political growth." The victory over Spain and the annexation of the Philippines had brought an East Asia in "transformation" and China in particular within the ambit of American concern. In a view shaped by early reading of such missionary publicists as W. A. P. Martin and S. Wells Williams and by later contact with the missionary organizer and fund-raiser John R. Mott, Wilson saw China as ancient, long unchanged, yet now "plastic" in the hands of "strong and capable Westerners."[52]

The idea of civilizing China and the concept of commercial expansion occupied niches in Wilson's mind that were fairly distinct and of unequal importance. The exigencies of the presidential campaign forced from Wilson his first statement (in February 1912) on the need "to take possession of foreign markets" to absorb surplus manufactured goods. (Once the campaign ended, however, he lost his interest in the problem of overproduction as quickly as practical politics had forced it on him.) By contrast, China was to Wilson not a market but rather fertile soil where Americans might help self-government and Christianity take root. He felt, he wrote to Sun Yat-sen, "the strongest sympathy with every movement which looks towards giving the people . . . of China the liberty for which they have so long been yearning and preparing themselves." Consistent with this view, he heaped praise not on merchants but on those, such as representatives of the YMCA, who had already done much to carry forward China's "political revolution."[53]

Wilson's special sense of the United States as China's guide was reflected in those he singled out to assist him in making China policy. From the start he was wary of a State Department tainted by a decade and a half of Republican control and recently identified with the discredited cooperative policy. The appointment of the party stalwart William Jennings Bryan was the first step toward putting things right. In making appointments for China, Wilson and Bryan fully agreed on the need to send "men of *pronounced Christian character.*" To fill the legation they considered chiefly educators and missionaries. Turned down by Harvard's retired president, Charles W. Eliot, and by Mott, they settled finally in June 1913 on Paul S. Reinsch, a hard-driving University of Wisconsin political science professor long interested in the development of Latin America and the "Orient." The appointment secured for Wilson a man of pious family background, while satisfying Reinsch's desire for a chance to turn his hand to diplomacy.[54]

No sooner had Wilson occupied the White House than he set to work tailoring a China policy to the measure of his own keenly felt "desire to help China." He withdrew official support from the American Group, effectively ending American participation in the international financial consortium formally set up two years earlier. This officially sanctioned trust violated China's integrity and impeded the exercise of benevolent American influence, so Wilson and Bryan concluded all within the first weeks of the new administration. Wilson was careful to emphasize in his surprise public announcement of March 18—made without a by-your-leave to Wall Street, the State Department, or the other powers—that the administration favored American commercial enterprise in China. But he would not countenance blatant government intervention in its behalf. Bryan at once congratulated Wilson for a stand that had "won the lasting gratitude of China" and forced the other powers "to become rivals for her friendship." Freed from the constraints of the cooperative policy, Wilson went on to bestow diplomatic recognition on China as a sign of encouragement and support.[55]

Wilson had laid claim to an exceptional position as the "friend and exemplar" of that "great nation now struggling to its feet as a conscious, self-governing people." But what was Wilson to do when, despite recognition and opening American money markets, the powers continued to extend their sinister influence, extracting special privileges from the Chinese government and frustrating the progressive aspirations of the Chinese people? Confronted with the same difficult question that had troubled American policy makers since the late 1890s, Wilson was compelled to search for some means to protect China and fulfill the special American responsibility for China's fate. A satisfactory solution was to prove elusive.

Wilson's professions of friendship, backed by his break with the consortium, were seductively appealing to a government in Peking peppered with the names of those already inclined to look favorably on the United States because of their American education or because of their experience in Manchuria.[56] The lingering image of the United States as the least dangerous of the powers and a promising source of capital and technology strengthened the appeal. That view had persisted within Ch'ing officialdom even in its last desperate year.[57] And it shaped the outlook of some of the leaders of the early Republic. For example, Hsü Shih-ch'ang, destined to head the cabinet between May 1914 and October 1915 and to become president in 1918, lent his name to a book, published in 1920, which looked at the United States as a developmental model and implied that Americans needed the China market as badly as China needed American support.[58]

But above all else, it was the financial threat from the consortium and the political designs of the powers that made renewed American expressions of good will particularly welcome. Following the fall of the dynasty, the British, Japanese, and Russians had intensified their penetration of Tibet, Sinkiang, Mongolia, and Manchuria while consolidating their positions in China proper. Yüan himself was haunted at this time by the recurrent anxiety of Chinese policy makers since the late 1890s—how "to escape the fate of dismemberment." In mid-December 1913 he would urge his political associates forward with the reminder "that our neighbors are interested spectators and unless we set speedily to work, others may take the task in hand in our stead. When our finances are under alien supervision and our territories apportioned into spheres of influence, the fate of Vietnam and Korea will be upon us, and it will be too late for repentence."[59]

The key task was to find a way to turn expressions of sympathy from the Wilson administration into some tangible benefit to China. Yüan deployed the familiar array of devices to put his government in the best possible light in Washington. He advertised his reliance on American-trained Chinese and in 1913 and 1914 hired as publicists such Americans as W. W. Rockhill, Jeremiah Jenks, and Frank J. Goodnow. He ostentatiously played up to the mission movement in China. He praised the American record of helpfulness and friendship. And he repeatedly offered assurances of his own commitment to liberty, law, educational and economic development, and respect for foreign rights.[60]

By the fall of 1913 Yüan had secured a major loan from the consortium essential to keep his government temporarily afloat, put down an armed revolt by his political opponents, and appointed a "cabinet of talents" headed by Hsiung Hsi-ling. Yüan could now turn to consider with his associates the problems of Japanese and Russian territorial threats in Mongolia and Manchuria, Japanese support for political dissidents in the South, growing Japanese naval strength, and another foreign loan to replenish the treasury. They soon arrived at one all-encompassing—and familiar—solution to these problems: entice Americans into development projects that would bring financial relief (in the form of some $60 million in loans) and at the same time break down spheres of influence. Though the projects would be entrusted to private enterprise, Peking offered them with the expectation of informal commitments from Washington to act, perhaps along with Germany and Britain, in defense of the open door. Accordingly, Yüan reactivated in late 1913 three projects involving Bethlehem Steel, Standard Oil, and the Red Cross—all conceived before the revolution but arrested by the ensuing political turmoil.[61]

Shortly after its appointment in September, the new cabinet, including Prime Minister Hsiung himself, Minister of Communications Chou Tzu-ch'i, Minister of Foreign Affairs Sun Pao-ch'i, and Minister of Justice Liang Ch'i-ch'ao, began to sound out the American legation. China, they explained in a series of interviews, was in trouble and urgently needed American assistance. The American chargé, E. T. Williams, was cool to their appeal, but the arrival of Minister Reinsch in November changed all that. He quickly endorsed "active assistance" to a China struggling for economic development and against the pernicious spheres of influence. By early 1914 he had put China's case before his superiors in Washington, and in Peking he had pushed to a conclusion preliminary agreements for all three of Yüan's major projects.[62]

In the by now well worn pattern, Chinese overtures endorsed by enthusiastic American diplomats failed to move Washington to meaningful support. There was no lack of sympathy there. Wilson and Bryan had found in Yüan's appeal for mission support in the spring of 1913 heartening evidence of a tremendous spiritual revolution, and they joined their colleagues in the Cabinet in a discussion of the role of prayer in American China policy. Moreover, Wilson had provided moral support in the form of diplomatic recognition, and he had emphasized to Reinsch on the eve of his departure for Peking the importance of the United States continuing to encourage China while holding out educational opportunities and setting a good political example. But Wilson and Bryan shared an aversion to dollar diplomacy and power politics. These views had led to the repudiation of the consortium and shaped the restrictive guidelines sent the legation in September 1913 on support for American economic interests in China. Now, despite Reinsch's pressure, those same views held Wilson back from a signal of support or encouragement for the projects Yüan had laid out to tempt the Americans.[63]

Developing doubts about a regime unable to maintain order or even the semblance of constitutionalism may have further prejudiced Washington's response. In the spring of 1913 while contemplating recognition, Wilson and Bryan had become aware of the political divisions and instability which plagued the early republic, and Wilson became convinced that "China isn't a homogeneous unit." News that summer of Yüan's suppression of revolution in seven provinces and his decision in October to outlaw the opposition Nationalist party struck Bryan as too close an imitation of the objectionable methods of the strongman Huerta in Mexico. To Wilson he wrote in November of "temper[ing] our expectations somewhat of rapid progress in the Orient." Yüan's dissolution of the assembly in February 1914 further fed the sense of disillusion. By 1915 Wilson would become firm in his judgment that China was "inchoate" as a nation.[64]

The outbreak of the World War in August 1914 drove the Yüan regime to improvise a second appeal to the United States. The war at once removed Germany as an actor in Chinese affairs and distracted the Entente powers, leaving Japan a free field. Yüan's cabinet, since reshuffled to set Hsü Shih-ch'ang at its head and to include Liang Tun-yen, was desperate to restrain the Japanese. It at once declared its neutrality and then invited the United States and Japan to join China in steps to prevent the European conflict from spilling over into Asia. Japan was, as Peking feared, intent on seizing the German concession of Kiaochow in Shantung province and so refused. Still hoping to head the Japanese off, Yüan's government proposed to Germany that she hand over her Shantung concession (which seemed lost anyway) to China or the United States. At the same time, it asked Washington for diplomatic support and an increase in American forces in China to counterbalance the now dominant Japanese. Finally, in early October, with Japan having entered the war under the terms of her alliance with Britain and now advancing into Shantung, it once again solicited Wilson's good offices to restrict Japanese military activity.[65]

This attempt by Peking to wrap itself in the mantle of neutrality and thereby forestall the Japanese at first elicited little reaction from the American government. Wilson was deep in grief over his wife's death. Bryan, left in charge of American policy, agreed at once to neutralization of the foreign-dominated treaty ports, but he shelved the broader (and for the Chinese, central) question of United States support for the neutrality of China's territory and waters. Any doubts about where the United States stood became clear when Wilson and Bryan contended that the United States as a neutral country had no right to take a stand one way or the other on the impending attack on the German concessions, though they expressed the pious hope that Japan would seek "no territorial aggrandizement in China" and would eventually arrange a restoration to China.[66]

Following these Sino-American exchanges in late 1913 and again in late 1914, Peking's ardor for American backing cooled perceptibly. Yüan had ample grounds for concluding by fall 1914 that Wilson, whatever his rhetoric might promise, was no more likely than Roosevelt had been earlier to take a stand with China against Japan. His new estimate became evident the following January when Japan presented him with a list of twenty-one demands establishing her preponderant position in China and recognizing her spheres of influence in Fukien, Shantung, eastern Mongolia, and southern Manchuria. In fighting what he interpreted as an attempt by Japan to make China a second Korea, he now looked for help from the British rather than Reinsch's legation. An appraisal by Ts'ao Ju-lin, a senior member of the foreign office, reflected the pessimism in Peking as the talks with Japan came to a close in May. China could not stand alone against Japan, and there was no third power to turn to for support. The Americans were good only for "hollow" declarations, Tsao contended, and in any case felt no real concern for China.[67] Two years later an invitation issued by Reinsch for China to follow the United States into war with Germany elicited more of

223   China's Defense and the Open Door

the same. Though some Chinese remained in the grip of the old infatuation with the United States, those impressed by past American failures to help China were by then more numerous. The announcement later in 1917 of a Japanese-American understanding (the Lansing-Ishii agreement) supported the view of the skeptics. Dolefully, the minister in Washington, Ku Wei-chün, concluded that American diplomacy had done great harm to China.[68]

At the very time Chinese interest in the United States was declining, Wilson was beginning to regard Japanese ambitions with alarm. The twenty-one demands did the most to upset the President's earlier optimistic assumption that a benevolent, hands-off policy coupled with the rhetoric of friendship and the work of the open door constituency was sufficient to carry forward China's development and ensure her independence. By 1919 Wilson and his aides in their search for a better China policy would have explored virtually the full range of open door tactics handed down by their predecessors.

Bryan was guided through the twenty-one demands crisis by a devotion to peace that overshadowed the principles of the open door. To the President, to the Chinese, and to the Japanese, he had repeatedly insisted right down to the conclusion of the crisis in May that all sides should show neighborliness and patience. "As Japan and China must remain neighbors," he soberly advised Wilson in late March 1915, "it is of vital importance that they should be neighborly, and a neighborly spirit cannot be expected if Japan demands too much, or if China concedes too little." Two months after Japan had presented her demands to China, Bryan felt he could safely conclude, "It is evident that each country is suspicious of the other." When neighborliness failed, the only alternative Bryan seemed to entertain was to hope that the problem would go away. He professed relief to discover that some of the Japanese demands were only requests, which he thought the Japanese unlikely to press. He entertained for a time the idea of ceding Manchuria to Japan in return for assured security for the rest of China. Finally in March in the first formal American reaction, he did a balancing act, making the vague concession that "territorial contiguity created special relations" for Japan in Shantung, South Manchuria, and East Mongolia, while reasserting the sanctity of the open door (carefully left undefined.)[69]

Wilson followed the twenty-one demands crisis with interest. From the start he ruled out direct involvement for fear it "would very likely provoke the jealousy and excite the hostility of Japan, which would first be manifested against China herself." He later approved as sufficiently "weighty and conclusive" Bryan's March stand on special relations and the open door. Soon, however, reports reaching him from Reinsch made him "very uneasy" and led him to take a firmer stand in defense of the open door. In early May, after China had already bowed to a Japanese ultimatum, Wilson put Tokyo on notice that the United States would not recognize any part of the resulting Sino-Japanese agreement that violated American treaty rights or China's integrity. (Gone now were references to Japan's "special relations" with China.) Thus Wilson rescued the open door and signified his preference for note-writing in the style of John Hay.[70]

Robert Lansing, along with Wilson's personal adviser Colonel Edward M. House, spoke within the administration for the power-oriented approach to the China problem formerly advocated by Theodore Roosevelt. Despite the existence of navy war plans designating Japan as the most likely Pacific foe, the Wilson administration was not prepared before 1914 to face up to a conflict. The distractions of the World War and the issue of neutrality being played out in the Atlantic made such a contest even less likely later. If the use of force was not in the cards, why not strike a bargain with Japan? In March 1915 during the twenty-one demands crisis, Lansing (then the State Department Counselor) suggested to Bryan acquiescing in Japanese territorial expansion on the Asian mainland in exchange for Japanese acceptance of the American position on the still smoldering immigration question and a promise of respect for the commercial open door. Lansing had already reached the conclusion in response to Chinese overtures at the outbreak of the war that it would be "quixotic in the extreme to allow the question of China's territorial integrity to entangle the United States in international difficulties." As Bryan's successor as Secretary of State, he held to that judgment, repudiating Reinsch's 1917 invitation to China to enter the war. He did not intend to allow the Chinese "to play a little international politics" at the expense of the United States.[71]

Wilson could go only so far with Lansing. The President shared Lansing's reluctance to get involved with the Chinese. Peking's willingness to come into the war under American aegis early in 1917 was, Wilson thought, ill-advised though "singularly generous and enlightened." He could not, however, agree to a bargain with Japan at China's expense. Wilson was still prey to a strong paternalism that made him want "to help China and . . . shield her against the selfishness of her neighbor." When Lansing used the visit of Japanese envoy Ishii Kikujiro to argue through Colonel House for making a deal with Japan, Wilson stiffened. He was willing to allow no more than another of those vaguely worded public accords balancing the "open door" against Japan's "special interests," and he insisted that the Japanese understand that the "special interests" conceded by the United States were not political.[72]

Having rejected lines of policy preferred by Bryan and Lansing and recognizing the futility of note-writing, Wilson was at last forced to come up with his own answer to the China problem. He began moving toward the methods of dollar diplomacy, an approach encouraged by Reinsch and by Lansing's State Department. Like Knox, Wilson had learned the difficulty of opposing Japan alone, and was coming to appreciate the need for international cooperation to meet China's financial needs and put a stop to Japanese political loans. But to attract Wall Street, Wilson had to reverse himself and promise government guarantees (getting, in return, a promise that the financiers would broaden their membership base). Though the American Group came back into existence as an auxiliary of policy in 1918, it was of no immediate use. Negotiations with the other powers over a reconstituted international consortium dragged on into 1920 and were concluded only after Japan had secured general agreement to remove Mongolia and Manchuria from the consortium's purview.[73]

At the Paris peace conference—with the consortium still in limbo and his alternatives exhausted—Wilson reluctantly shifted to a policy of accommodation toward Japan.[74] Though he had talked of a peace settlement that made no room for war booty, Wilson had to accept Japan's continued occupation of the former German Shantung concessions to keep Japan in the conference and in the projected League of Nations. News of Wilson's retreat struck in Peking like a thunderbolt, sending crashing the high expectations of postwar justice that Wilson's pronouncements had aroused in China. Reinsch resigned, distraught over Wilson's failure to take direct aim at the system of "localized preferences" by which the powers and especially Japan diminished China's integrity, violated the open door, and poisoned East Asian international relations. The Chinese foreign office, for its part, now wondered how it was to resist Japan as long as "the support promised China by Europe and America is only lip-service."

But Wilson could not see his way either to square his practice with his professions or to abandon the open door ideal as pretense.

Instead, he clung to the slim hope that the League and world public opinion would provide the antidote to aggression in China. In the fall of 1919, as he spoke around the country in support of the League, he contended that his policy toward China was still guided by sympathy for its "great, thoughtful, ancient, interesting, helpless people" and by a desire to assist her in putting a stop to exploitation under the old and unjust policy of concessions. Thus the open door ideology—reformist and paternalistic—persisted even though diplomatic tactics repeatedly proved unequal to its realization.

Chapter Seven

Exclusion Stands the Test, 1898–1914

AMERICAN EXCLUSION POLICY entered a new, more restrictive phase in 1898 as the Bureau of Immigration fell under the control of Sinophobes from the ranks of labor. The increasingly stringent regulations imposed by the Bureau revived the debate over exclusion in the United States and finally provoked a boycott movement along the China coast. Thus in the first years of the twentieth century the immigration question once again assumed a central position in Chinese-American relations and provided more evidence that the developing relationship between the two countries contained as much room for a special enmity as it did for a special friendship.

The Contest Renewed

Sinophobia was triumphant in the United States by the end of the nineteenth century. Mob violence and a sustained political drive had successively carried local, state, and the federal government, and secured a sharp drop in the new arrivals and a gradual reduction in the Chinese population in the United States as a whole. Local harassment and attacks continued into the twentieth century, but with legitimate channels of protest wide open, local incidents were neither so frequent nor so explosive as earlier.[1]

The last important step for exclusionists to take in order to consolidate their position was to bend the Bureau of Immigration to their purposes. As a step toward centralizing control of immigration policy, Congress had created the Bureau in 1891, and in 1900 gave it full administrative charge of Chinese immigration.[2] In the exercise of its newly won powers the Bureau was directed by Terence V. Powderly. The son of an Irish immigrant, Powderly had become a leader in the Knights of Labor in the 1880s and predictably an exponent of strict controls on the Chinese. His reward for supporting William McKinley's presidential ambitions was his appointment in 1898 to head the Bureau.

Before his departure in 1902, Powderly had a campaign well underway that was ostensibly merely intended to rationalize and institutionalize exclusion by establishing uniform rules and applying them nationwide. But operating with wide discretion in interpreting congressional legislation and court rulings, the Bureau under Powderly and his successors was in effect to create a new layer of exclusion measures by administrative fiat. To ensure the exclusion of all laborers, the Bureau sought to stretch its net as widely as possible, even if it meant catching some of the exempt class. Supported generously by Congress, the Bureau posted a close watch at the main ports of entry, mounted patrols along the Mexican and Canadian borders, fought through the courts appeals against its rulings, and attended to the details of deportations.[3]

The treatment the new regime accorded arriving Chinese was as intimidating, arbitrary, and abusive as its regulations were stringent. For most newcomers the Bureau arranged on Angel Island a daunting, Kafkaesque drama which they had to play through successfully in order finally to set foot on American soil.[4] The stakes in this contest of wits were particularly high for those who gambled all their resources on getting into the United States and who, if refused entry, would have to return home in desperate financial straits. To make sure no one escaped its scrutiny, the Bureau defined as Chinese anyone with even a fraction of Chinese blood, regardless of country of origin or nationality. Immigration inspectors, habitually overbearing and insulting even to traveling Chinese officials, isolated the newcomers from friends and made legal and in some cases medical assistance difficult to obtain. With the aid of the Bureau's own doctors and Chinese interpreters, the inspectors conducted close examinations with the aim of discovering some grounds for rejection. Any symptom, however minor or improbable, of glaucoma, hookworm, or tuberculosis was reason enough for refusing entry. Any technical flaws in the visa obtained from American consuls in China before departure served equally well. Even a minor discrepancy between Chinese and English sections of the visa or the omission of some piece of incidental information was sufficient grounds for rejection.

All else failing, inspectors resorted to long, detailed and, if necessary, repeated interrogations to trap the immigrants in inconsistencies or contradictions in their personal histories that could be used as evidence that they were laborers impersonating one of the exempt class and hence subject to rejection. Callouses on the hands might be enough to establish a man as a laborer, his claims and written evidence to the contrary notwithstanding. Businessmen who admitted carrying bulky merchandise about their stores on occasion or students intending to work part-time to help support themselves were classified as laborers and sent back home. Chinese travelers, entitled by treaty to move across the United States, found the way blocked by Bureau officials fearful that these visitors might stay on illegally and take up work. Another group within the exempt class, the wives and dependent children of Chinese merchants, encountered frequent rejection by inspectors suspicious that the applicants were in reality prostitutes or young laborers. For example, the Bureau in 1904 rejected one in five of these merchant dependents on precisely those grounds.

Those ultimately turned away at the port had the right of administrative appeal to Washington. But the process was cumbersome, slow and, worst of all, seldom successful because the inspector's report carried more weight than the testimony of Chinese, who were regarded throughout the Bureau as an unscrupulous lot. An appeal to the courts, which were sometimes critical of the Bureau's not overnice procedures, offered the only remaining hope of getting around an unfavorable Bureau ruling, at least for those who could bear the legal costs.

The Bureau's battle against "coolie" labor carried it beyond the main ports of entry and into Chinatowns and work camps, where it conducted raids in search of illegal entrants with scant regard for legal niceties or the rights of American citizens of Chinese descent. The mass arrest of over two hundred Chinese in Boston in October 1903 illustrated these abuses on a grand scale. The Bureau was equally capable of offensive behavior as it monitored Chinese traveling about the United States. An inspector with the slightest suspicion might with impunity stop, interrogate, insult, and even detain a Chinese, whether American citizen, Chinese official, student, businessman, or plain traveler. The better to secure its control, the Bureau implemented a system of identification certificates in 1905.

During the Powderly years exclusion policy was even extended to the territories brought under United States control in the wake of the Spanish-American War. By making exclusion part of colonial policy in Hawaii, the Philippines, and Cuba, Sinophobes made certain that coolie labor would not enter the United States by the back door. Exclusion laws were applied at once by American authorities in the newly annexed territories and later sanctioned by Congress.

In Cuba, American military authorities in the last days of the occupation in 1902 brought forward a suggestion, duly adopted, that the new constitution incorporate exclusion provisions.

Despite the failure of previous complaints and protests, Chinese in the United States were driven once again to oppose exclusion, now in a new, less violent yet still dangerous and insulting guise. The marshal of that opposition was Wu T'ing-fang. Wu came as minister to the United States in 1897, at the age of fifty-five, and remained until 1902. He was, like most of his predecessors, from the Canton delta region (Hsin-hui district). Wu's stand against exclusion derived in part from his sensitivity to mistreatment of his fellow provincials, but it also owed much to a sense of nationalism offended by the humiliating American treatment of Chinese as a second-class people and China as a second-class nation. Once in the United States, Wu gave top priority to the criticisms and complaints against exclusion that Chinese immigrants and residents began bringing to him. His good command of English enabled him to point out the abuses suffered by Chinese immigrants to gatherings of influential Americans and in the popular press, while at the same time he acted as a gadfly to the federal bureaucracy, which he stung with his formal protests.

By the fall of 1900 the pattern of the new, administratively imposed restrictions had become clear enough to Wu that he began a sustained counterattack, the first shot in a decade-long renewal of the Chinese battle against exclusion. The argument Wu employed was to become central to the Chinese position: Chinese were being treated "not as subjects of a friendly power lawfully seeking the benefits of treaty privileges, but as suspected criminals. . . ." Wu concentrated his fire on the Bureau's blatant prejudice, its capricious application of its rules, and its unjustified restrictions in the key areas of transit rights, the definition of exempt classes, and reentry procedures. He decried its random harassment of Chinese residents in the United States, its plans for a system of identification certificates, and its support for extending exclusion to recently occupied overseas territories. Wu demanded that immigration regulations be brought into line with treaty provisions and that Chinese be accorded the same free access to the United States and the same freedom from harassment expected as a matter of course by Americans in China and most foreigners in the United States. Exclusion, Wu warned Americans, would in the long run drive Chinese merchants to take their trade elsewhere, while students would go to Europe and Japan where they knew they would be welcome. If neither a sense of justice nor a concern for self-interest would move Americans, then (Wu hinted) China would refuse to renew the immigration treaty due to expire in 1904 and resort to retaliation, perhaps by impeding entry to China of various classes of Americans.[5]

Wu's warnings, while a matter of supreme indifference to immigration officials, did strike a responsive chord among members of the open door constituency. By the late 1890s they had come to occupy a position that might best be described as moderately exclusionist. Some form of restriction, they then realized, was a political inevitability and perhaps even desirable in light of the difficulties the West had encountered in assimilating the Chinese and the public disorder their presence had occasioned. But American missionaries, businessmen, and diplomats also realized that their open door to China might well hinge on the treatment of Chinese in the United States. Diplomats had been among the first to sense the dangers and warn—discreetly when in service, more forthrightly when out—against pushing exclusion so far that it undermined respect for treaty obligations and endangered the American position in China. "If the Americans in China had suffered one tithe of the wrongs that the Chinese have endured within the United States since 1855," S. Wells Williams observed from the legation as far back as 1868, "there would certainly have been a war on account of it." American missionaries and merchants in China had also petitioned for better treatment for Chinese immigrants, while at home shipping and textile interests, the business press, and missionary boards and periodicals had carried the cause beginning in the 1870s.[6]

Now with the Bureau of Immigration arousing Chinese ire, the open door constituency went to work to modify exclusion. The American Asiatic Association took the lead. Its secretary, John Foord, concentrated his energies on drawing the connection between the American need for trade with China ("by far the greatest unexploited market in the world") and the competitive disadvantage under which exclusion put that trade. Other business groups, including even commercial organizations on the Pacific Coast, joined in supporting his case. As Chinese anger mounted, missionary publicists chimed in with complaints that tales of American injustice estranged their students and stirred up widespread ill-will that threatened to set back years of patient work. How could the United States hope "to teach one-third of the human race the value of life and how to live," wondered a petition from Americans in Canton, and at the same time discriminate against "the students who are to be the moulders of the new China"? Old hands in the diplomatic service, such as Chester Holcombe and George Seward, continued to speak out in public, while Minister Conger, the consuls in China, and the State Department's own China expert, W. W. Rockhill, privately deplored the effect of exclusion on American interests in China.[7]

Early in 1902 the open door constituency got its first chance to show its muscle as it squared off against an exclusionist drive to secure congressional endorsement of the Powderly policy. An alliance of labor (Samuel Gomper's American Federation of Labor together with the Knights of Labor and other independent unions), West Coast congressional delegations backed by their state legislatures, and various patriotic societies devoted to general immigration restrictions combined to demand that Congress write into law the Bureau's key rulings and affirm the application of exclusion to the new American possessions in the Pacific. Foord and other representatives of the open door constituency countered by putting their case before congressional committees in January and February. The first trial of strength ended in a near draw. Congress withheld its stamp of approval from Bureau proceedings, but it did not repudiate or in any way jeopardize the Bureau's broad and still expanding interpretation of exclusion, and to give Sinophobes some cheer, Congress did endorse the earlier decision extending exclusion to Hawaii and the Philippines.[8]

More agitated over continued exclusion than either Wu T'ing-fang or the open door constituency were the 90,000 Chinese still in the United States. Their memories of old grievances continued to rankle at the same time that new restrictions posed a threat to their self-interest and pride. Lee Chew, a resident of New York City who had climbed the economic ladder from servant to laundryman to general merchandiser, identified the two memorable features of his twenty-five years in the United States—hard work and the repeated abuse he had taken from Americans. Fong F. Sec, who had arrived in 1882 and later returned to China to become the chief English editor for a major press, painfully recalled the humiliations he had endured—of being "spat upon, kicked, stoned, and forced to run for my life time and again just because I was Chinese." Now as the outlines of the Powderly policy (with its ultimate goal seemingly nothing less than the elimination of the Chinese from the country) became clear, merchant leaders in the United States moved to renew the fight to secure their residence and preserve their ties to home. Alarmed Chinatowns began to lay the groundwork for a protest movement, raising funds (for example, the small El Paso community of seven hundred alone contributed about $1,500) and appealing for support back in China with tales of humiliation and injustice. In 1901 the first such appeal, initiated by a comprador employed by an American steamship company, went out with 56,000 signatures on it. "There are all kinds of cruelties that we Chinese have personally seen and physically suffered." The appeal rehearsed the long list of discriminatory measures which Chinese faced daily in restaurants, hotels, hospitals, schools, and matters of real estate, and wondered whether this sad state of affairs could be changed.[9]

Chinese elsewhere overseas were becoming equally restive. For the 100,000 economically powerful and culturally unassimilated Chinese in the Philippines, annexation by the United States had proven a mixed blessing, improving their economic security and political status but also bringing restrictions that severely strained their economic and social ties to their province of origin, Fukien. After spasmodic protests, the Chinese in the Philippines—led by prominent merchants—finally organized against exclusion in March 1903. Chinese in Hawaii, who also found themselves suddenly burdened by new disabilities following annexation, were even more resentful after authorities in Honolulu had (intentionally, it seemed to them) burned down the entire Chinese quarter while trying to contain an outbreak of plague. Compensation for the resulting property loss was delayed and only partial. From Chinese merchants in Central and South America came still more complaints that the exclusion policy, as it was applied by immigration officials in San Francisco, disrupted their trade through that port.[10]

The onerous treatment of Chinese participating in the 1904 St. Louis exhibition further dramatized the abuses that attended the Powderly policy. After accepting a formal American invitation to the exhibition, the Chinese government discovered that the Bureau of Immigration had devised a special set of regulations for Chinese merchant exhibitors and their staffs, placing them under close restraints during their stay. The *Pei-yang kuan-pao*, a government paper published under Yüan Shih-k'ai's supervision, sarcastically took note of "this hypocritical pretense of friendly intercourse and surpassing kindness!" It recommended that participating Chinese merchants give the project a second thought. After threatening to withdraw, Peking secured a promise from the Roosevelt administration to liberalize the regulations. But in practice the Bureau went ahead and imposed impediments and inflicted insults that caused the Chinese press to comment unfavorably on how Americans treated guests, at least when they were Chinese.[11]

In January 1904 China rang up the curtain on the long-simmering confrontation over immigration. Against the background of increasing public and official agitation, the central government decided to denounce the 1894 immigration treaty, due to expire the following December, and indicated that the United States would have to make some changes in existing policy toward Chinese immigrants if there were to be any new treaty. Once the Chinese position became public in March, exclusionists in Congress seized this "provocation" as an opportunity to divorce exclusion policy formally and entirely from treaty provisions. The open door constituency mustered a counterattack. But outnumbered by anti-Chinese forces in Congress and unable to delay action, it finally had to settle for removing only some of the more objectionable features from exclusionist proposals. Late in April Congress set existing exclusion policy on a base of domestic legislation of indefinite duration.

At the same time, attempts by diplomats to find a way out of the impasse got nowhere. Peking would not reverse itself, as the U.S. government was pressuring it to do, and accept a simple renewal of the old treaty for another ten years. With the United States legation by June giving up hope of preserving the status quo, talks shifted to Washington, where Wu's successor, Liang Ch'eng, presented a revised draft treaty in August. But neither it nor an American counterdraft brought the two sides any closer to a settlement.[12]

The Chinese government's position on the terms of a new treaty at this time closely followed Wu's during his tenure as minister in the United States. It called for a repeal of some of the more offensive measures introduced under the Powderly regime and a relaxation of exclusion regulations in the Philippines and Hawaii. But as talks with the United States proceeded into 1905, Chinese officials, perhaps seeing little chance for an accord, stiffened their position, putting increasing emphasis on the principles of reciprocity and equality.[13] For Peking reciprocity meant the application to various classes of Americans in China of the same restrictions applied by the United States to Chinese residents and newcomers, while equality meant that Chinese in the United States would have the same rights as immigrants from other countries, and at a minimum would be subject to restrictions no more severe than those applied to the Japanese and other Asians. This new emphasis in Peking's approach was probably less intended to safeguard overseas Chinese than to protect China's national honor in the face of an insulting and apparently inflexible American position.

Behind this more adamant official position and the accompanying popular cry for some form of retaliation against the United States lay the haunting fear that China, trapped in a Darwinian world of fierce competition, was in deep peril. To many the empire, which had since 1894 suffered one diplomatic and military reversal after another, seemed near collapse and the Chinese race humiliated and close to subjugation. "Our people are still sleeping drowsily," bemoaned one observer. "Indeed the people are heart-stricken and the nation dispirited." While highlighting the alarming state China had fallen into, the immigration issue—by evoking a common sense of outrage—had also presented an opportunity for Chinese to begin to redeem themselves. "If this great humiliation to our four hundred million people does not lead us to assist the Cantonese, then we are blind to our basis as a nation and have forgotten our sense of righteousness as a race."[14]

This appeal and others like it had their effect. A little over a year after Peking had taken its stand in favor of treaty revision, public agitation in China against American policy was in full swing. In May 1905 Chinese in the United States, Hawaii, and the Philippines intensified their appeals to the government in Peking and to fellow provincials in Canton, Hong Kong, Fukien, and Shanghai. In response the Shanghai Chamber of Commerce met on the tenth to propose a national anti-American boycott to begin in two months unless exclusion terms were modified. Though primarily aimed at American goods and firms, the proposed measures also called for employees of American firms to walk out and students to stay out of schools with American teachers. In a mass meeting held on the twenty-seventh Canton accepted the Shanghai program. After the assigned grace period had passed with Chinese demands still unmet, the boycott went into effect, first in Shanghai on July 20, and three days later in Canton.

As an instrument of retaliation, the boycott was peculiarly attractive. A threat to the profitable trade of the Americans, known as a commercial people with a strong interest in China, seemed a promising means of securing redress. The boycott would also stand as a patriotic gesture demonstrating popular support for the stand the government had taken in defense of overseas Chinese. Finally, the boycott carried with it few risks or burdens. It was not provocative, it involved no coercion of Americans. Because government officials were not directly involved, there were no justifiable grounds for official U.S. protest. To avert financial hardship, participating merchants were allowed two months to sell off their American stocks and locate substitute sources of supply—Russian and Sumatran kerosene, Australian flour, Japanese textiles, and Chinese-manufactured cigarettes.[15]

By resorting to the boycott in 1905 Chinese were putting a traditional weapon of local politics at the service of a nationalistic foreign policy. Merchant organizations had long used group solidarity and control of trade as an instrument of protest (for example, against some specific injustice done by officials or foreign residents). Now in 1905 the boycott was extended to deal with an issue of broad concern, and though merchants through their guilds and chambers of commerce took a leading part in initiating the movement, virtually every major class of educated, politically active Chinese living along the China coast joined in. Students, energetic and enthusiastic, took to the streets to build up public awareness and support. They gave talks, posted handbills, distributed anti-American literature, and gave instruction on how to identify American goods. But the participation as well of women's organizations, workers, officials, gentry, modern intellectuals, journalists, and even political dissidents is testimony to the growing strength and expanding appeal of Chinese nationalism in these years.

The protest movement quickly spread to cities and towns up and down the coast and to a lesser degree to population centers in the interior. In each locale resentment against the United States was sustained by a richly detailed indictment—in posters, newspapers, novels, and even songs—of American misdeeds. But exclusion had wounded so many individuals that firsthand testimony was always to be had. Those with personal grievances (and sometimes a desire to displace American trade) were often at the forefront of local agitation. For example, in Shanghai the most prominent of the boycott leaders was Tseng Chu, a native of Fukien (the region peculiarly affected by the application of exclusion to the Philippines) and the head of the Fukien merchant association in that city. Also in Shanghai, in the single most dramatic act of protest that was to occur during the entire boycott, Feng Hsia-wei, a Cantonese who had been denied access to the United States to study, committed suicide on the steps of the American consulate. In Newchwang in Manchuria the most vocal advocate of retaliation—an associate of a British import firm that stood to gain from ousting American kerosene—had himself once, while passing through the United States, been detained and placed in irons. In Amoy a merchant born in the Philippines of mixed parentage became the prime mover in the protest against laws that had hurt those like himself with ties to both the Islands and the Chinese mainland. And so the examples could be multiplied.[16]

Specific irritants sharpened the boycott sentiment in the Canton delta. Long accumulating resentment against abuse of fellow provincials who had gone abroad formed a bedrock of anti-American feeling. To it was added unhappiness over the American sale of the Hankow-Canton railway to Belgian interests, the subject of condemnation by several hundred gentry and merchants meeting in Canton in late 1904. The drowning in September of that year of the comprador Ho Ts'ai-yen, pushed off a bridge by a group of drunken American sailors on shore leave, had further inflamed local opinion. The local consul general, a Sino-American court of inquiry, and ministers Conger and Rockhill all thought American guilt established beyond a shadow of a doubt. But when the navy insisted on shielding its men, Hay went along, refusing even to admit wrongdoing. Finally, in June 1905 the State Department, hoping to undercut the pending boycott, arranged to pay $1,500 to Ho's widow "as an act of friendly good will to China." For Cantonese the incident gave further reason to believe that only united protests could force the United States to right injustices.[17]

News of the protest was spread outward from Canton by newspapers, boycott literature, and agents sent by the boycott societies to organize rallies and in some cases branch committees. Sze Yup, the four delta districts that most of the Chinese then in the United States counted as home, provided the warmest reception, but even areas where emigration to the United States was uncommon gave their support. Passengers on small boats traveling the delta waterways were seen exchanging literature and discussing the exclusion problem, while monks in a remote monastery fifty miles from Canton revealed a keen interest as they plied a visiting American with questions.

Overseas Chinese communities, whether directly affected by exclusion or not, lent their support to the boycott. Chinese in the United States, who had collected $6,300 in supporting funds even before the boycott had begun, continued to send money—in excess of the needs of the movement. Chinese in the Philippines and Hawaii also raised funds, and those in the Philippines joined the mainland boycott of American goods. Several hundred merchants in Singapore organized in support of Shanghai's call for a boycott and cabled their support to Peking. Penang quickly followed suit. In Bangkok 3,000 Chinese residents met and resolved to instruct their agents in Singapore and Hong Kong to stop shipments of American goods. By early August the boycott had also spread to overseas Chinese communities in Saigon; the Japanese ports of Yokohama, Nagasaki, and Kobe; Victoria in British Columbia; and Havana.[18]

Though officials took no direct part in inaugurating the boycott and in the main kept their distance from its activities, many were nevertheless strongly sympathetic. Within the capital, the foreign office shielded the boycott against American diplomatic pressure through much of the summer. Its head, Prince Ch'ing, backed by a new generation of Canton-born, foreign-educated foreign affairs experts, emphasized the role of American misdeeds in arousing public opinion, and he defended the right of Chinese to take their trade wherever they wished. Leading provincial officials also welcomed the boycott as a sign that the poor morale of the people was on the mend. For some it inspired hope for the future. "If the people are not exhausted," observed Lin Shao-nien, the governor of Kwangsi, "the country cannot be reduced." In Canton itself governor general Ts'en Ch'un-hsüan made no secret that he wished the movement well, while Yüan Shih-k'ai and Chang Chih-tung privately appealed to Roosevelt to take a more lenient attitude on the immigration question.[19]

The boycott movement, initiated in May and put into effect against American trade in July, was by the fall in trouble everywhere but in south coastal China. By early 1906 it was losing force even in Canton—without even its minimal demands met. The boycott's failure was in part the result of confusion over the precise changes in American policy that it was to effect. Not surprisingly, the complexity of American exclusion policy, compounded as it was of treaty, congressional, and administrative measures as well as court rulings and state and local laws and ordinances, proved difficult for the protesters to grasp. Thus, for example, the July 20 Shanghai rally calling the boycott into force demanded "the repeal of the Chinese exclusion treaty," apparently unaware that even without the treaty (which Peking had in any case already terminated) exclusion would continue essentially unchanged thanks to the 1904 congressional legislation.[20]

To make matters worse, the participants differed among themselves over goals. Some, more concerned with symbolic protest than with practical achievements, insisted on nothing less than the overturn of the entire structure of exclusion so that all Chinese, laborers included, could enter the United States without hindrance or insult and reside there in peace. For others ready to settle for more modest gains, the boycott aimed at achieving no more than better treatment for the classes (chiefly students and businessmen) previously exempted from exclusion by the treaties of the 1880s and 1890s. In general, student participants inclined toward the more sweeping demands, while businessmen took a more moderate stance.

The ultimate ineffectiveness of the boycott can also be traced to its failure to build true nationwide support or organization. Begun as a spontaneous movement feeding on deeply felt grievances, the boycott remained for the most part confined to the Canton delta region and the treaty ports. Even there its support was limited to the relatively well-educated and to those with a personal interest in easing the American restrictions against immigrants. The movement in each locale went its own way, maintaining—chiefly through established merchant organizations—only spasmodic and informal contacts with other centers of activity. No central committee ever emerged to coordinate the movement up and down the coast. That no effort was even made suggests that perhaps the leaders of the boycott believed their goals might be easily achieved, that they were reluctant to provoke official concern by organizing outside preexisting charitable and commercial organizations, or that they were as much divided by regional differences as they were united by outraged nationalism.

The boycott also failed because its advocates had badly misjudged likely foreign reaction. They overestimated the importance of the China trade to the United States and the eagerness of the other powers to displace American trade, and at the same time underestimated the political and popular support exclusion enjoyed in the United States. As it turned out, exclusionists were unmoved by the commercial damage the boycott did, while the American government, which did indeed worry about commercial losses, reacted not by making concessions on immigration but by turning the pressure on Peking to halt the protest. American officials in China rejected claims that the boycott was strictly a nonofficial movement and feared that it might develop into a wider and more violent antiforeign campaign. So too did their colleagues representing Britain, Japan, Germany, and Portugal. Worried in addition that a precedent was being set for some future attack on their own trade, they threw their weight against the boycott in their respective spheres of influence—in Hong Kong, Newchwang, Tsingtao, and Macao.[21]

Under this mix of unfavorable circumstances the alliances on which the boycott movement had been built began to crack. The government's tacit support for this popular cause was the first to go. Even as planning for the boycott had begun in the spring, Yüan

Shih-k'ai had taken the lead in trying to persuade the central government to suppress it. By July pressure from Yüan and the American legation had led the foreign office to caution provincial officials against encouraging the boycott and to advise them to exercise a restraining influence wherever possible. It warned that the agitation might impede the ongoing immigration talks with the United States and, even worse, might contain the seeds of sedition. Finally, at the end of August a strongly worded imperial edict appeared, calling for a complete halt to the boycott.[22]

In Canton, governor general Ts'en now moved carefully to contain and then dampen the protest. A native of adjoining Kwangsi and a reform-minded nationalist, Ts'en had through July and August played a crucial role as buffer between the capital and provincial opinion. At the same time he had fended off the protests of the local American consul general. Ts'en seems to have shared to a degree Peking's concern that the boycott not get out of hand, yet he also realized that a sudden attempt at suppression might provoke the very outburst that the government feared. Ts'en sought to mollify boycott leaders by publicizing the steps Roosevelt had taken to liberalize exclusion regulations, and late in August he asked them to suspend their activities until December, when the American Congress was supposed to act on China's complaints. (Boycott leaders parried the request but they did at least become more discreet in their activities.) Even after the appearance at the end of the month of the imperial edict, interpreted locally as a counsel of caution and not as an absolute prohibition, Ts'en continued to defend the right of merchants to refuse to handle American goods.

Finally, in September, in this atmosphere of official ambivalence, boycott leaders fell to feuding. When three students put up some particularly offensive posters, two merchant leaders revealed their names to the American consul general. The resulting consular protest identifying the offending parties led to their arrest by Ts'en and set off a furor in the local boycott movement over this betrayal from within. Again, in December, the movement split when American businessmen in Canton and Hong Kong solicited a list of specific changes the boycott aimed at effecting. Some of the moderate leaders of the boycott from Canton and Hong Kong met with the Americans on December 3 and conceded continued exclusion of unskilled laborers in exchange for the free admission of all others. These terms were in turn repudiated by the Canton organization, which had not been consulted and which favored nothing less than total repeal of exclusion.

The movement in Canton dissipated as it continued into 1906. In January Ts'en issued his strongest warning against public agitation. At the same time he became embroiled with merchants and gentry in a dispute over control of the Kwangtung section of the Hankow-Canton railway (recently recovered from the United States), diverting public attention from the boycott. Students, still the most strongly committed to the anti-American protest, used every opportunity to draw attention back to the immigration question. The return of the body of Feng Hsia-wei, the Shanghai martyr, to Canton for burial in July was one. The second came in September, when the students arrested a year earlier were released from jail and in a public ceremony received, in recognition of their sacrifice, scholarships to study in Japan. That same month, following an incident in which a drunken American sailor knocked a Chinese into a creek, they chided their countrymen for supinely allowing the Americans to "maltreat us while we are in our own country as they do while we are in America." (They singled out Chinese merchants for particular obloquy as slaves to foreign interests.) Finally, in December 1906 when the U.S. Congress reconvened, the students succeeded in bringing out a thousand or more demonstrators into the streets to demand changes in immigration policy. But popular enthusiasm proved evanescent and the political atmosphere inimical to renewed protest. In December the new governor general, Chou Fu, banned further public gatherings, gagged the press, and seized boycott funds. When one of the three recently released students had the temerity to protest, Chou threw him back into jail.

Roosevelt's Finesse

The boycott of 1905, the last serious Chinese challenge to exclusion, had failed to effect any fundamental change in U.S. policy. But as long as the protest had lasted, it had at least forced Theodore Roosevelt and those close to him to face up to the injustices and inequities of exclusion.[23] An observant Chinese with an eye for irony might well have looked back on this reaction to the boycott and concluded facetiously but truthfully that Americans responded only to a forceful policy.

Roosevelt himself had in the 1890s supported exclusion of the "Chinaman" as necessary to put a stop to cheap labor "ruinous to the white race." Within his first year in the White House, Roosevelt had endorsed the Powderly policy. His first annual message to Congress, in December 1901, called for the strengthening of regulations to make exclusion of Chinese laborers "entirely effective" or, as he phrased it a few months later, "ever more stringent." When Congress got down to specifics in 1902, Roosevelt threw his support to the strictest of some seventeen different bills introduced in the House, and then signed the resulting measure without hesitation or complaint. In May of that year he appointed as Powderly's successor as the Commissioner of Immigration Frank P. Sargent, an associate of Samuel Gompers and a Republican loyalist in the ranks of labor.[24]

China's decision in early 1904 not to renew the immigration treaty finally forced from Roosevelt his first expression of concern over treatment of merchants and students coming to the United States. He advised the Secretary of Commerce and Labor, who had jurisdiction over the Bureau of Immigration, to "do everything to prevent harshness being done to merchants and students," while still ensuring that laborers not sneak in "by any fraud or evasion." Otherwise Roosevelt regarded the immigration question as a matter of no particular urgency. He did not want even "an allusion" to it in the Republican campaign platform. "We wish to honestly enforce our laws, but we are doing that anyhow."[25] At the same time he allowed the State Department to proceed in a leisurely manner in its negotiations for a new treaty. By dragging the talks with China on beyond the November election and by keeping them low-keyed, Roosevelt kept out of the race an issue which was potentially divisive for Republicans. Even after his reelection Roosevelt guarded against any unpleasant surprises coming out of the treaty talks by giving the Bureau of Immigration a virtual veto over the terms offered the Chinese. Finally, after the talks in Washington had stalled in January 1905, Roosevelt let three months pass before deciding to transfer them and the chief American negotiator, W. W. Rockhill, to Peking, perhaps on the hope that the Chinese foreign office might prove more accommodating than the Chinese minister in Washington had been.

Only in May and June 1905 with the boycott imminent and the open door constituency aroused did Roosevelt finally give his full attention to the immigration problem. With remarkable political adroitness he now dealt with the diverse and often contradictory pressures that had come to bear on him. To exclusionists he repeatedly offered assurances that he was still adamantly opposed to Chinese laborers, whether skilled or unskilled, coming to this country. "We have one race problem on our hands and we don't want another." At the same time Roosevelt expressed to visiting representatives of the open door constituency solicitude for their interests in China, and agreed with them that students, merchants, and others of the exempt class should receive a more courteous welcome. He confessed that he had only recently discovered that the Bureau of Immigration "acted with utmost harshness."[26]

The reform of the Bureau was essential, Roosevelt recognized, if he was to reconcile the exclusion of laborers with decent treatment for the exempt classes. In mid-May 1905, shortly after Shanghai had proposed the boycott, the President took the first step toward that goal, chiding the Bureau for its "insolence against Chinese gentlemen." On June 24, after thrashing the issue out in Cabinet the day before, he ordered immigration and consular officials to apply existing regulations "without harshness" and "more scrupulously" or risk losing their jobs. The number of Chinese now living in the United States had fallen to 85,000, Roosevelt noted, and only a "wholly insignificant" number of new residents would gain entry by fraud. At stake on the other hand, Roosevelt contended, were the interests of the nation and humanity. By mid-August, after the boycott had begun, Roosevelt himself was talking about a major shake-up of the immigration service. He put Sargent on notice: The Bureau had to put aside its old "erroneous policy" and discriminate more carefully between laborers on the one hand and students, merchants, and travelers on the other. The latter were not to be "annoyed or interfered with in any way."[27]

Roosevelt followed in October and November with a public endorsement of the open door constituency's call for changes in the exclusion laws. "As a people we have talked much of the open door in China," he told Congress when it convened in December, "but we cannot expect to receive equity unless we do equity. We cannot ask the Chinese to do to us what we are unwilling to do to them. . . ." Roosevelt was careful, however to avoid specific proposals that might embroil him in controversy. He made clear to representatives of the open door constituency that they—and not he—would have to move Congress to ameliorate exclusion. As well ask the lamb to reason with the lion. Congress was still the stronghold of exclusionist sentiment, and no reform measures—whether embodied in a bill or in a treaty—could get through intact, as the legislative tests of 1902 and 1904 had already demonstrated. To make sure that Roosevelt did not forget the political facts and become mesmerized by the campaign by the open door constituency, exclusionists redoubled their efforts. Samuel Gompers, unshaken in his conviction that the Chinese "menace the progress, the economic and social standing" of American workers, organized a labor delegation to the White House in July, again the following December, and once more in March.[28]

Roosevelt doubtless hoped that his attempts between May and December 1905 to temper exclusion would calm the Chinese. In China Rockhill and the American consuls had tried in May and June to reinforce this impression of American reasonableness, emphasizing the President's sympathetic attitude and the favorable prospects for congressional action and treaty revision. But to Roosevelt's surprise and perhaps chagrin these efforts had neither neutralized Chinese resentment nor arrested the boycott movement. Chinese obstinance now began to provoke a man keenly sensitive to that "curious tendency on the part of the Chinese to interpret as weakness anything that we do in the direction of treating them well."[29]

This conception of the Chinese character had from the beginning of the crisis compelled Roosevelt to couple the promise of amelioration of exclusion with a display of toughness toward the boycott. In June, at the same time he administered his warning to immigration and consular officials against mistreatment of Chinese, Roosevelt had also urged the foreign service to put a stop to the protest movement. Rockhill's protests culminated in August with an announcement that the United States would hold the Chinese government responsible for the losses suffered by American trade. At the same time Rockhill broke off treaty talks and demanded punishment for Tseng Chu, the boycott leader in Shanghai. Consuls along the coast used their influence with local officials to drive the boycott out of the press and off the streets and in general promote a restoration of American commerce. Roosevelt now assured Rockhill that "a stiff tone" was necessary because the Chinese "despise weakness even more than they prize justice." In early September Roosevelt instructed Secretary of War Taft, then on his tour of the Far East, to tell Chinese officials in Canton that "we cannot submit to what is now being done by them." When an imperial edict calling a halt to boycott activity proved only partially effective, Roosevelt charged that the Chinese were trying to hoodwink him. Not only did officials, particularly in Canton, still seem to tolerate the boycott but at the same time they were pressing to recover the Hankow-Canton railway from American concessionaires. "I think we shall have to speak pretty sharply to the Chinese," Roosevelt now observed.[30]

In November the murder of five Americans at a mission station in Kwangtung (at Lien-chou, near the Hunan border) helped convert Roosevelt's growing annoyance into truculence. Despite the Canton boycott committee's immediate denial, Roosevelt drew a causal connection between the inflammatory anti-American campaign and the brutal killings. Reports from China of spreading popular unrest and a crescendo of complaints from American merchants over declining trade added to Roosevelt's determination to act.[31] He now chose to make clear to the Chinese that they could not—with impunity—repeat the 1900 rampage against foreigners or continue to harm American interests. In an early draft of his annual message to Congress he complained, "Not only the Chinese people but the Chinese Government have behaved badly in connection with this boycott, and a boycott is not something to which we can submit." On November 15, he ordered an increase in American forces off the China coast, and two weeks later elements of a strengthened Asiatic fleet began gathering in Canton harbor. In December Roosevelt set in motion preparations for a joint army-navy expedition against the city, to include at least 15,000 troops. He did not want the Chinese to mistake his concessions on exclusion for weakness. By April he was satisfied the Chinese had gotten the message.[32]

Meanwhile, the exclusion reform effort continued into 1906. Roosevelt appointed a commission early in the year to investigate the Chinese service of the Bureau of Immigration. His appointments—James B. Reynolds, a Chicago businessman critical of the Bureau and the damage it was doing to American educational influence and trade in China, and Jeremiah Jenks, a Cornell economist equally devoted to cultivating American influence in China—suggest that he meant business. In October the President replaced the Secretary of Commerce and Labor, Victor Metcalf, a California politician who had openly defended the Bureau against the criticisms Roosevelt and others had lodged against it, with Oscar Straus, himself an immigrant and identified with neither labor nor West Coast nativism. Under presidential prodding in 1906 and in the wake of the commission report in early 1907, the Bureau dropped or modified some of its more objectionable rules, and deportations in 1906 fell to half the previous year's count. The American Asiatic Association responded to Roosevelt's endorsement of some form of legislative change by preparing a compromise bill conceding to exclusionists a broad definition of laborer but otherwise removing the restrictions imposed on the Chinese by the Powderly policy. Foord once again orchestrated support, bringing spokesmen for chambers of commerce, textile associations, mission boards, steamship companies, and the National Association of Manufacturers to Washington in March 1906 to testify. Meanwhile the State Department renewed negotiations with the Chinese over a new treaty.[33]

But as the Chinese protest died, so too did the commitment in Washington to making changes favorable to the Chinese. Exclusionists kept the American Asiatic Association's bill off the floor of the House, and by May 1906 the open door constituency had, for the third time in recent years, to concede defeat in Congress. At the same time, the search for a diplomatic solution to the immigration dispute was bogging down. While the Chinese still insisted that any treaty embody the principle of reciprocity, the Bureau of Immigration continued to oppose even facilitating entry for the exempt classes, and Secretary of State Root's adviser on Chinese affairs, Charles Denby, Jr., warned that to make any concessions at all would be a mistake. The Chinese would regard them as "only so much ground gained and will continue to connive at every trick of which an unscrupulous people can conceive to still further evade the restrictions." By mid-1906 the slender resources of diplomacy had been exhausted. Even the Bureau of Immigration's shift toward greater moderation, secured by repeated ad hoc interventions by Roosevelt and his agents, proved tactical and short-lived. The Bureau retained its administrative autonomy and its Sinophobic biases. Even at the height of criticism the Bureau's inspectors remained capable of the most improbable affronts to Chinese.[34]

By 1908, Roosevelt's last full year in office, the Bureau had returned to business as usual, applying its regulations in the old restrictive and offensive manner and provoking a renewal of Chinese protests.[35] Newcomers to the continental United States and its insular possessions, as well as residents returning after a visit to China, faced the same set of daunting and time-consuming procedures, described aptly by one traveler as "more intricate than a spider's web." The process began before leaving China and ended one, two, or more weeks after reaching the United States. It included a trachoma test, which became after 1906 a byword among Chinese for arbitrariness and extortion. It still isolated immigrants in close and uncomfortable quarters, while suspicious and often hostile immigration inspectors still subjected them to close scrutiny in hopes of finding some grounds for rejection. Visiting students, businessmen, and even officials still ran the real risk of arrest and deportation. Even Chinese residents, the Six Companies complained to Roosevelt in early 1909, "are dragged from their hearths, confined in prison without bail, held incommunicado, denied the advice of counsel, and even refused the right to consult their own medical advisors. All these things are done without any accusation of any kind being lodged against them, and it is done by officers directly under your control. . . ."[36] In addition, Chinese residents still faced hostile state and municipal authorities. For example, in San Francisco the city fathers discriminated against Chinese in the public schools, and, after the 1906 earthquake and fire, had tried to expel them from the old, well-situated Chinatown quarter in order to make way for whites.

For the Chinese in the United States it was clear that Roosevelt's reforms of 1905 and 1906 had not drawn the sting from exclusion and that protests directed toward Washington were in vain. Once more they turned to South China for support, and once more their appeals stirred talk of retaliation, beginning in 1907 and hitting a peak of intensity in 1909 and 1910. In Canton the reform-oriented Self-Government Society threw its influence behind boycott sentiment. Merchants in Hong Kong and Canton again publicly recounted their "bitter experiences" in trying to get into the United States to conduct business. Chinese merchants in Manila and nearby Fukien ports still attacked as "oppressive" and "beyond the bounds of reason" the wall of exclusion which had disrupted the once flourishing trade between the Islands and the Chinese mainland.

But no longer were the political conditions as propitious to retaliation as they had been in 1905. The major obstacle was the deepening difficulties of the Ch'ing dynasty. The death of the Empress Dowager and the Emperor in 1908 had diminished the court's prestige and stability, while burgeoning revolutionary activity made the central government increasingly intolerant of popular political movements whether for domestic reform or a stronger foreign policy. In this tense and uncertain atmosphere a renewed anti-American protest movement was certain to fare poorly. The government in the better days of 1905 managed only a weak-kneed stand against exclusion; nothing better could be expected in its current defensive state. And while emotions ran high in the Canton region, elsewhere along the coast the immigration issue was overshadowed by the crisis of legitimacy through which the Ch'ing was passing.[37]

However, just the talk of a new boycott and expressions of resentment against American exclusion policy were enough to keep the open door constituency in the field calling for changes. In 1910 a worried merchant group in San Francisco organized a committee, chaired by Robert Dollar, to investigate immigration procedures at that port. Its report indicted the procedures as "unreasonable," and observed that for any arriving Chinese "to answer the questions correctly was an impossibility." That same year the New York Chamber of Commerce and the Cotton Goods Export Association protested the decision to close the Canadian border to the Chinese. The next year the Associated Chambers of Commerce of the Pacific Coast and the American Association of China (an affiliate of the American Asiatic Association) warned that treatment by immigration officials was "a cause of humiliation and continuous suppressed resentment" among Chinese. After an investigation in 1912 Seattle businessmen described the handling of new arrivals by immigration authorities as "harsh" and "barbaric." As late as 1914 the San Francisco Chamber of Commerce was still protesting to Washington against the "humiliating and barbarous treatment" that Chinese were subjected to.

Missionaries once again warned that a loss of cultural influence was part of the price the United States paid for its exclusion policy. Obstacles thrown in the way of arriving Chinese students forced some to return home and deterred others from even trying to come to the United States. Those that gained entry had fixed in their minds an abiding impression of American anti-Chinese prejudice. Reports from the foreign service echoed the warnings of businessmen and missionaries. Consuls along the southern coast repeatedly noted the financial cost and personal pain that the exclusion policy inflicted on Chinese and the ill-feeling that resulted. "I cannot believe," the consul in Hong Kong wrote back in July 1908, that "the humane men at the seat of government realize what suffering is caused, what rankling sense of injustice is kindled among the Chinese. . . ."[38] From Peking Minister Paul Reinsch complained that exclusion had provoked "a great and growing dissatisfaction" and "constitutes the most serious danger threatening the good relations between China and the United States."[39]

The Bureau of Immigration countered these renewed protests with the same arguments it had been using successfully since embarking on the Powderly policy: The Chinese were an unscrupulous people who resorted to every imaginable form of evasion to get into the country. Laborers studied assiduously to impersonate one of the exempt classes such as businessmen, the minor sons of businessmen, and students. They bought forged documents or certificates of reentry from other laborers who had returned to China with no intention of going back to the United States. They made false claims to American citizenship, a deception made possible by the destruction of birth records in San Francisco during the 1906 earthquake and fire. They bribed doctors, American consuls, and immigration inspectors. They enlisted the help of sailors, relatives, or white adventurers in smuggling them in from Jamaica, Cuba, Mexico, Canada, or Hong Kong. They took advantage of the laxity and venality of American consuls in screening initial immigrant applications, the connivance of Chinese officials, and the preoccupation of American courts with legal technicalities. To its critics in Congress and its supervisors in the executive branch, the Bureau offered only the stark alternatives of continued vigorous enforcement of the law or letting down the bars to a flood of undesirable Chinese simply to satisfy a small band of self-interested businessmen and deluded humanitarians.[40]

The continuing complaints against exclusion, carrying now no credible threat of retaliation, left "the humane men" in Washington unmoved. "We have behaved," Roosevelt announced in Congress in 1908, "and are behaving, towards other nations, as in private life an honorable man would behave toward his fellows." The death of Frank Sargent in the last months of the Roosevelt administration offered the President an opportunity to give the Bureau more enlightened direction, but instead he ensured continuity in its methods and program by filling the post with Daniel J. O'Keefe, a Republican stalwart from the American Federation of Labor. Presidents Taft and Wilson took no more interest in the Chinese problem than had Roosevelt in his last two years in office. Indeed, the 1912 presidential campaign compelled Wilson to repudiate his view of the Chinese immigrants as "intelligent and extremely useful" and to accept instead the exclusionist position that (in Wilson's own words) "oriental coolieism" was "a most serious industrial menace." The State Department—whether under Root, Knox, or Bryan—was inert on exclusion, doing no more than dutifully passing on to the Bureau the numerous complaints against it, while by contrast the consul general in Canton was vigilant against the renewal of organized protest, hastening at any hint of a new boycott to deliver to local officials a lecture on law and order.[41]

Revolutionary Currents in the Chinese Community

By the early twentieth century American exclusion policy seemed to have set a death sentence on Chinatown. Chinese in the United States, numbering 90,000 in 1900, would fall to a low of 61,000 in

1920. The men who had made up the great immigrant waves of the 1850s to 1870s and now constituted a large portion of Chinatown's bachelor society were in the twilight of their lives. At the same time, the gradual appearance of a native-born generation (in 1900 roughly one in ten of the Chinese population) threatened to dilute the cultural identity of the community. That new generation's tendency to go its own way was reflected in the establishment in 1895 of its own formal association, the Native Sons of the Golden State (later renamed the Chinese American Citizens Alliance to reflect its national scope), and in the growing incidence of intermarriage with whites and of residence away from Chinatown.[42]

Chinese in the United States struggled for a livelihood. Most had been isolated from the mainstream of the American economy since the intense anti-Chinese agitation of the latter part of the nineteenth century. Many had to find work within the confines of Chinatown—in the restaurants and shops that made up its increasingly tourist-oriented economy. Those working outside Chinatown had to find niches where they did not face white opposition, did not have to engage in personal interaction as approximate social equals, did not need large amounts of capital, and could turn to good account their willingness to work hard at a low rate of return. Chinese continued to work in laundries (25 percent in 1920), as servants (18 percent), in truck farming and other kinds of agricultural work (11 percent), or in various kinds of manufacture such as clothing or cigars (9 percent). This limited economic opportunity accompanied by social prejudice, impeded upward social mobility. For example, as late as 1920 there were only 462 Chinese classified as professionals (only 1 percent of the 45,600 Chinese then listed as gainfully employed).[43]

Assailed and isolated by the host society while increasingly cut off from China, the Chinese in the United States tried all the harder to maintain their cultural identity. The education of a new generation took top priority. Schools multiplied after the turn of the century, and tutors and schoolmasters—generally classically trained, with some official degree-holders—were brought in from China. San Francisco with its large concentration of Chinese, had led the way. The first community school appeared there sometime after 1884, supplementing the work of tutors employed by the affluent few and the private schools restricted to those who could pay four to five dollars a month (about one-sixth of a laborer's income). After the 1906 disaster the community school came under the supervision of the Six Companies, in whose new headquarters it was thereafter located. The school, with its exacting classical curriculum, was accessible to all. The tuition was low (fifty cents per month), and the hours of instruction arranged conveniently at the end of a busy day, generally from 5:00 until 8:00 in the evening. After the turn of the century, schools began to appear in other Chinatowns as well, sometimes under the patronage of the Chinese minister, the local consul, a reform or revolutionary political organization, or a Christian church.[44]

A Chinese community fighting against heavy odds to maintain itself listened with interest to the appeals of political exiles offering succor. Reform nationalists were the first to attract support.[45] They called for a cautious program of change in China, to be effected within the existing imperial system. But this excited Chinatown less than their insistence on a vigorous foreign policy that included defense of China's nationals abroad. To underline their concern for overseas Chinese, reform nationalists took a prominent role in the campaign against U.S. policy. Through their newspapers in Honolulu they promoted the boycott idea, and in the continental United States they organized an early campaign to petition Peking against renewal of the exclusion treaty. The two leaders in the reform camp, K'ang Yu-wei and his lieutenant, Liang Ch'i-ch'ao, were particularly attractive figures. In part it was due to their Canton delta origins and their standing as scholars and well connected political activists. Their outspokenness on an issue of consuming interest to Chinese in the United States added considerably to their popularity.

Liang himself began to criticize exclusion laws while visiting the Hawaiian Chinese in 1900 following the destruction of the Honolulu Chinatown by American health authorities and the imposition of the exclusion laws to the Islands. Later, at once encouraged by and eager to encourage the rise of a new national spirit in China, he emphasized the importance of the boycott as a chance for the government to strengthen itself by cultivating popular support. The people's spirit was like a stump of a great fallen tree, which "if helped along would flourish, if abused would be finished."

K'ang also took a stand against exclusion. To eliminate the stigma American policy attached to Chinese immigrants, changes would have to be made so that they were at least as well treated as arriving Japanese. Visiting scholars and students were equally deserving of better treatment and freer access. Once the boycott movement began in China, K'ang personally carried these concerns to Roosevelt, twice in June 1905 and again in January 1906 after Congress had failed to enact new legislation. "Americans have been wont to condemn Russian cruelty toward the Jews. How much more humane has been America's treatment of the Chinese?" Sounding much like Wu T'ing-fang earlier, K'ang warned that "when the sentiment of nationality shall have attained full development, a united Chinese nation will seek to assert its rights and avenge its wrongs." When he lost hope in both Roosevelt and Congress, K'ang took his protest to the American public.

In 1899 K'ang had organized his China Reform Association (known in China as the Protect the Emperor Society—Pao Huang Hui). The next year he sent aides to the United States to establish party branches, the first of which appeared in Honolulu, San Francisco, Portland, Seattle, and Boston. The party flourished, much of its success due to the popularity of its position on exclusion. When Liang Ch'i-ch'ao toured the country in 1903 he found over seventy branches embracing, nominally at least, virtually the entire adult population of the major Chinatowns (according to one account by a party historian). Leaders of the community became association officers. In San Francisco T'ang Ch'iung-ch'ang, a prominent newspaper editor and leader of the Triad Society (Chih Kung T'ang), became active as the local association's secretary, while in New York City a wealthy Americanized Chinese, Chao Wan-sheng (also known as Chu Mon Sing and Joseph M. Singleton), took the lead in party affairs. In the major population centers the association engaged heavily in political education with a strong stress on nationalism. It organized schools (the one in Honolulu, to take an example, had 700–800 students by 1911), and it operated its own newspapers in New York, San Francisco, and Honolulu. K'ang supervised the association's affairs, sought new converts to the cause of constitutional monarchy, and saw to the investment of the ample funds collected by the network of party branches.

The failure of the boycott brought an important shift in Chinatown politics. Disillusionment with the Ch'ing turned Chinese in the United States away from the reformed monarchy advocated by K'ang and his association and toward the revolutionary alternative with its promise of a new democratic China, strong and united enough to defend their interests. The chief beneficiary of this shift in political sentiment was Sun Yat-sen, a native of Hsiang-shan district in the delta and the revolutionary leader most active among overseas Chinese. Impatient for immediate relief, Chinese in the United States had found scant appeal earlier in Sun's argument that China required total political overhaul, including foremost the overthrow of the dynasty, before there could be any hope of an effective foreign policy or successful defense of the interests of overseas Chinese. The collapse of the boycott, however, bore Sun out. The central government had suppressed the boycott and stood in the way of any revival, while its representatives in the United States still failed to give Chinese effective protection and support against an increasingly oppressive immigration policy.

Sun had gotten off to an early but slow start with overseas Chinese. He had organized his first revolutionary group, the Society to Revive China (Hsing Chung Hui) in Hawaii in 1894 and broadened it on his return to Hawaii in 1895 following the failure of a rising in Canton. But its members, most of them poor, could not provide the funds he needed, so Sun proceeded to the United States in 1896. A cross-country tour revealed even less interest in his cause there than the well-to-do, more conservative Chinese in Hawaii had shown. Nowhere was his call for revolution against the Manchus and for China's reconstruction on democratic lines well received, not even among the secret societies (causing Sun to lament that these inheritors of a tradition of anti-Manchu nationalism had degenerated into mere mutual aid societies).[46]

A second visit in 1904 went only slightly better, even though Sun was by then fairly well known for his revolutionary exploits and his ability as an orator. His strategy was to gain entrée to the secret societies and through them to secure from their mass membership contributions to support his own activities. To that end he himself became a society member while in Hawaii during the first stage of his tour. But his anti-Manchu rhetoric and his call for a revolution to pave the way for a republic evoked little enthusiasm in the United States. Only a small circle of Christian converts, modern students, and others touched by Western ideas proved responsive, and they had little money and a considerable reluctance in many cases to enroll in Sun's party and risk retaliation against themselves or relatives in China should their disaffection become known to the Ch'ing. Sun left New York for Europe dispirited and empty-handed. The $4,000 he had collected in San Francisco on his arrival had dwindled away during his fruitless five-month tour of the United States.[47]

Despite his pressing need for funds, Sun was not prepared to cater to the Chinese in the United States by making room in his program for an attack on exclusion. During his 1904 tour, a time of widespread complaint by Chinese against exclusion, Sun had ignored the issue. A year later, when the boycott began, Sun was at best indifferent. He quietly accepted the efforts of some of his followers to use the boycott to foment revolution and draw attention to the cowardice and ineptitude of the Manchus. But his own view was that the boycott or any related activities, such as appealing to the American public or attacking exclusion, diverted attention from the chief task at hand, the seizure of power in China.[48]

Yet in spite of his own calculated silence on exclusion Sun was, on his several visits to the United States between 1909 and 1911, to find an audience increasingly receptive to his claims that the Manchus were the enemies of the Chinese. A Chinese community that once had ignored the revolutionary literature was now reading with interest such indictments of the Manchus as Tsou Jung's *The Revolutionary Army*, a stirring popular pamphlet published in China in 1903 and circulated by Sun in the United States in 1904. Tsou, an eighteen-year-old Szechuanese, had complained, "Our fellow countrymen settled abroad are humiliated by foreigners in ways which they would not tolerate towards bird or beast. Yet the Manchu government remains politically blind and deaf to this." These same charges of injustice and neglect were now becoming part of the rhetoric of Chinese in the United States themselves as they appealed to their countrymen for support. The Chinese in San Francisco wrote home to Canton in 1909 and 1910, bewailing "the weakness of the government and its inability to protect overseas Chinese" and blaming the failure of the boycott on "a monarchical government." Years later men whose memories reached back to those days recalled, "At that time . . . we Chinese seemed to be without a country." Another looking back explained, "Our people in this country had been pushed around so much they wanted some way, sort of revenge, to get themselves a better way to survive."[49]

By 1909 the tide had clearly begun to turn against the reformers and in favor of Sun. That year he traveled from the East Coast cross-country to San Francisco, and then continued on to Honolulu, setting up at each stop new branches of his party, known in the United States as the Chinese Nationalist League (or, in Chinese, the United League—T'ung Meng Hui). Finally in June 1911 Sun effected a merger with the secret societies. Revolutionaries were inducted into secret society membership, and the societies made financial contributions to the revolutionary cause. By then Sun's party was solidly on its feet. In San Francisco (designated its North American headquarters) as well as in New York and Honolulu, the party had in operation schools, newspapers, and youth groups. Through them it pressed the anti-Manchu cause and neutralized residual reformist sentiment. Now Sun's rallies drew hundreds. More important, donations flowed in that helped underwrite the last years of revolutionary struggle. For example, one-half of the $77,000 used in the abortive Canton revolt in April 1911 came from the pockets of Chinese in the United States and Canada. By late 1911 Sun and his secret society allies had, by one account, collected $144,000.[50]

The Chinese government and its diplomatic agents understandably regarded these currents of reform and revolutionary influence in the Chinese community with growing alarm. Minister Liang Ch'eng warned Peking in 1903 that "these traitors"—Sun, K'ang, and Liang—were "stirring up the overseas Chinese and collecting millions of dollars. . . ." The Washington legation and the consulates in San Francisco, Honolulu, and New York sought to isolate political subversives from the Chinese community. Diplomats branded them as bad men with whom it was dangerous to have any associations. In public warnings the character for Sun's given name was rewritten so that it meant "the defiler," and K'ang was vilified as a traitor who had defied the Empress Dowager and led the Emperor astray. On occasion consuls tried to check the influence of the reformers by threatening their relatives living in Kwangtung with punishment. Later, faced with a rising revolutionary threat, they joined with reform leaders in seeking the deportation of Sun's adherents and in keeping closer tabs on Chinese residents.[51] The Ch'ing foreign service also threw traps and obstacles, often in cooperation with the U.S. Bureau of Immigration, in the way of peregrinating political troublemakers.[52]

Chinese officials balanced their campaign against political subversives with an effort at cultivating the loyalty of the Chinese in the United States.[53] Peking reiterated the policy of protection that it had first formally proclaimed in the 1890s. In 1904 and 1905, during the exclusion treaty controversy and the early phases of the boycott, it had actually tried to translate its rhetoric into action. In 1906, following the earthquake and fire that destroyed the San Francisco Chinatown, Peking had sent relief funds, while the minister himself displayed his concern by visiting the scene of the disaster. In 1909 the government promulgated its first nationality law, which made children born abroad of Chinese parents subjects of the empire with a right both to return to China and to enjoy protection as residents in foreign countries. Within the United States the minister and his consular subordinates still tried to extend their mantle of protection.

With protection went supervision over Chinatown and patronage of its affairs. In this regard the official still played the role of magistrate, trying to prevent disorder among his people, threatening the indolent with deportation, superintending morals, contributing to the construction of schools and, of course, inculcating loyalty to the dynasty. But these official intrusions into the life of the community only served to irritate its members and to increase the disaffection produced by Peking's ultimate betrayal of the boycott. No longer after 1905 could the foreign service convincingly argue the government's devotion to the welfare of overseas Chinese or embarrass those such as Sun with accusations of having used "wildly disloyal language." Increasingly isolated and discredited, the legation was powerless to hold back the growing revolutionary influence.

For Chinese in the United States the triumph of the revolutionary cause in 1912 restored their hope. They had contributed generously to see the dynasty overthrown and had used their influence in the United States on behalf of the insurgents. An appeal made in October 1911 in the name of four-fifths of the Chinese in the United States, British Columbia, and Mexico asked the Taft administration not to intervene in favor of the imperialists. The same month 5,000 Chinese in Honolulu rallied to make the same demand and to praise the revolution as "a national movement which aims to establish in the place of a government corrupt, effete and absolute one that will be representative of the nation. . . ." Later the leaders of the Chinese community lobbied to secure American diplomatic recognition for the new republican government. Some Chinese returned home to serve; others collected funds for the financially struggling new nation and purchased the airplanes and paid for the training of pilots that were to constitute the core of China's air force. Within Chinatowns reformers and revolutionaries put aside their former rivalry and for a time at least joined in support of a government in Peking which seemed to contain room for both. In this atmosphere of political harmony even older divisions within the community which had spawned internal conflict for several decades were papered over by mediation, first on the West Coast in 1912 and the following year in the East.[54]

The hopes that Chinese in the United States invested in the new republican China were soon dashed. It proved if anything weaker and more distracted by internal problems than the old regime. In 1913 Sun collided with President Yüan Shih-k'ai. Members of Sun's party, expelled from the government, put up a brief armed resistance and then fled abroad to resume the revolutionary struggle. Sun himself, once again in exile, reorganized his forces into the Revolutionary Party (Ko Ming Tang) and resumed his struggle for power. In the United States the revolutionaries, who continued to operate under the name of the Chinese Nationalist League (registered as a humanitarian organization to forestall any U.S. government interference in its activities), resumed their appeal for money. And once more the Chinese in the United States gave generously. By one account, contributions between 1913 and 1916 came to $270,000. Party leaders, recognizing the richness of their North American preserve, jealously guarded it against political interlopers, whether old reformist elements now allied with Yüan or envoys from Yüan himself coming to appeal for support in the United States. Sun's men did not hesitate to use strong-arm tactics, including breaking up public meetings and gunning down opposition leaders. On occasion they engaged in political lobbying in Washington, as in 1914 when they sought to dissuade the Wilson administration from offering the $50 to $100 million in financial assistance that Yüan was rumored to have asked for to prop up his regime.[55]

The Revolutionary party's unity and popularity suffered a severe blow in 1915 in the wake of Japan's twenty-one demands advanced against China. Resentment against Japanese imperialism ran high in the Chinese community, and many sympathetic to Yüan's resistance once more dipped into their pockets to support their homeland in this moment of peril. The revolutionaries, however, were badly split. Sun, the single-minded revolutionary, would no more give Yüan rest (even if he thereby inadvertently aided Japan) than he would have cooperated earlier with K'ang Yu-wei to modify exclusion. His seemingly unpatriotic position drew criticism even from his close associates. Sun's headquarters in San Francisco went into open revolt and, to prove its patriotism, organized a "dare to die corps" to fight against the Japanese. Revolutionaries wishing to put the anti-Yüan struggle aside to form a popular anti-Japanese front found a leader in Huang Hsing, himself a prominent revolutionary who had come to the United States in 1914 as a political refugee. Though he now appealed to Sun "not to get rid of a tiger to let in a wolf," Sun held firm, even at the price of diminished support in the United States.[56] By mid-decade, with the Republic still weak and the revolutionaries in disarray, Chinese in the United States had exhausted their appeals against discrimination and harassment. Now they would simply have to endure.

Chapter Eight

American Reform and Chinese Nationalism, 1900–1914

REFORM IDEAS ROSE to prominence in early twentieth century China and came to occupy a central place in Sino-American relations at that time. Interested Americans and politically conscious Chinese agreed at least in general terms on the importance and nature of change and the value of foreign models. But on one key point—the role that foreigners in general and Americans in particular were to play—they found themselves sharply at odds. The unprecedented appeal of reform thought on the American side after 1900 extended beyond the mission movement and the foreign service to policy makers in Washington and tied all to a conception of the United States as guide and patron of a "modernizing" China. These pretensions were, however, to run up against a Chinese conception of reform heavily tinctured with nationalist preoccupations. Already in these years Chinese nationalists challenged and in some cases disrupted American activities. Would-be American reformers responded with expressions of puzzlement, dismay, and even anger that were to remain a central feature of American attitudes toward China for decades to come.

Nationalist Images of the United States

Chinese nationalism is conventionally traced back to the 1890s, when a string of foreign policy disasters aroused a progressively wider circle of intellectuals, officials, students, and treaty port merchants over China's deepening crisis. Nationalists agreed on the need to make China economically and militarily strong and to safeguard territory and sovereignty rights, but they divided over the best path to follow to achieve those ends. That division—between nationalists of a reform and those of a revolutionary persuasion—in turn did much to shape their respective outlook on the United States.[1]

Reformers and revolutionaries alike began with the same core set of images of the United States, right out of Wei Yüan and Hsü Chi-yü. Chinese continued to look admiringly back to the early years of the United States as a golden age dominated by its inspirational figure, George Washington. With more than a touch of envy, Chinese still read about a heroic generation of Americans who had laid the foundation for national political stability, security, technological and economic growth, and expansion to the natural limits of the continent. To Wei and Hsü's vision had been added by the turn of the century an awareness—promoted above all by the outcome of the Spanish-American War—that the United States had become a major industrial and international power committed as never before to a search for markets and hence to a more expansionist Pacific policy.

> The United States in the last few years has shifted from Monroe-ism to imperialism and from agricultural operations to industrial operations. In the past several years goods for export have gradually increased. . . . Although formerly American influence in general flourished on the eastern shore of the Pacific, recently it has hastened from the eastern shore to the western shore. Thus Hawaii was occupied, the Philippines acquired, the axis of trade between the United States and Asia seized, and American influence extended to East Asia.

Comments by McKinley on the dangers of overproduction and by Roosevelt on Pacific destiny, Hay's proclamation of the open door, the sudden American interest in Manchuria, the rise of American naval power in the Pacific, and the Panama canal project serving American commercial and strategic aims in the Pacific constituted the growing stack of evidence for this generally accepted picture of maturing American power and commercial vitality.[2]

From this shared picture of a glorious American past and a recent surge of active involvement in Asia, Chinese revolutionaries and reformers drew different conclusions consistent with their divergent approaches to China's current crisis. The revolutionary began with the overriding conviction that corrupt and inept Manchu rule was the cause of China's weakness. The Ch'ing dynasty had not only oppressed the Chinese people but obstructed reform at home and capitulated to the powers in order to secure its own survival. The key to China's salvation was, then, the overthrow of this alien regime. Revolutionary nationalists contended—in the face of considerable doubt on the part of reform nationalists—that the powers would not exploit revolutionary turmoil in China or attempt to suppress it as long as revolutionary forces posed no challenge to foreign treaty rights. Indeed, so the revolutionary position went, the powers realized that their intervention would provoke an irresistible popular fury and that in any case they had much to gain by a revolution and the consequent emergence of a stronger China no longer suspicious of the world and no longer the source of East Asian international rivalry.

The tendency in the revolutionary camp to play down the foreign threat in order to move the revolution along was represented in its most extreme form by Sun Yat-sen. The United States in particular, he and others of similar persuasion argued, had especially strong reasons to welcome a new democratic regime in China that was committed to reform, well-disposed to foreign economic enterprise, and strong enough to remove China as a bone of contention between Washington and Tokyo. Americans would surely recognize in the revolution against the Manchus an effort like their own in the 1770s and 1780s, in which resolute, self-sacrificing leaders had organized against an ineffective and oppressive ruling class, achieved independence, and established the foundation for a strong modern republic. Conversely, Chinese should recognize in the United States a model for their own development. Sun, who had repeatedly visited the United States in the decade before the 1911 revolution, described it as "the land of riches," admirable for its political order, economic efficiency, and technological achievement. He saw no cause for concern over its sudden thrust across the Pacific. Hawaii, where he had lived for a time in 1879 and 1883 and subsequently returned after its annexation, served Sun as an illustration of the vigor of progressive American influence. Even in the American annexation of the Philippines Sun could find grounds for encouragement, for it created closer bonds between the United States and China by ensuring American industry more secure access to the China market.[3]

Other revolutionaries, less devoted than Sun to securing foreign backing and less able to fix single-mindedly on the anti-Manchu crusade, could not turn a blind eye to the American record of racial injustice, evident in the enslavement of the blacks and abuse of the Chinese. But otherwise they too put the best face on the recent pattern of American expansion. They welcomed the new

American open door policy as a check on the more aggressive, partition-minded powers. Even the boycott, so one article in 1906 strained to argue, offered evidence of American good will: the United States had not resorted to force; Roosevelt had spoken out in favor of better treatment of Chinese; and Congress had at least considered nondiscriminatory treatment of Chinese. The American colonial venture, put in the proper light, could also offer encouragement. In Cuba the United States had supported the cause of revolution and independence. In the Philippines American forces had brought Aguinaldo back into the country to help defeat the Spanish and to set up his own government.[4]

It was not, however, the views of Sun and other prominent revolutionaries but rather those of reform nationalists that were to define the dominant Chinese attitude toward the United States and Americans in China, even after the success of the 1911 revolution. Reform nationalism influenced the outlook of ranking officials serving under both the Ch'ing and the early republic as well as the provincial elite. They shared a fundamental pessimism about China's imperiled position in the world. They saw the international environment in Darwinian terms: brute power determined relations among fiercely competing states. In that environment extinction for a weak China at the hands of the dominant powers was a frightening possibility. Far from saving China, a revolution might, they warned, prove its undoing. An examination of the French Revolution and recent memories of the Boxers underlined the dangers of popular excesses. The powers would seize the first sign of antiforeign violence as a pretext for armed intervention and for posting new demands perhaps more sweeping than in 1900. Even if the revolution managed to keep to its anti-Manchu purpose, its ultimate goal of creating a stronger China might still alarm the powers and lead them to intervene to safeguard their own long-term interests. The reformers' solution for China's unhappy plight was to work within the existing imperial political structure. They put their faith in moderate, nondisruptive constitutional reform; the development of the military, commerce, and industry; expanded opportunities for practical education; and such social reforms as opium suppression. Here was the safest and surest means to unite and strengthen the country against the foreign threat.

That threat, as the reformers conceived it, took two forms—the grab for spheres of influence in anticipation of a possible territorial division of China, and the pervasive foreign penetration of China's economic and cultural life exercised mainly through the mechanism of the unequal treaties. To a considerable degree the reformers' view of the United States depended on which of these two facets of the foreign threat was uppermost in their concerns.

Officials charged with general supervision of foreign affairs and those in frontier areas understandably tended to give precedence to staving off the loss of territory and thus relied heavily on a makeshift balance of power strategy to ward off new incursions, neutralize old ones, and win time for reforms to take effect. Their devotion to the recovery of sovereign rights lost in the unequal treaties, while real enough in principle, was often attenuated in practice by a reluctance to compound China's problems with the powers, especially those that might be counted friendly to the maintenance of China's territorial integrity. Among this group of reform nationalists, the United States increasingly figured as the most eligible ally, one whose dependence on foreign markets and investment opportunities might be manipulated to China's advantage. The architects of a balance of power policy from Chang Chih-tung, through Yüan Shih-k'ai and his aides, to Hsi-liang, and back to Yüan again were ready to hand over special concessions to Americans to secure their government's diplomatic support.[5]

The alternative tendency of many reform nationalists was generally to deplore China's dependence on any power for protection and emphasize the need for greater strength and self-reliance (though at moments of great diplomatic crisis even they would sometimes endorse the pursuit of foreign allies). They preferred a foreign policy that aimed at neutralizing the privileged foreign presence in China, and hence tended to see the United States as an adversary, unyielding in defense of the unequal treaties, always on the lookout for new economic opportunities to exploit, and prone to treat Chinese with contempt. The United States thus deserved, reform nationalists of this persuasion contended, to be counted in, not outside of, the camp of powers threatening China.

These critics watched the United States and marshaled their evidence.[6] Under McKinley they saw the United States cutting its last ties to the golden age of Washington. Impelled by industrialization, Americans had joined the struggle for foreign markets and succumbed to the lure of territorial aggrandizement and colonial domination. The annexation of the Philippines supplied the first sure sign that the United States was bent on "invading the East" and harbored "malicious designs" on China. This American domination over another Asian people long retained for Chinese a special emotional charge as a warning against American aims. A popular periodical, the *Tung-fang tsa-chih* (Far Eastern magazine) concluded a full-length article in 1909 on atrocities in the Philippines with the bitter observation that Americans seemed to believe freedom was for whites only. A Hunan paper trotted out the same tale five years later replete with all the classic ingredients of the earlier antimission literature—an immoral white woman, a fiendish Western doctor, the indiscriminate use of American kerosene as a weapon of terror, and a heroic Filipino general mercilessly tortured. "You who fear becoming slaves in a conquered country," the paper grimly enjoined its readers, "take notice."

Reform intellectuals had ample materials still closer to home (where, after all, American actions and intentions meant the most) from which to construct their cautionary tales. They characterized Hay's open door policy not as a timely contribution to China's salvation but rather as foreign aggression in a new guise. In their view, the open door policy had succeeded in obstructing partition only because the powers did not yet wish to take on the cost of direct rule. While Hay's diplomacy tried to preserve for his late arriving countrymen a chance to penetrate the China market, the U.S. navy set about strengthening itself the better to enforce American claims in China. The United States had made clear (as they saw it) that it was prepared to use its power, at least against Chinese, in 1900 when it had joined the allied intervention and again five years later when Roosevelt, in a display of contempt for the Chinese as a second-class race, had rejected the protests against exclusion and browbeat Peking into crushing the boycott. The denial to Chinese of even minimal opportunity and respect in the United States at the same time that the United States was demanding greater opportunity in China was galling to Chinese and seemed to them to give the lie to American professions of benevolence. In the last years of the decade the danger from the United States loomed more menacingly as the Taft administration intensified the American drive to secure a share of the China market. Washington's determination to win a seat for American capitalists at the feast the European financiers had spread, the Manchurian neutralization proposal promising to replace one kind of outside control for another, and the attempt to impose an American financial adviser all sustained the view that the American advance would come at the cost of Chinese economic rights.

Liang Ch'i-ch'ao, the single most influential and prolific of the reform intellectuals, did much to popularize this view of the American menace. He made his discovery of the United States in 1890 in a predictable way—through a reading of Hsü Chi-yü. But whatever of Hsü's emphasis on American exceptionalism he may have picked up was soon overwhelmed by his concern with Western aggression and the Westerners' pervasive attitudes of condescension and contempt. By belittling China, they prepared the way to "destroy one's country and enslave one's people in the name of righteousness." By 1899 Liang had come to differentiate between "visible dismemberment" and "invisible dismemberment" and to minimize the much touted difference between commercial powers such as the United States and the territorially aggressive ones. Both infringed on Chinese sovereignty, one by subtly depriving Chinese of effective control, the other by using brute force. By 1902 Liang was describing the dangers to China that the United States in particular posed as she put aside isolationism in favor of an Asia-first policy and threw herself into the international struggle for survival.[7] He regarded the American conquest of the Philippines with keen interest. China was tied to the Islands not just by trade and migration but as an Asian state with a shared desire to achieve democracy and national independence against foreign domination. The liberation of the Philippines from American control would constitute an important step in turning Western power away from Asia. Under American control, on the other hand, the Philippines would serve as a base for a continued American Pacific drive westward, with China's southeastern coastal provinces the next target. Liang warned his countrymen still entranced by the myth of the American revolution that the militancy of the other powers and China's passivity had emboldened the Americans, who now stood ready to strike on the first sign of partition.[8]

During his seven-month visit to the United States in 1903 Liang discovered the wellsprings of this recent, abrupt American conversion to expansionism. All around him he saw evidence of the power and prosperity created by the dynamic American economy. But he also noted that long-term growth had given rise in the past decade to the problem of overproduction, which in turn had called forth "the great spirit of the twentieth century," the industrial trust. To deal with surplus product the Morgans of the United States had centralized economic control, and when that proved insufficient, had begun reaching abroad for new markets. At once fascinated and concerned by this interrelated trend toward economic concentration at home and economic imperialism abroad, Liang secured brief interviews with Morgan and Roosevelt, the embodiments of this new American power. Looking ahead, Liang predicted that economic concentration would proceed apace. As the trusts wove an ever more intricate network of communications, transport, and trade about the globe, national boundaries would dissolve and national sovereignty would evaporate. Already the United States had made Latin America its commercial preserve under the banner of the Monroe doctrine. Now Americans were directing their attention to China, the most promising open market left in the world. They would secure their control of a share of that market through informal means if possible, but they would resort to a policy of outright partition in case of internal instability or threatening actions by the other powers.[9]

Much of Liang's subsequent commentary was informed by the general conclusions he had reached by 1903. To a 1904 American proposal for foreign-directed reform of China's currency, Liang responded with a warning that this would be a step toward handing over control of the entire government, as had happened in Egypt and Korea. It "would be not our good fortune but our ruination." Immediately after the Russo-Japanese War he eschewed endorsing American involvement in Manchuria (preferring instead neutralization of the region as the most promising device for curbing Japanese influence). And in 1909 and 1910 Liang and others writing in the *Kuo-feng pao* (National spirit), a journal edited by Liang himself, followed Washington's investment drive with growing agitation. Liang repeated his earlier warnings against taking the honeyed words of the Americans at face value. China's barbarian managers who did would discover to their chagrin that the United States was disinclined to clash with the other powers when she could obtain a share of the China market by the cheaper and surer expedient of joining with them against China. Liang criticized Hsi-liang's railway loan as a self-defeating political initiative. Rather than evicting the Russians and Japanese (as Chinese policy makers wished), the Americans were more likely to settle for a share of influence in Manchuria alongside the Russians and Japanese. Knox's subsequent proposal for putting Manchuria's railways under international control bore Liang's assessment out. Such an arrangement would substitute six-power control of Manchuria for Russo-Japanese domination with no gain for China. To preserve China's independence and save the Chinese from becoming "the world's most despised people," Liang recommended that the government avoid large foreign loans—even when arranged with the United States on the mistaken notion that it thereby obtained diplomatic backing against Japan, Russia, and their allies. Liang watched with dismay as the Regent's government did precisely the opposite, burdening China with the Hukuang and the large so-called currency reform loans.[10]

By 1911 many reformers had defected to the side of the revolution, driven there by the Regent's cynical manipulation of political reform and his weak foreign policy. In the years that followed they were to find their earlier predictions about the dangers of revolution amply fulfilled. The revolution of 1911 did indeed unleash centrifugal forces and political factionalism that left the central government weak and distracted and created new opportunities for foreign penetration and domination. Yet the dream of securing China's territory from outside control and grounding her relations with other countries on the principles of equality and reciprocity lived on. And so too did the reformers' ambivalent view of the United States as the safest of the powers—yet still an imperialist.

Diplomats and the Reform Ideal

While the Chinese began to employ the vocabulary of nationalism against foreign domination, American diplomats stood by a broad program of reform that presupposed a major, even guiding foreign influence. Washington, which earlier had largely limited itself to guaranteeing Americans in China their privileged position, now reinforced the reformist impulse by lending support to specific projects directed toward China's overhaul. American officials in China were sensitive to a degree to the strength of the new currents of Chinese nationalism, but their attachment to the ideal of foreign-directed reform, a fundamental and widely accepted facet of the open door ideology, stirred misgivings about the direction the Chinese were taking. These misgivings weighed even more heavily on thinking in Washington, rendered it insensitive to the significance of Chinese nationalism, and acted as a further brake against the few concessions the foreign service was prepared to recommend.[11]

The experience of W. W. Rockhill, the only nonmissionary expert to rise to some eminence from the ranks of the foreign service, is instructive on the strength of the reform ideal. Reserved and scholarly, Rockhill had gone to China in 1884 to begin his apprenticeship as a diplomat and to seek out the horizons and the opportunities for experience lacking in the United States. "Rather a cycle of Cathay or Korea than fifty years of America for me." By the late 1890s he had come to view China's reform as "essential to the maintenance of peace." Otherwise, "partition and subjection to foreign rule are but questions of a little time."[12]

The Boxer crisis, while highlighting Peking's inability to manage its own affairs, also created (as Rockhill saw it) an unprecedented opportunity for forceful foreign intervention on behalf of long-overdue reforms. As the American representative at the post-Boxer negotiations in Peking in 1900–1901, Rockhill, urged on by Hay, proceeded to lay out the American reform program as fully and systematically as it ever would be. Containing something for each of the groups within the open door constituency, the program in effect set the agenda for a decade.[13]

The legation's own special concerns were reflected in Rockhill's efforts to save China's foreign office from the inefficiency and indecision which had brought it into disrepute among foreign diplomats. A revamped and more prestigious institution, staffed by men of "rank, influence, or strength," would function as a strong voice in behalf of moderate and progressive policies within the councils of the government.[14] For the business community, Rockhill sought to remove impediments to trade and to create a stable and uniform fiscal system. At the top of the long list of commercial provisions that he carried into the Peking conference stood his demand for an end to the "oppressive" internal transit tax (likin) on imports and exports and the opening of all parts of the empire to foreign business and residence, both measures long considered key to opening the vast but elusive interior market to American trade.[15] Even for missionaries, whose work he regarded as subordinate to the great tasks of China's economic and political development, Rockhill was willing to make an effort, adhering to the lines of the advanced missionary policy laid down only a few years earlier.[16]

In the fall of 1901, at the end of the prolonged and difficult negotiations, Rockhill headed back to Washington, gloomy over China's "very dark" prospects. In particular he was concerned over the incompetence of the Chinese government. He saw it resorting to the familiar tactics of evasion and delay and failing to accept the direction and assistance of the United States, "the one power which seeks to act justly. . . ." The Empress Dowager was the most serious obstacle to change, but generally, Rockhill lamented, "there is no life, energy or patriotism throughout the whole governing class." Only "strong outside pressure" would move China to put her house in order. But the powers, divided against each other, had just proven that they could not agree on a general program.[17]

The China to which Rockhill returned as minister in 1905 was approaching the high tide of reform nationalism. In 1901 the central government had announced a broad range of its own reforms and had subsequently hastened them forward.[18] Rockhill quickly grasped the significance of the new tone of Chinese national life and initiated a steady stream of reports intended to alert policymakers to these changes. Among the Chinese people, whom he had as late as 1904 described as "devoted to their individual interests and devoid of public spirit," he now detected "a national spirit." And the same Chinese government that he had in 1904 labeled "weak and corrupt" and even on his arrival condemned for "indecision and a determination to drift with any current" was now, he reported, "irrevocably committed to a vast scheme of national progressive reforms. . . ."[19]

Not that all was suddenly well. The "reformed" foreign office had, Rockhill lamented, fallen back into its "extremely irritating and childish" ways, while the rise within its ranks of such nationalists as T'ang Shao-i and Wu T'ing-fang made it more exigent on such issues as rights recovery and Chinese immigration. Instead of a sound economic policy—which he regarded as crucial to China's stability and survival—he saw only bumbling and feeble attempts at economic reform and "blindness" to the rules of political economy. Likin still obstructed commercial development, and meaningful currency reform was a dim prospect. Rockhill also worried that popular nationalism, "usually misinformed," might get out of hand and revert to the old crude and dangerous antiforeignism under the direction of "malcontents and agitators for political reform" in the provinces and in intellectual circles. For that reason he did not like the 1905 anti-American boycott—"foolish and lawless," "a dangerous precedent," tainted by "anti-foreign" feeling—nor its leaders— "a set of silly Chinese" motivated by a vague feeling of discontent and restlessness. Finally, Rockhill worried that government officials, in their impatience to escape their place of inferiority in the world, might run the program of reform right off the track. "China no longer requires to be pushed ahead, but rather to be restrained," he would write when in these grey moods.[20]

Yet despite all these doubts and irritants, Rockhill still on balance regarded Chinese reform nationalism hopefully, and even where advocates of "China for the Chinese" threatened American interests, he urged his government to follow a conciliatory policy.

Recognizing the strength of the popular protest during the 1905 boycott, Rockhill cautioned against forcing Peking to move against treaty port nationalists and in the process undermine its own authority. Again, in the case of the Chinese drive to cancel the American China Development Company's railway concession, he advised Roosevelt against fighting a strong and largely justified popular protest.[21] In yet a third case in 1906 in which China sought to extend its jurisdiction over foreigners in newly opened ports in Manchuria, Rockhill showed a willingness to follow a flexible reading of American treaty rights in order to accommodate Chinese efforts to reclaim lost sovereignty. He expressed his sympathy "with the Chinese desire to be masters in their own country," and recommended going along with them in this instance as well.[22]

Washington was, however, unreceptive to Rockhill's call for a policy more responsive to the changed situation in China. Roosevelt, who had paid tribute to Rockhill by giving him the Peking post, would not compromise his conception of the civilizing mission and would not tolerate "native" agitation. Thus he grew angry when he felt the Chinese were unfairly wresting the Hankow-Canton line from American concessionaires and accused Rockhill of "a complete misapprehension of the facts." He again rejected Rockhill's call for moderation during the boycott, which he vigorously opposed as an insolent and inexcusable campaign against the United States.[23]

Root, guided by his lawyer's faith in the sanctity of treaty rights, was similarly unmoved. He condemned the popular boycott organizations as "unlawful combinations in restraint of the free commerce stipulated by treaties." (This somewhat forced reading of the treaties suggests that reverence for international law did not preclude occasionally stretching it into more satisfactory shape.) He rejected Rockhill's balanced approach to the open port issue. To bow to the "aggressive tendencies" of the Chinese would only encourage them at some future date to attempt "to deprive foreigners of all advantages which they have legitimately acquired."[24]

Lesser lights in the foreign affairs bureaucracy also rejected Rockhill's views on accommodation. They viewed the recent restlessness in China as a passing, if dangerous, phase which should not be allowed to disrupt American influence and enterprise. Rockhill's call for conciliation convinced them that he was not an expert to follow, but a Sinophile to ignore—a man whose overexposure to Chinese ways had dulled his zeal for American ideals.[25] They countered his influence by simply amplifying his critique of Chinese officialdom. Wu T'ing-fang and T'ang Shao-i, to take the two who raised bureaucratic hackles the most, were denigrated as political lightweights, "extreme" in their views, and "at heart deeply antiforeign."[26]

On the other hand, where Rockhill gave vent to lingering doubts about Chinese-directed reform measures and personally pushed reform American-style, he found Washington and his foreign service colleagues supportive—and the Chinese resistant. This occurred in the case of education—considered the royal road to securing a stable and liberal order and remolding presumably plastic Chinese. In "filling the chinese [sic] mind with American ideas as he [sic] studies," the consul in Nanking (himself not particularly well lettered) saw hope for establishing "the most lasting influenc [sic] for America." Foreign service advocates of American education also wanted to offset Japanese prestige, at its height in 1905 with some 8,000 Chinese already studying in Japan in contrast to a mere 130 then in the United States. Denby, Jr., in the Shanghai consulate saw the alternatives sharply drawn—between Chinese students becoming imbued with "political yearnings" under the Japanese or, under American auspices, becoming "intelligent advocates of the prudent and timely, rather than the revolutionary introduction of reform." The Tientsin consul appealed to the "national duty" to replace the hysterical influence of the Japanese with the stabilizing influence of the Americans. Besides, he added, education was a good "commercial investment."[27]

Rockhill, who shared this enthusiasm for education "on modern lines," made his pet project the application of the U.S. Boxer indemnity surplus to education. When he discovered that Yüan Shih-k'ai and his associates had their own plans for using the returned funds for economic development in Manchuria, Rockhill threatened to block the return of the money (though he conceded it was rightfully theirs), and he warned Washington both against the "perfectly impracticable" "wildcat scheme" of the Chinese and against its chief proponent, T'ang Shao-i ("densely ignorant on all financial questions, and of political economy I doubt if he . . . know[s] even the name"). Predictably Root and Roosevelt both supported Rockhill on linking the return of the indemnity to education, and Root's advisers on Asian affairs gleefully prepared to knock down the Chinese alternative plan when T'ang arrived in Washington to present it. "The return of the indemnity should be used to make China do some of the things we want," Huntington Wilson informed Root. "Otherwise I fear her gratitude will be quite empty."[28]

On another reform issue, opium suppression, Rockhill and Washington also marched together. True to form, the minister regarded an end to the opium evil "as an indispensable first step to the moral regeneration of the race." But he warned that Chinese plans to deal with the drug problem were not only overambitious but probably a screen for expanding the domestic opium crop from which the government, in turn, would derive a growing and welcome revenue. Roosevelt decided at this point that the United States should take the lead in an international campaign against opium, with China as the initial focus. Accordingly, the State Department issued a call in mid-1907 for a conference to meet in Shanghai. Influenced by Rockhill's doubts about Chinese backsliding, it sought an advance pledge from Peking to take "genuine radical measures" against domestically produced opium. Rockhill and Washington were also united in their confidence that the United States was in a unique position to advance this reform. In a remarkable mix of untruth and illogic, Denby, Jr., explained that the initiative had fallen to the United States because it "has kept its hands clean" in the China opium traffic and because it controlled in the Philippines "a race akin to those most affected by that traffic."[29]

After 1909, the year of Rockhill's departure, the legation had to entertain a new feature of the China scene, growing revolutionary unrest and political instability. The increasing frequency of attempts against the dynasty, provincial resistance to central authority (both before and after the 1911 revolution), and the armed trials of strength in the early republic revived the legation's fears, dating back to the missionary incidents of the 1890s and the Boxer trauma, of "native unrest." China "might burst into a sudden flame of violence which would sweep over the entire country."[30] The legation's preoccupation with disorder was in turn to attenuate its commitment to reform and to give priority to the pursuit of stability, which was of course the prerequisite for progress in China. The crucial step was to find some strongman, preferably one with reformist credentials, to act as the guarantor of stability.

After the Boxer debacle Yüan Shih-k'ai rose to fill that position. By 1908 the foreign service had come to regard him (in Rockhill's words) as the "chief influence for order, stability and progress in the government."[31] Although Yüan fell from power early in 1909, the dynasty's need of his services in combatting the revolutionary upheaval of 1911 restored him not only to the political limelight but also to his old place as the legation's favorite. William J. Calhoun, a Chicago lawyer appointed minister the year before, warned that beyond Yüan there was "nothing in sight but anarchy." Although initially doubtful of China's readiness for republican government, Calhoun was reassured by Yüan's dominant presence once the republic was set up early in 1912. His government, whatever its deficiencies in living up to its democratic pretensions, deserved American support as "the only government in sight." Calhoun's successors, chargé E. T. Williams and Minister Paul Reinsch, agreed with this assessment. Though Yüan would in fact make a mockery of democratic politics, assassinate the opposition, and restore worship of Confucius, the legation continued into 1914 to picture him as the necessary force for "order and authority" in a country beset by foreign aggression, domestic unrest, and financial and administrative confusion.[32]

The legation's concern with disorder at this time inclined it toward a restrained commitment to reform and a flexible reading of American treaty rights, particularly where accommodation might serve to strengthen the authority of the central government. Not long after his arrival Calhoun advised against pressing costly new projects on China, however desirable from a reform point of view, because the resulting new taxes "might easily provoke further expressions of anti-foreign feeling." Nor was it politic, Calhoun later argued, to obstruct the search for new sources of revenue on the part of the financially hard-pressed central and provincial governments. He also stressed, as Rockhill once had, that the "modern educated men" of China, especially those who had studied abroad, were opposed to the extraterritorial status foreigners enjoyed and "possessed of the idea that the sovereignty of China must be recognized and accepted as the only administrative authority in the country." Reinsch, for his part, was optimistic about the prospects for "a new era in China's development," particularly if disinterested Americans helped in education, trade, and finance. But like Calhoun, he was concerned about the dangers of a broad and inflexible reading of the treaties. Opposition to the Chinese in such matters as internal taxation, he warned in March 1914, "stands in the way of the development of an efficient and adequate government."[33]

The concrete concessions recommended by Calhoun and Reinsch were to find no more favor in Washington than had Rockhill's earlier. The State Department was not prepared to back-pedal on reform. Knox relentlessly pursued the old economic program—currency reform and the elimination of likin and other obstacles to foreign trade—while continuing American support for the international effort to suppress opium.[34] Moreover, many in the State Department feared that concessions to the Chinese on treaty rights would endanger American interests and influence. "China for the Chinese" was fine, Taft had once observed, as long as it did not interfere with order, trade, progress, or foreign rights. Finally, the old paternalistic and pessimistic view of the Chinese made retreat on either reform or treaty rights unthinkable. Knox's advisers on China regarded their "Yellow frinds [sic]" as pawns in the great game of international diplomacy. They had no patience with Sinologists ("brains and no body") such as Rockhill who studied backward China as if conditions there had any real bearing on the great practical questions of the moment. The State Department's China experts thought Smith's *Chinese Characteristics* had said it all a decade earlier: The Chinese were supremely self-interested, passive, and unpatriotic. They "yield nothing to reason and everything to fear." The arrival of Bryan as Secretary of State and E. T. Williams as the State Department's chief China expert changed the tone of the analysis but not the essential commitment to holding treaties sacrosanct and the Chinese suspect. Thus while China changed and the legation struggled to respond to it, Washington kept policy moving along in a well-worn rut.[35]

The Politicized China Market

In the legation's view, U.S. economic enterprise occupied an important, arguably even the central, role as a force for progressive change in China. But business needed a free field if it was to make its contribution to China's development (not to mention American prosperity). For American diplomats the treaties of the 1840s and after, which had forced the Chinese to concede an ever broader horizon for foreign trade and investment, were a welcome intrusion of politics into the trade scene. By contrast, European financial diplomacy in China in the mid-1890s and after proved an alarming development. So too did the reaction of Chinese nationalists, whose fears of foreign economic exploitation and subjugation precipitated sustained resistance on a variety of fronts. The American response to this increasingly politicized China market was curiously divided. On the one hand, American policy makers, compromising their free trade principles, began actively to intervene in economic affairs—both to encourage business and to remove barriers in China that seemed to stand in its way. On the other hand, the investors and traders for whom the government ostensibly labored displayed either a limited interest in the China market or a level-headed capacity to adapt to it rather than an inclination to crusade to reform it.

For economic expansionists in the foreign service and the State Department, the data that came in between 1901 and 1913 suggested that American trade and investment were in trouble. Though trade doubled and investments (exclusive of mission property) increased from $20 million to $49 million, these increases were not enough to meet the high expectations that carried over from the boom of the previous decade. Nor did they disguise the disturbing fact that in relative terms Americans were at best holding their own and, in some respects, actually losing ground.[36] The collapse of two of the great hopes of the economic upturn of the 1890s underscored the general malaise thought to be afflicting American business in China. The promising cotton goods trade entered a long decline after the Russo-Japanese War, falling from 35 percent of the total Chinese market for foreign cotton goods in 1905 to 5.7 percent in 1907 (and by 1929 virtually disappearing from the China market).[37] No less a letdown was the loss of the Hankow-Canton railway project as it passed first to Belgian control and then, in 1905, back to China. Other investment projects went nowhere. Neither Japan nor Russia would sell E. H. Harriman the Manchurian railway lines he wanted. While the 1907 financial crisis in the United States nullified one of his construction agreements with Chinese officials in the region, his death in 1909 left a second agreement in jeopardy. By 1911 Americans could boast of having built all of 29 miles of rail (the total result of the American China Development Company's seven years of activity) out of China's total of 5,771 miles. The marked disparity between talk and concrete achievement was as evident in mining as railway enterprise. Of the 62 mines in which foreigners had an interest around the turn of the century, Americans were represented in only 4, and for only a short time at that.[38]

Some in the foreign service, and to a lesser extent commercial pressure groups, argued that these setbacks were due mainly to foreign political interference in the free market. Already in the 1890s Charles Denby and E. H. Conger had worried that the powers would deny Americans the rich opportunity China held out. After the Boxer crisis, Conger attacked Russian trade discrimination, which was likely "to seriously hamper, if not destroy, foreign trade" in Manchuria, and his warnings were echoed in the alarmist reports churned out by the consul in Newchwang. Throughout the Boxer crisis and the Russian occupation of Manchuria, the American Asiatic Association (backed by its Shanghai affiliate), the southern cotton textile industry, and the New York export houses continued to speak out for export interests, while the Association's vocal secretary, John Foord, kept in touch with McKinley, Roosevelt, and the State Department.[39] But after the Russo-Japanese War, organized exponents of the China trade grew quieter as they began to recognize that China was a poor market for most American finished goods and that Japan was a larger and more promising one. It now fell to the foreign service to keep the myth of the China market alive.[40] The consular service in Manchuria proved a particularly fertile breeding ground for eager economic expansionists. After the Russo-Japanese War, Willard Straight at Mukden and like-minded consuls nearby bombarded the State Department with dire warnings about the Japanese drive to dominate the Manchurian market.[41]

The response in Washington to these complaints of discrimination and the concomitant calls for more government support for the China trade varied from administration to administration. Roosevelt in his recommendations to Congress between 1901 and 1907 went as far as adopting some of the general demands of the commercial pressure groups—for reform of the U.S. consular service, a trans-Pacific cable, aid to American shipping, and creation of a foreign trade bureau. He was also quick to meet Chinese attacks on American economic interests (notably during the 1905 boycott and the struggle for control of the Hankow-Canton railway concession). But at the same time the Roosevelt administration restrained the legation with orders not even to appear to give official support to Americans seeking economic opportunities. And in dealing with the alleged threat from Russia and, later, from Japan, the Roosevelt administration was similarly cautious.[42] The Taft administration shifted to a distinctly more aggressive economic foreign policy, guided by a firm belief in the reality of the China market and by a preoccupation with Japan (and not China) as the chief threat to American economic interests. But, after four eventful years, the incoming Wilson administration repudiated both the philosophy and the instruments of dollar diplomacy and kept putting its scruples in the way of a score of investment schemes that Minister Reinsch promoted with an unflagging enthusiasm that would have put the best of Knox's dollar diplomats to shame.

Overall the policy makers who in one way or another might be designated economic expansionists—Huntington Wilson, Knox,

Reinsch—left behind a barren record. Knox, who was in the best position to overcome the timidity of investors and the sluggishness of exports, failed to boost the financial stake in China beyond a mere $7.5 million or to revitalize American exports as he had hoped. The obvious question, posed by Minister Calhoun in 1912, was: "What difference does it make whether the 'door' is open or shut if we are not disposed to go in or out of it, even when it is open?" Knox and the others offered no answer and soldiered on, beguiled by a fundamentally flawed understanding of the patterns of international trade and investment that they wished to manipulate.[43]

The cotton goods trade offers a key example of this official naiveté about the strategies that guided export industries.[44] Knox and other economic expansionists attributed the decline of the cotton trade to a discriminatory policy undertaken by the Japanese government, and failed to understand the complex dynamics of the export trade and their own inability to control an essentially economic, not political phenomenon.

American textile mills, which had taken the cheaper end of the China textile market from the British, were now having to surrender it to the Japanese. The golden age of American cotton goods in China drew to a close just as it had arrived—because of economic factors. American producers started out with a major advantage over their Japanese rivals—an immediately available and large supply of first-class raw cotton (usually accounting for somewhat more than half of basic production costs). But American wages were several times higher than in Japan, and although American workers remained more productive, Japanese labor narrowed the gap as it gained experience and as the rapidly developing industry progressively introduced after 1905 a new generation of advanced, Japanese-designed machinery. Moreover, Japanese producers and exporters, unlike the Americans, effected savings by coordinating their marketing activities. The result, then, was a dramatic Japanese price advantage in China—in the range of 20 to 30 percent—and consequently increasing dominance of the market.

The American cotton goods setback must also be understood in terms of a failure to cultivate the China market. To win sales in China the Japanese regulated supply to meet demand, aggressively marketed their products, eliminated costly middlemen, and gladly sold on credit. The American industry, on the other hand, aimed primarily for the home market. In part because of intense rivalry for this home market, American producers failed to achieve some of the industry-wide cooperation evident in Japan. And with intense domestic competition unabated, they were naturally reluctant to risk diverting capital, to engage in costly small-scale production of goods tailored to meet Chinese tastes, or to start up marketing operations in China. Only one of the major American producers opened an office in China to keep in touch with the market; the rest relied on a system of distribution through American, British, or Japanese export agents which proved inefficient as well as expensive. The agents were often inattentive to proper shipping methods, while goods, once in China, frequently passed through the hands of several intermediaries before reaching the ultimate consumer. To make matters worse, producers on occasion accepted foreign orders when domestic demand was low only to cancel them when it revived. No wonder then that this foreign trade dwindled. Clearly for Americans, though not for the Japanese, domestic opportunities simply overshadowed foreign possibilities.[45]

When American policy makers turned to the problem of getting American capital committed to China for the long haul, they displayed not only the ignorance evident in their approach to exports but also a willfulness that is puzzling. Since the late 1890s they had well known—or at least had been repeatedly told—that China did not offer the security and returns to compete with the American market.[46] Indeed, there is nothing to indicate that Roosevelt, Knox, and Reinsch, or Conger and Denby earlier, thought China a good place to put their own money. Yet they expected Morgan and others to take an interest in investment in China, perhaps on the judgment that the moguls of Wall Street had enough money to afford patriotism. This divergence of perspective between Washington and Wall Street over the essential ends of investment fixed the two in an uneasy relationship marked by repeated efforts on each side to use the other.

The interests of some of the financiers had a speculative bent that ill-served Washington and the legation's desire to build up long-term influence. Such was the case with the American China Development Company, whose principals were careful to screen their proceedings from Washington.[47] In September 1898, only five months after winning the Hankow-Canton concession and even before the first survey team reached China, the American stockholders in the enterprise were already entertaining a takeover bid from Leopold II. By November 1900 a third of the shares had passed into his hands, and by February 1904 he controlled five of seven seats on the company's board of directors. Shareholders may have been shaken by the death in December 1898 of Calvin Brice (reputedly the driving force behind the enterprise at its inception), financial instability associated with the Boer War, and the Boxer crisis itself. Leopold's offer to take shares at nearly twice their original value gave nervous investors an attractive way out of the doubtful investment; they did not hesitate to seize it. Leopold's American collaborators, led by none other than J. P. Morgan, discreetly went along with Leopold's quiet takeover, not wishing Washington to know their part in turning this investment prize over to another power, or Peking to become aware of the violation of the concession's nontransfer clause.

Only after some delay and much publicity did an awareness of these developments penetrate the consciousness of the men in Washington. By 1901 all the other major foreign offices had confirmation of the takeover, rumors were circulating widely in China, the *Journal of Commerce* had published an accurate report, and Rockhill and Conger had provided notice. But Hay persisted until 1904 in taking company assurances at face value. At last, the dispatch of a Belgian to manage railway work in China, loose talk by the American he displaced, a *London Times* article, and finally an alarming report from Conger proved too much for Hay's credulity. For a time he ceased defending the company. But after receiving assurances in May 1904 from Charles A. Whittier, a wealthy Bostonian whom Leopold had put in charge of the company, Hay returned to the defense—to prevent "an act of spoilation" by the Chinese, who now wanted the concession cancelled.

To head off the Chinese campaign and with the urging of the State Department, the Morgan group proceeded to restore American control. But the Chinese government, unappeased, persisted, offering to buy out the stockholders for $6.75 million ($3 million for less than thirty miles of track built, and the balance to compensate for loss of rights). To this Roosevelt responded by asking Morgan to stand fast against the sale in the interest of preserving American prestige and commercial influence and to trust Washington to hold off the Chinese. But when the stockholders at last voted in late August, they failed to measure up to Roosevelt's patriotic standards. Having once sold out to Leopold at a fat profit, they saw no reason to decline an equally attractive offer for the troubled concession from the Chinese. In the end Roosevelt conceded that Morgan and his associates had acted in their own best interest, but he persisted in condemning the original management for allowing the Belgian takeover, never realizing that it was Morgan himself who had negotiated its terms.

E. H. Harriman followed the long-term investment strategy favored by Washington, but he too preferred to keep Washington at arm's length, in large measure because what he was after was not immediate profit but power that he did not wish to share. As one of the original shareholders in the Hankow-Canton concession he had opposed Belgian control. He had later turned to Manchuria, where he sought to forge a new link in his grandiose global transport system. Already his trains spanned the North American continent, and his Pacific Mail as well as Japanese lines covered the next step across the ocean to Japan and the Asian mainland. The acquisition of trackage across Siberia and on to the Baltic and a trans-Atlantic shipping line would complete the circle. Harriman was not trying to maximize his capital return in the short run, nor could he hope to live to see his plans realized. What he sought was an outlet for his instincts for empire-building, and for that reason it is difficult to imagine Harriman settling into harness with the same government that had by its trust-busting shut off his opportunities at home. Yet settle into harness Harriman eventually did, along with Morgan and the others brought into Knox's American Group. Perhaps by 1909 Harriman had come to the conclusion that his Manchurian railway stood a chance only with government backing.[48]

At any rate, Harriman's death shortly after coming to terms with the Taft administration left Morgan and the remaining financiers of the American Group to test the utility of the alliance. Though they professed a patriotic concern for "the maintenance of American prestige in the far East" (and not "pecuniary gain"), after a year of working with Knox they were complaining that patriotism, along with whatever credits with Washington they banked for another day, was involving them in long and profitless negotiations and straining relations with valued European colleagues. Taft responded by describing Wall Street's leaders collectively as "the biggest ass that I have ever run across." The rift in the Washington–Wall Street alliance, gradually deepening through the Taft years, became complete when Wilson decreed a divorce—to the apparent satisfaction of both parties.[49]

Reinsch's efforts to reconstitute on an informal basis the alliance between finance and diplomacy illustrate the dictum that a failure to learn from the past is the best guarantee of repeating it. Reinsch tried to push American enterprise and investment forward as a stabilizing and developmental force in China, but business held back, its earlier doubts confirmed by the 1911 revolution and political and financial instability in Peking. As incapable of learning from his own personal experience as from that of preceding dollar diplomats, Reinsch continued to search with all the luck of a Diogenes for a financier with a sufficiently enlightened view of the American role in Asia.[50]

American diplomats and policy makers understood Chinese economic nationalists no better than they did American businessmen. Officials tended to regard economic nationalism (a term they never used) as an intolerable atavism or a passing delirium that Chinese were subject to. The powers, they felt, had only to exercise a strong steadying and restraining hand to insulate China's development and foreign enterprise from this form of disruption.[51] These officials would have come closer to the mark had they regarded Chinese interference with foreign trade and investment not as an aberration peculiar to a "backward people" but rather as comparable in many ways to the spirit of protectionism that gripped even "civilized" countries.

To its citizens it was clear that China labored under the peculiar burden of the unequal treaties guaranteeing foreigners fixed low tariffs, exemption from some taxes, and freedom from Chinese legal control. To right the balance, scholar-officials looked to treaty revision to restore impaired sovereign rights and regain formal control over the most dynamic sectors of the economy. But until that day arrived they pursued a variety of strategies to block, blunt, or exclude foreign economic enterprise. To offset the foreign presence, the central government started taking steps in the 1890s to promote competing Chinese industry and commerce.[52] Where open competition alone was not enough to beat back foreign influence, economic nationalists resorted to buying out foreign concessions and harassing and obstructing foreign firms.

The campaign of economic nationalists repeatedly impinged on American interests. For example, American cotton goods encountered competition from the Chinese textile industry, which concentrated in the area of American strength—low quality fabric— and which shared with the Japanese distinct advantages in labor costs and proximity and sensitivity to the market.[53] But nationalists generally regarded not foreign imports, but foreign investments— popularly associated with political intrigues, spheres of influence, and economic exploitation—as the major economic threat to China.

Mining interests were the first category of American investment to feel the pinch. Official mining regulations, first issued in 1898 and repeatedly revised thereafter, required that foreigners accept joint operations (with the minimum Chinese share increased from 30 percent in 1898 to 50 percent in 1914), and they provided for an increasing Chinese role in administration of the mines, official review of foreign mining loans, and taxation on mining operations. Violators stood to lose their concessions.[54] The Roosevelt, Taft, and Wilson administrations indicted these regulations as an "absolutely unreasonable" obstacle thrown in the way of foreign enterprise, as a blow to China's development inflicted by ignorant officials oblivious to the need for intelligent foreign management, and in extremity as a blatant violation of treaty rights.[55] But a steady barrage of criticism and protest did not prevent economic nationalists by 1910 from completely eliminating American mining interests, once concentrated in North China and Manchuria.[56]

Opponents of foreign investment scored a second victory in ousting Americans from the Hankow-Canton railroad. Public protest against the Belgian takeover appeared in Hunan (subsequently the focus of agitation) in May 1903. A year later, after Leopold had shown his hand, the movement became a broadly popular one, spurred on by Chang Chih-tung, the governor general of Hunan-Hupei and the former sponsor of the American concession, and backed by other provincial and several senior metropolitan officials. The gentry, merchants, and students of the three provinces through which the railway was to run and the new patriotic press charged that the sellout to the Belgians was a serious betrayal of China, for it put a strategic line under French and Russian diplomatic influence. Some feared central China would go the way of Manchuria, where a major railway under the Russians had proven deadly to Chinese control. Others resented the inept and cavalier performance of the company's American employees.[57] By September 1904 the protestors had forced a halt to work on the line, and by November had convinced the central government to give Chang carte blanche to settle the dispute. Although by then control of the railway was back in American hands, Chang rejected continued American involvement of any sort on the grounds that current popular feeling, as intense (he shrewdly contended) as the old anti-Chinese agitation in the United States, gave him no choice. He then went on to arrange the repurchase, rejecting the antagonistic course that some had proposed of simply voiding the concession.[58]

Initiatives in dollar diplomacy after 1909 revived Chinese opposition to American investments. The growing agitation felt by Liang Ch'i-ch'ao and other reformers soon came to be widely shared. By 1910 and 1911 officials responsible for China's foreign policy were bemoaning the broadening American accord with the consortium powers over control of foreign loans. Revolutionaries, toward whom Liang in his disenchantment with the Regent's government had moved, in turn endorsed Liang's views of the dangers of China falling (in Huang Hsing's words) to "invisible financial control by foreign powers," and they echoed Liang's charge that the United States had embraced (again according to Huang) the cause of "Chinese independence and integrity in order to monopolize foreign loans." In the provinces nationalists, fearful like Liang of foreign economic penetration, attacked Peking for selling out to the powers. The Hukuang railway loan, of which the Taft administration had claimed a share, sparked provincial resistance that was to contribute to the fall of the dynasty, while a concession to search for oil in Shensi province, granted to Standard Oil in 1914, inflamed opinion there, in Hunan, and in three other provinces.[59]

In contrast to the failure of investment schemes freighted with political possibilities attractive to Washington but alarming to Chinese, two American market-oriented multinational corporations enjoyed a singular success.[60] They proved that even in a politicized China market a good and growing business was to be had—without resort to the active diplomatic support that American economic expansionists advocated and that Chinese economic nationalists feared. The two firms, Standard Oil and British-American Tobacco (BAT), depended, to begin with, on a competitive product delivered by a well-financed, aggressive, and innovative marketing system. Beyond that, they devoted themselves to patient accommodation to foreign conditions and tastes (rather than a self-conscious dedication to transforming them), avoidance of collision with economic nationalists, and maintenance of a prudent distance from the ministrations of solicitous American officials with other ends in view. Rather than taking for granted the possibly exaggerated estimates of China market enthusiasts, these firms came to their own conclusions, relying heavily for intelligence not on consular reports but on their own agents in the field.

At the heart of Standard's successful strategy was its marketing system, built up between 1905 and 1925 at a cost of $43 million and consisting of six major regional offices, eighteen subagencies, and a widespread distributing network composed of Standard's own Chinese agents and local Chinese merchants. While the agents handled sales from company warehouses in the interior, the merchants, selected by Standard and guaranteed by other wealthy merchants, sold to the ultimate consumer on a commission basis. Standard relied on these Chinese merchants much as the missionaries did on their Chinese assistants in order to build and serve far-flung interests. Through this marketing system passed over half the kerosene sold in China in the decade down to 1910, amounting that year to 96 million gallons, or roughly 15 percent of Standard's sales outside North America and 2.5 percent of Standard's annual domestic refinery capacity in those years. Standard made further substantial headway in its sales in China before the World War disrupted its supply lines, and then resumed its sales momentum in the 1920s.[61]

When BAT entered the China market at the turn of the century, it began almost at once to repeat the pattern of well-financed and aggressive marketing that Standard had arrived at through a decade and a half of piecemeal efforts. The vision and initiative for the enterprise had come from James B. Duke, head of the American Tobacco Company, and James A. Thomas, his lieutenant in China. They had decided that China was a promising market, making up in size what it lacked in per capita wealth, whereas in the United States, Duke had concluded, he had little prospect of expansion. He turned his projected China venture to good account by forming an alliance in 1902 with the British firm, the Imperial Tobacco Company, his chief rival in foreign markets. Organized into what was known as BAT, they were to operate together on a global scale with China its major market. Duke held two-thirds of its stock, and for its first two decades he directed its worldwide operations. Its agents, well-paid and well-trained foreigners, operated across the empire. Working through its Chinese employees and merchant associates, BAT employed innovative advertising techniques and pushed its sales along traditional Chinese channels of trade. BAT was, in addition, to take its manufacturing operations right into China, where taxes and wages were low and major rivals nowhere in sight, and to induce Chinese peasants to produce quality tobacco to meet the company's production needs. By 1915 BAT had investments in China totaling $16.6 million and had become one of its two largest industrial employers. Initial production in 1902 of 1.25 billion cigarettes had climbed by 1916 to 12 billion. On sales of $20.75 million in that latter year BAT cleared a profit of $3.75 million. For BAT as for Standard, the China market was no empty myth.[62]

The cases of Standard and BAT support the proposition that economic nationalists were less aroused by market-oriented firms than by concession hunters with real or imagined political designs. The main threat to the BAT came from the Nanyang Brothers Tobacco Company. This Chinese firm got its start during the anti-

American boycott of 1905 and emerged as a serious competitor during World War I. The BAT fought back with price wars, with propaganda playing up Nanyang's Japanese ties, and with an attractive merger offer. Nanyang held out against a foreign takeover on the theory that "it is better to be a chicken's beak than an ox's buttocks," but it was never able to overcome the BAT's economic advantage and shake its commanding position. BAT encountered no other serious opposition. A number of small Chinese cigarette firms sprang up alongside Nanyang at the time of the boycott, but BAT muscled them easily aside. BAT's sales campaigns—a mix of advertising and the distribution of samples—provoked sporadic local hostility, but the frequently made charges by nationalists that foreigners offered cigarettes as a substitute for opium failed to deter those taken by the cheap pleasures of tobacco.

When Standard first arrived on the China scene, it encountered some resistance from Chinese officals sympathetic to complaints that imported kerosene was displacing native vegetable oils as the principal illuminant. In fact, the 1890s and 1900s saw the Chinese erecting an imposing combination of obstacles to foreign kerosene—municipal prohibitions against its use, boycotts by merchant guilds, opposition to the establishment of storage tanks, and illegal taxation. To neutralize this challenge from economic nationalists at the local level, Standard mainly depended on kerosene's price advantage over vegetable oil and the better light the foreign product gave.[63] But where Standard did take direct action, it resorted less and less to diplomatic support and instead increasingly sought its own solutions. For example, when critics singled out its bulk storage tanks not only as a substantial fire hazard but also as a prominent symbol of foreign intrusion, Standard agents generally appear to have bided their time until local passions cooled and then worked out with local authorities some mutually satisfactory agreement. To the marvelous variety of official imposts applied to kerosene, Standard appears simply to have often acquiesced, in part because prolonged negotiations hindered company business and imperiled good will, in part because its Chinese merchant associates were vulnerable to local pressure, especially the official threat of imprisonment.[64]

A paradoxical relationship (distasteful to American reformers in the open door tradition) had come to exist between Standard Oil and BAT on the one hand and China on the other: to conquer the China market the firms had to a degree to surrender to it. This meant not only that their agents had to learn the language and customs of the people, that their products and advertisements had to appeal to Chinese tastes, and that their goods had to travel along established Chinese marketing patterns and through the hands of Chinese merchants. It meant also that they had to be solicitous of "public opinion." This included cultivating good will by contributing flood and famine relief, sponsoring agricultural schools, and waiving indemnity claims for company losses inflicted by "banditry." At the very least they had to avoid unnecessary antagonism, a delicate task as they engaged in fending off both Chinese competitors and the Chinese government's occasional sallies against their privileged position.[65]

The ironic result of a decade and a half of government effort to support American interests in the China market was a line of policy out of touch with those it claimed to serve. Commercial expansionists cried political interference against a cotton goods trade in difficulty for economic reasons. They tried with predictably disappointing results to persuade American financiers to stake out a parcel of China's economy for the future—despite Wall Street's doubts about investing in China and growing Chinese resistance to their doing so. And finally they played only a negligible part in the success story of two U.S. enterprises that had made a reality of the China market dream.[66] Yet Washington (including ultimately the Wilson administration) was no more ready to turn its back on economic foreign policy than it was prepared to repudiate its allied hopes for China's development and American influence.

Missions and a New China

A mission movement still in search of an effective strategy of penetration began in the early twentieth century to digest the lesson familiar to China traders from Russell and Company to Standard Oil—to prosper in China required some accommodation to Chinese conditions. The lesson was hard for missionaries to absorb. While businessmen could profit from China as it was, missionaries had as their central goal China's transformation. But the frustrating slowness of proselytizing and the savaging the missions took in 1900 drove the movement to change its tactics. It diverted an increasing part of its energy and resources from the saving of individual Chinese souls to a program of good works and above all to a patient, broadly-based educational effort. By easing the irritations created by evangelism in the interior, the mission movement hoped to assume a strong and permanent influence in a new China. This hope, however, did not take account of the nationalist resentment against missionary paternalism and against foreign control of mission institutions ostensibly at the service of China.

It was the Boxers who administered the shattering blow to mission security necessary for reassembling the pieces of the mission movement in a new pattern. The disaster that struck in the summer of 1900 disrupted work, destroyed property throughout the empire, and left 32 Americans (including 8 children) dead along with 157 Protestants from other countries. In Pao-ting alone 13 other Americans fell alongside Pitkin and Gould. Everywhere uncounted numbers of converts were executed. The American Board mission operating out of Peking lost 170 converts in the countryside (53 of 65 members of one congregation), while out of a group of 240 Chinese Presbyterians near T'ung-chou, 171 were killed. The survivors who emerged from hiding at the end of the summer were destitute, demoralized, leaderless, and in some cases fearful of returning home.[67]

Once relief arrived in August, those surviving missionaries still able to function set to work securing the funds and provisions urgently needed by their scattered and homeless flock. With foreign troops in the ascendance, it proved a surprisingly easy and straightforward task. In Peking, T'ung-chou, and Pao-ting missionaries seized housing as they wished, sometimes buying it at bargain prices from Chinese compromised by Boxer connections. They extorted foodstuffs and auctioned off loot to foreign memento-seekers. The bolder toured the countryside, under military escort where possible, demanding that village elders or magistrates settle losses suffered by converts or face retribution.[68]

But merely to settle accounts and restore the status quo after "this awful insurrection against foreign benevolence" (to use the words of one missionary) was not enough. Reverend D. Z. Sheffield, the missionary-educator from T'ung-chou, gave voice to the pervasive conviction among his colleagues that China had arrived at a crucial juncture and that only a policy of force would set her in the right direction. In July, on his way back to China from home leave, he had wanted the powers to establish a protectorate over China and impose reform. "The Chinese must be taught as never before the power of outside nations." In August after the siege, Sheffield's vision brightened with the hope that "this upheaval will prove to have been the plowing in preparation for wider sowing and fuller harvests." By November he observed confidently that "a New China is to rise out of all this sorrow and confusion, and our work is to proceed with more rapid progress and along broader lines." In keeping with Sheffield's views, American missionaries in Peking not only reminded Washington immediately after the siege of its duty to arrange the usual indemnities and assurances for the future, but beyond that urged cooperation with Britain in imposing on China educational and legal reform.[69]

American missionaries had gone into the crisis already resentful of Washington's tendency to subordinate their needs to those of commercial interests. The failure of mission warnings to stir Washington to action against the Boxers deepened their sensitivity to this neglect and fueled fears through the winter of 1900–1901 that Washington would follow the old pattern of minimizing "our vast interests at stake." Those fears were soon confirmed by an official policy that left converts and mission helpers in the lurch (by ignoring past losses and the need for future protection) and that failed to adopt a sufficiently punitive policy toward Chinese officials and areas where American mission losses were especially heavy. Reverend W. S. Ament, soon to become the focus of a controversy over "missionary looting," put the key proposition clearly: "The American policy of treating Asiatic nations as though they were civilized human beings does not meet with the approval of those who know the situation. In the Philippines as here, kind treatment is taken for weakness, and to punish less than the circumstances call for only enhances the evil which you seek to cure." A prominent North China missionary, Elwood Tewksbury, reached for the same comparison in calling for armed might to bring backward Asians into line. "American policy is criminally weak both in the Philippines and China. Forgiveness of crime—moral suasion—demands an educated intelligence to understand, and this does not exist plentifully among the heathen natives here!"[70]

By degrees it became clear to the China missionaries that not just the McKinley administration and its agents in China but also to a significant degree the mission boards and public opinion as well were out of sympathy with them. Before the Boxer year was out the *New York Times, Arena, Nation,* and *San Francisco Call* had criticized the missions' dependence on force. The issue finally gained national attention when the *New York Sun*'s correspondent in Peking sent in a dispatch describing looting by Ament and his followers. (Ament had allegedly exacted $58,000 from forty villages in the vicinity of Peking.) Reading of Ament's misconduct, inauspiciously on Christmas Eve, Mark Twain immediately fashioned this information into a devastating indictment of "the Blessings-of-Civilization Trust." He disingenuously asked, "Is it, perhaps, possible that there are two kinds of Civilization—one for home consumption and one for the heathen market?" Ament and other missionaries such as Gilbert Reid and W. A. P. Martin defended "looting" as an expedient that men who led "lives of uprightness, honesty, and generosity" had to resort to to get missions through the post-Boxer emergency. If these men seemed too harsh on the Chinese, it would have to be written off to their "spirit of magnificent enterprise and the enthusiasm of conquests in new regions." Twain, unimpressed, proceeded to congratulate the missionaries for their revision of the Ten Commandments: *"Thou shalt not steal—except when it is the custom of the country."*[71]

Though leaders of the mission boards could in private call for the "most careful inquisition and righteous punishment" of Chinese responsible for the recent "atrocity," their public pronouncements revealed their awareness that an aggressive mission policy would fail of popular support and thus imperil the funds and recruits that made any long-range effort possible. Meeting in an emergency session in late September 1900, the representatives of seventeen major missionary societies predictably eulogized the martyred—"the soil of China has been forever consecrated by the blood of God's saints"—and promised to "seize the coming strategic opportunity to win China for Christ." But they also frankly conceded that "at home the expediency of the whole missionary enterprise is being challenged" and sources of funds were closing off. Aggressiveness was needed not in China but at home to rekindle popular support for an already debt-ridden mission movement. For China the new watchword was prudence. Only by "sensible" conduct could the China missionaries avoid embittering the Chinese and giving ammunition to their domestic critics.[72]

Confronted by adverse publicity, by the home boards' counsel of caution, and ultimately by the danger of losing the indispensable base of support in the United States, the initial sense of missionary outrage gave way to a mood of reappraisal. Ament himself found on his home leave in 1901 that pastors in Michigan viewed the mission enterprise as an "alien element." Shocked, he quietly bowed to Judson Smith's injunction of silence on the issue of loot. Missionaries in Peking, who had earlier called Twain's charges libelous, now in April conceded such errors on the part of converts as looting, extortion, recantation, and acts of revenge. Missionaries were also moved by a newly dawning appreciation of the inherent risks of an aggressive evangelism in the interior. Not even a more forceful U.S. government policy could protect isolated mission stations from hostile neighbors who might in several days of riot undo decades of patient and courageous work. This new sensitivity to the counterproductive nature of compulsion and to the dangers of deracinated converts at odds with those about them carried missionaries toward the logical conclusion that some degree of coexistence with China was essential to the security and growth of their cause.[73]

A keener sense of the perils of evangelism went hand in hand with a heightened appreciation for education as the key to winning China, perhaps slower but also surer and safer. Advocates of education, long vocal in the foreign service, gradually won a respected place for themselves in the mission movement. The career of W. A. P. Martin, one of an Anglo-American band of educational freelancers, at once epitomized the missionary search for an effective instrument of conversion and the growing interest in education as that instrument. Although Martin began in China doing evangelical work in Ningpo, he later moved into diplomacy, and finally—on the basis of his Chinese scholarship and his open if not precisely sympathetic attitude toward Chinese culture—he gained an appointment in 1869 as head of the government school for translators in Peking. As adviser in foreign affairs and translator of important Western works (preeminently his 1864 version of Wheaton's *Elements of International Law*), Martin acted on the more promising strategy that to capture Peking for Christianity was to capture the empire and that to win the elite was to win the masses, a view that followed logically from the widespread missionary conviction that popular resistance rested on a bedrock of gentry-official hostility.[74]

The devotion of this group of missionaries to secular education was strengthened by the tendency, common among foreigners, to see the Chinese as children, who under proper tutelage could be carried to a modern—and implicitly a Christian—perspective. Isaac Taylor Headland, a missionary teaching under Martin in Peking, caught the essentials of this view in describing in early 1900 the Kwang-hsü Emperor's gradual conversion to Western ways and reform, first through exposure to toys and inventions, then through the Bible and religious tracts, and finally through scholarly works on science, political economy, and international law. His progress recapitulated the process all Chinese would have to go through.

Once men of intelligence and power were brought by these stages to a full grasp of the spiritual and material superiority of the West, then their influence could be turned, in the words of Gilbert Reid, "unto the evangelization of their fellow-countrymen."[75]

In the nineteenth century the educational freelancers had occupied only a marginal place in the mission movement. Calvin Mateer and other educationalists had contended that to neglect the full range of modern learning would be in effect to renounce the moral mission influence over the Chinese destined to overthrow conservatism, develop their country, and modernize its government policies. But the 1877 general conference of China missionaries rejected this view, and the 1890 conference conceded only an auxiliary role to schools.[76]

The old arguments of the educators took on more force in the post-Boxer period. Mission training of China's youth, not yet entrapped by tradition, would in time secure China's cultural transformation without foreign intrusion or resulting native resentment. Such an effort would also demonstrate the mission movement's altruism and good will and thereby silence its Chinese critics. On a psychological level the shift to a gradual and long-range process of remolding Chinese culture and away from an intense effort at winning individual Chinese had the marked advantage of freeing missionaries from the pressure and frustrations of counting souls. Moreover, the educators could claim to be in step with the social gospel, a coherent doctrinal position in the United States by the 1890s and increasingly influential in the church. A mission program premised on the need for social transformation as a precondition for individual conversion would have greater appeal at home, no small matter for board secretaries struggling to keep up staff and budgets. The new emphasis accorded with the progressive faith in the potency of education. If proper training could remake immigrant youth in American cities, why not the youth of China? Finally, the proponents of secular education had the advantage of being able to point to sound institutional foundations on which they could build. By 1914 less than half of Protestant missionaries would be engaged in direct evangelism, and American missionaries would have staked out education as their special preserve.[77]

The new educational emphasis brought changes in mission work on several fronts. The focus of mission effort shifted from the countryside to the city. There future Chinese leaders were to be recruited; trained in the subjects necessary to prepare graduates to assume a role in government, civic affairs, and business; and turned back to their urban environment where the hope for modernized China was strongest. Further, English came into increasing use (in some schools for all subjects but Chinese studies)—because of the interest of urbanized Chinese as well as the educators' lack of fluency and the inadequacy of texts in Chinese. Finally, the increasing concentration on education put a premium on attracting American college students to the missionary field. Since young Chinese were thought to be most susceptible to the example and appeal of their American contemporaries, the mission movement had to reach out for the "brightest and best of our countrymen." The Student Volunteer Movement, which had brought young Pitkin and other college-educated men to China in the 1890s, now turned to meet this educational need, alone supplying 2,500 young Americans by 1919. So too did the Young People's Mission, established in 1902 as part of the boards' post-Boxer campaign for support. These trends toward urban service and recruitment of well-educated youth expressed themselves in the YMCA. Soon after its establishment in the United States, it extended its operations to China, first Tientsin in 1895 and then Shanghai in 1900. In the post-Boxer decade it expanded both its offices and its staff of recent college graduates devoted to urban reform and social services. It sponsored literacy campaigns, attacked problems of public health, worked with opium addicts, financed social surveys, and provided recreation and residences.[78]

Just as the missionaries sought new approaches in China to avoid provocation, so too did the Chinese seek to ease the earlier overt and often violent hostility to missions. Nineteen hundred had taught Peking as well as missionaries the dangers of confrontation, and thereafter the Chinese government made a concerted effort to prevent any recurrence of anti-Christian activity. The result was an unprecedented level of security for all missionaries and above all for evangelists still working in the interior. Where attacks did occur, Chinese officials moved quickly to suppress them and to reach a settlement of accounts with the injured missionaries.[79]

The responsiveness of Chinese to the new educational and humanitarian orientation of the mission movement was equally heartening. The intense interest which urban youth and intellectuals took in Western learning, if not religion, led them to treat missionaries, at least in their guise as educators, with a new-found respect. Enrollment in mission schools after 1900 burgeoned. Students at the college level, only 10 in 1873 and 164 in 1900, flocked in thereafter so that by 1915 thirty-three institutions boasted an enrollment of over

2,000 men and women (to increase by another 1,500 over the next decade). Soon the government itself was implementing programs that coincided with reforms long favored by missionaries and that seemed to promise greater openness in the future to the mission message. The government drastically revamped the educational program, sending students to study abroad, introducing a new curriculum in the schools, and in 1905 abolishing the old exam system (to the delight of missionary observers). Footbinding, long a symbol to missionaries of the brutal subordination of women in Chinese society, came under official criticism, and by 1906 Chinese had displaced foreigners as the leaders of the anti-footbinding societies. Finally, the Chinese government took up opium suppression, much to the gratification of a mission movement long hostile to the imported drug.[80]

The change in expectations which Chinese-sponsored reform brought about among missionaries is reflected in the changing views of one of their most influential spokesman, Arthur H. Smith. The antimissionary violence of the late 1890s had served only to inject into his writings, already critical and condescending toward the Chinese, a new note of pessimism about China's future and impatience with American policy. In the wake of the Boxer Uprising, he had registered all the familiar warnings against a weak China policy and defended his colleagues against charges of looting. But by 1907, when his popular *China and America To-day* and his mission handbook *The Uplift of China* appeared in print, Smith had made a volte-face more remarkable even than Rockhill's at the same time. He saw China at last advancing on a wide range of fronts—from penal reform to railway construction—and displaying "the germs of real patriotism." The Chinese people suddenly acquired a "galaxy of race traits" which "outfitted [them] for the future as no other [people] now is, or perhaps ever has been," while the Chinese government appeared a work of genius, "adroitly contrived to combine stability with flexibility, apparent absolutism and essential democracy." This pioneering evangelist and former advocate of force now urged policymakers to take a sympathetic attitude toward Chinese nationalism. He berated his countrymen for their racial intolerence, their ignorance of China, and their shortsighted neglect of opportunities to build up their influence. And he held up to them a "long catalogue of our crimes against China," including prominently the persecution of Chinese immigrants, before which "the worst Boxer atrocities in China" paled in comparison.[81]

Mission hopes for a new democratic and Christian China rose

to unprecedented (but only briefly sustained) heights with the outbreak of the 1911 revolution. "Now the people are awake, unprogressive Manchu rulers have been given notice to leave." Of all American observers, missionaries responded earliest and most fervently. They took credit for laying the groundwork for the modern China now arising before them and looked for greater opportunities to contribute to "the new civilization." China's new leaders, they noted with satisfaction, included Christians such as Sun Yat-sen trained in mission schools and on familiar terms with missionaries, while Yüan Shih-k'ai—hailed by Arthur J. Brown as "the ablest living Chinese" and still remembered for his suppression of the Boxers and protection of missionaries in Shantung—promised toleration and appealed for Christian prayers for the new republic. In their enthusiasm missionaries and mission bodies supplemented prayer with old-fashioned lobbying in Washington to secure prompt diplomatic recognition and to put a halt to the vicious and exploitative policy of dollar diplomacy.[82]

The decade and a half after the Boxer rising marked the golden age of the China mission movement. Evangelists enjoyed unprecedented safety, and the educational enterprise prospered. In 1915 between 2,000 and 2,500 Americans served in China under the sponsorship of over forty different societies. These Americans had in their charge property worth something in the range of $25 million, directed some 9,600 Chinese assistants, superintended a native church with 120,000 members, and provided schooling for 93,000 Chinese at all levels. By 1916 American mission societies were sending funds to China at an annual rate of about $3.5 million. This expanded and intrinsically more costly mission program required a larger and more specialized supporting staff at home to do the accounting, raise funds, organize campaigns and conferences, prepare and distribute publications, and maintain liaison with a proliferating number of mission boards and umbrella organizations. The new mission bureaucracy produced its own statesmen, such as John R. Mott and Sherwood Eddy; its own rhetoric, less militant and more attuned to the business vocabulary of careful planning and efficient operations; and its own special instrument, the Laymen's Missionary Movement (established in 1907), for tapping the coffers of businessmen.[83]

However, serious problems lay ahead, the most deadly of which was that posed by Chinese nationalism. Missionaries in the early twentieth century were coming to regard this movement with hope. At the same time they could neither restrain their paternalism nor abandon their intrusive role in Chinese life sufficiently to neutralize the antagonism that nationalists felt toward foreign penetration of all sorts—and that foreshadowed the eventual demise of all foreign mission work in China.

Some mission statesmen, such as Arthur J. Brown of the Presbyterian mission board, argued in broad terms for conciliation with "the growing Asiatic spirit of self-consciousness and independence," and demonstrated an emerging awareness of the risks of political intervention. At the September 1900 inter-board conference, Brown had vowed that Presbyterians would not call on the state for protection "unless absolutely necessary." In 1904, after an on-the-scene investigation in China, he publicly argued that missionary intrusion into legal cases involving converts was counterproductive, that attempts by missionaries to guide diplomats better informed on the "delicate" situation in China were misguided, and that cries for diplomatic protection hurt the mission cause at home and hardened Chinese prejudices. Three years later his board accepted Brown's position discouraging (but not absolutely ruling out) legal interference or diplomatic appeals.[84]

Most of Brown's colleagues would not, however, go this far. The heads of the other boards had turned aside his watered-down call for renouncing state protection in 1900. When the 1903 mission conference in China embraced the principle of noninterference in Chinese judicial affairs, it did so only in keeping with the terms of the 1903 treaty; and even then missionaries still reserved the right to make representations where they thought converts suffered from local intolerance. At the major conference held in 1907 Calvin Mateer, head of the committee on the missionary and "public questions," accepted the case for restraint in matters of treaty rights and cultivation of Chinese good will already made by Brown, the head of his board. Yet at the same time Mateer conceded that only intervention by missionaries would secure converts from persecution. To refuse assistance would be "to sunder the tie that binds the convert to his teacher." The final resolution balancing the dangers of intervention in local affairs against the need to protect converts still left much to the individual missionary's discretion—and his ability to withstand the pressure his congregation exerted on him.[85]

Whatever the concessions to Chinese sensitivities that missionaries were prepared to make in theory, they in practice continued to conduct themselves in a way repugnant to Chinese nationalists. The missionaries themselves still hid behind the treaties and diplomatic protection. They still controlled the native church, guiding it spiritually, sustaining it financially (by supplying about two-thirds of the budget for most congregations), and protecting it against local abuse. The missionaries remained generals in command of their carefully subordinated army of native helpers. Those helpers were excluded from major mission gatherings down to 1907, and even into the 1920s they were held back from leading positions of authority for fear that the deep vein of selfishness and superstition thought lurking in the Chinese character might break through, overwhelm long Christian training, and undo the patient effort and sacrifice by generations of missionaries. The same sense of paternalism, caution, and foreign superiority also led missionaries to hold on to control of the new system of educational institutions. Trustees in the United States insisted that by providing the financial lifeblood for schools no less than churches they were entitled to define policy, thus reinforcing the grip of the local missionary-educators. Chinese faculty long remained second-class citizens, while Chinese administrators either functioned as junior associates or discovered their power was only nominal.[86]

The drift toward accommodation, already loaded with reservations and contradictions by the missionaries themselves, was further complicated by diplomats who vacillated between giving priority to preserving treaty rights (by resort to coercion if necessary) and preventing any missionary indiscretions that might lead to trouble. So at times when missionaries might favor concessions to the Chinese, diplomats might respond with a stirring defense of treaty rights as the legal foundation of American influence and throw back at the missionaries their own iron law about the Chinese only respecting force. On other occasions, when some missionary might blunder into trouble, his call for help might elicit, from a diplomat anxious to avoid needless altercation with the Chinese and doubtful of missionary prudence, a call for caution. Sometimes, when caught between what they perceived as objectionably aggressive mission claims and their own avowed devotion to a broad reading of treaty rights, the legation and State Department sought a way out by fuzzing the issue. Frequently this happened in the case of converts embroiled in legal trouble or of property disputes, often brought on by missionaries making speculative purchases on their own account or registering title under the name of a convert.[87]

The massacre at Lien-chou in Kwangtung in 1905, the major incident of the decade, shows mission policy at its most contradictory and irritating.[88] The case had all the classic signs of missionary imprudence: a long-smoldering property dispute, missionary interference in a local religious celebration for the second year in a row and, finally, once local passions were high, the incriminating discovery of a human fetus preserved in alcohol (presumably only an innocent if ill-chosen prop in a medical missionary's office). Five Presbyterian missionaries were killed and the mission buildings were destroyed. Though missionaries had brought the trouble on their own heads (as newspaper accounts at home pointed out), Brown publicly countered that the fault for the attack lay with bad elements among the Chinese, pervasive antiforeign feelings unconnected with mission work, and irritation throughout Kwangtung over American exclusion policy. He nonetheless held to his new conciliatory line by waiving his board's claim to any indemnity. The Canton consul agreed with the wisdom of the step, since he guessed that the money would come from the "wretchedly poor" of Lien-chou who bore no responsibility for the massacre. In that case, "the life of a missionary at Lienchou will not be a bed of roses."

The State Department and the legation would not, however, let the incident rest. They went on to advance sternly and without hesitation the full set of conventional demands appropriate to "aggravated cases" of this sort: indemnity ($39,000), punishment, and assurances for the future. For good measure they added the destruction of the local temple whose "pagan" celebrations had offended the missionaries. Unless the Chinese quickly put matters right, Rockhill warned, they must expect "the most unfortunate consequences." Not until a full year and a half after the incident did Washington get around to quietly urging missionaries in Kwangtung "to exercise the greatest possible tact and circumspection and to act with the most kind and broad-minded consideration for the people amongst whom [you] live at all times avoiding everything that might cause irritation, wound pride, or arouse prejudice or resentment."

The Chinese government had been trying since the 1860s to deal with this irritating and sometimes provocative foreign missionary presence.[89] The Tientsin massacre in 1870, the missionary incidents of the 1890s, and the recent Boxer fiasco had further impressed on officials the need to substitute for ad hoc local management of mission problems a general policy that would above all else secure authority over converts and block missionary interference in local affairs, thereby saving Peking from being again caught between popular protest and foreign pressure. Officials had also realized the need for regulations requiring registration of missionaries traveling in the interior on legitimate business, ruling out land purchases that might arouse popular anger or that lacked official sanction, and generally enjoining a respect in their outward, public behavior for Chinese institutions and customs. But each of the early attempts to impose a policy had failed of support among the powers while arousing those officials who favored the complete exclusion of mission influence in place of mere regulation.

In the post-Boxer period, officials resumed the campaign to wring from the powers an explicit statement ruling out missionary interference in judicial and administrative matters and acknowledging that converts were fully subject to Chinese law. The goal was partially realized, as far as the Anglo-American mission movement was concerned, in the 1903 treaties, but Chinese officials continued to seek ways to control miscreant converts and restrict the activities of foreign missionaries. The most effective way to achieve the latter goal was to strip missionaries of the mantle of immunity from Chinese authority provided by the unequal treaties. Wu T'ing-fang in an 1898 memorial had set the problem of controlling this determined and sometimes imprudent group in this larger context. Only by exercising jurisdiction over resident foreigners generally would it be possible to prevent reckless behavior and minimize tensions. Japan had regained control in the face of foreign objections, Wu observed, and he concluded that China could do the same through a ten- to twenty-year program of legal reform. Peking finally embraced Wu's proposal in 1903, when it began revisions of the judicial system and pressed him into the task.[90]

The rise of Chinese nationalism accelerated the campaign against missionary influence. It was in the schools, in a largely urban context, that the conflict between missions and Chinese nationalism began by the early twentieth century to assume its main focus. A government effort to neutralize mission educational influence—for want of direct control—represented the more active and effective Chinese challenge in these years. Textbooks prepared by missionaries disappeared from the curriculum, while the missionaries themselves, also once welcome in government schools, were excluded from them after 1900.[91] Peking simultaneously extended its own system of higher education to replace the defunct exam system, and sought to make its schools competitive with mission institutions. Government-sponsored colleges in Peking, Shanghai, and Tientsin were made tuition-free, unlike those run by missionaries, and graduates of government schools received preference in filling official posts. Missionary educators responded that their institutions were "doing better work than China will be able to do for herself for some years," and with Rockhill's backing tried to combat what they saw as an attempt "to substitute a Confucian for a Christian curriculum." To secure equal footing for their graduates who wished to enter government service, they were willing to incorporate some officially approved curricula, but their firm opposition to official inspection or an imposed general course of study left them at loggerheads with the education ministry in the last years of the Ch'ing and no less during the early Republic.[92]

The other facet of the pre-1914 challenge to mission schools was to be found in incipient student activism. Chinese students, who sought in mission schools a knowledge of the Western technology and political institutions thought necessary to save China, understandably expected mission institutions, as any other in China, to serve nationalist ends or suffer repudiation. Increasingly, then, mission schools—whose basic policies, essential funding, and senior staff came from abroad—occupied a problematic position. Already before 1914 students were agitating and sometimes going on strike to secure course changes, win free time during national holidays, and eliminate religious requirements such as mandatory chapel. Through the late teens and the twenties discontent with the mission enterprise would spread widely among the very students whom missionaries were counting on to save China. The battleground was no longer the countryside but the once secure cities where the cutting edge of uplift was supposed to have worked with greatest effect. Student activists now charged school administrators with trying to suppress patriotism and preserve an apolitical atmosphere on campus. They resurrected old charges that Christianity was an irrational doctrine and that the mission movement generally served as an instrument of foreign domination. By the 1920s missionaries, who only a decade earlier had glimpsed victory, were again on the defensive.[93]

Epilogue

The Special Relationship in Historical Perspective

THE IDEA OF a special Sino-American relationship has a long history. It was propagated in the nineteenth century by Americans in China and incorporated in the admiring view of the United States entertained by some Chinese. By the early twentieth century the concept of a special relationship was coming under heavy fire from Chinese nationalists, but among Americans the idea had become a staple of both popular and official rhetoric, to persist to mid-century and beyond. Through times of trial in China brought on by renewed foreign aggression or internal unrest, Americans held to the reassuring myth of a golden age of friendship engendered by altruistic American aid and rewarded by ample Chinese gratitude.

This view remained central to the message of distinguished popular experts on China, heirs to a tradition of reporting that went back to S. Wells Williams, Arthur H. Smith, and W. W. Rockhill. Pearl Buck, a missionary's daughter whose views were in public favor in the 1930s and 1940s, repeatedly invoked the myth in calling Americans to play their role of protector of the Chinese people against Japanese militarists and then Chinese Communists. Later, in the 1950s and 1960s, John K. Fairbank, distinguished academic and influential interpreter, celebrated the "traditional friendship" between the two countries (even as he deplored past American misunderstanding of China). His *The United States and China*, for at least two decades the standard general introduction to China and American China policy, contended that good works and good will had created a reservoir of "pro-American" feeling that the Communists (creators of a "monster" system filled with "profound evils") were trying to smother under propagandistic charges of past American interference and abuse.[1]

As a fixed ornament of China policy, the rhetoric of the special relationship flourished in official cold war apologetics. Even as the United States and China drifted apart in 1949 and 1950, Dean

Acheson pointed to an unsullied record of American assistance going back to the previous century ("our historic policy of friendship for China") and to the good will ordinary Chinese still felt for the United States. While Chinese and Americans killed each other in Korea, he contended that a "deep interest in and friendship for the Chinese people" was still a central pillar of China policy. Acheson's successors as Secretary of State continued to invoke the conventional formulas—the "long history of cooperative friendship" (John Foster Dulles) and "the historic ties of friendship" (Dean Rusk)—in their major statements on China policy well into the 1960s.[2]

If Sino-American contacts did indeed constitute a special relationship, it was surely in a sense far different from that of the common myth—with its intimation of some intuitive understanding between the two sides, out of which grew mutually beneficial contacts. A careful scrutiny of the fabled relationship suggests, on the contrary, that what was "special" was the degree to which two distinctly different and widely separated peoples became locked in conflict, the victims in some measure of their own misperceptions and myths.

Chinese traveling to the "mountain of gold" filled a need of the developing U.S. economy but also aroused an intense hostility. Although the immigrant was far less aggressive than Americans in China, his color and his resistance to assimilation spawned a powerful nativist movement. American nativists secured laws to block new immigration, expelled many Chinese residents, and left the remnants isolated and beleaguered by continuing popular violence and an imagery of Chinese inscrutability, filth, and vice. For these Chinese, American friendship took a peculiar form—that of opportunity denied, mixed with abuse and racial antipathy.

If Chinese nativists failed to match the accomplishments of their American counterparts, it was not for want of effort. A general hostility toward foreigners tended to focus more and more through the nineteenth century on the missionary as the most culturally intrusive and also the most vulnerable of the foreign invaders. Thereafter nationalists, casting international relations in terms of economic imperialism, began bringing businessmen, financiers, and their diplomatic agents within the circle of suspicion as well. But against popular violence and official harassment, against Boxers and advocates of rights recovery, the open door constituency generally managed to hold its ground, a triumph attributable not so much to its good intentions or good will as to the effectiveness of the powers in making good the claims of Americans and other foreigners to the treaty privileges wrested from China.

Commonly employed rhetoric and stereotypes offer a superficial measure of the hostility and tensions Chinese-American contacts gave rise to. Repeated American reference to the need to use force in dealing with China found its complement in a pervasive Chinese preoccupation with foreign abuse and intimidation. American nativists, no less than the authors of anti-Christian literature and nationalists alarmed by China's abasement, constructed a set of powerful and dehumanizing images, many of them drawn from earlier contacts with other peoples. Americans easily transferred the supposed incapacity and backwardness of blacks, Latinos, and Indians to the Chinese. Many Chinese facilely tarred Americans with the attributes of greed, perversity, and lack of reason long associated with other "barbarians."

Chinese immigrants and American missionaries, the two groups whose lives impinged most intimately on the other culture, evoked in the xenophobic imagination strikingly mirrored anxieties over sexual pollution, physical contamination, and the disintegration of the social and political fabric. The supposed proclivity of depraved missionary and immigrant alike to defy sexual taboos and to make use of drugs and potions to seduce unwary women and children and poison the community around them was a centerpiece of xenophobic literature. The mission compound no less than Chinatown was regarded as a hotbed of subversion. The despotism of Chinatown held newcomers back from assimilation, thus challenging American political and social ideals, while missionaries spread heterodoxy among the poor, the socially marginal, and the disaffected. These foreign enclaves became the lightning rods for the problems of the society. On the one side, Chinese served as scapegoats for the failure of California to provide the wealth its white settlers expected, while on the other missionaries drew down the wrath of a people plagued by hardship and unsettled by foreign aggression.

In the often tense dealings between Chinese and Americans, cultural intermediaries played a pivotal role, facilitating access to the other culture and interpreting it to their fellows. The voluntary associations in Chinatown reserved a place among their officers for someone who understood English, perhaps enjoyed some outside political contacts, and could recruit American lawyers and other assistance. American officials charged with control of Chinatown hired Chinese interpreters and informers. Chinese policy makers and their representatives abroad kept close at hand foreign-educated countrymen versed in international relations. The American foreign service had its own intermediaries in such China experts as W. W. Rockhill and the early missionary-diplomats, S. Wells Williams and Peter Parker. Missionaries recruited and trained native assistants to extend their work into the countryside, to make their writings intelligible to the Chinese, and to help in schools and hospitals. Those missionaries with a real command of the language emerged as interpreters in their own right, speaking to their colleagues and in time to Americans at home. Businessmen relied from the beginning on Chinese merchants to give them access to the treaty port trade and the interior market.

All these intermediaries occupied a potentially vulnerable, even perilous position. Foreign affairs experts such as Rockhill or the early Chinese barbarian managers who, if even only to a small degree, stood apart from one culture in order to better interpret the other provoked suspicions that they had "gone foreign." Such intermediaries as mission helpers, who were expected to change as well as interpret their own culture, played the most provocative role and became the most obvious scapegoats when tensions and frustrations grew overcharged. Even Chinese merchants, who could play their intermediary role without directly challenging basic cultural patterns, ran some risk of criticism by the very fact of their profitable association with foreigners.

Out of these cultural contacts a real China policy began to emerge in the United States in the 1890s. As Washington left behind the older views of China simply as an incidental market or the troublesome source of the most irritating of American immigration problems, it took up in their place the open door ideology with its commitment to China's reform and protection against aggression. The new outlook rested on only the flimsiest concrete economic or strategic justification and served the direct interest of a group of Americans too few to fill even a small city. Its appeal, otherwise inexplicable, has to be understood in terms of the vision that both drew from and fed back into the national fantasies of redemption and dominion. As the open door constituency and its domestic allies presented it, China—vast, populous, and teetering between renovation and collapse—held out boundless opportunity to the American expansionist impulse in all its guises.

The natural correlative to the tendency of American policy makers to exaggerate China's importance was the Chinese tendency to mistake the United States as a power whose presumed interest in China might be turned to China's advantage. This approach to the United States was firmly grounded in strategic necessity, evolving out of a growing desperation in the face of European and Japanese encroachments against China and her tribute states. The widespread influence exercised by the early writings of Wei Yüan and Hsü Chi-yü and the relatively quiescent part the United States played in nineteenth-century East Asian international politics made plausible the idea of American concern and benevolence. Now the assumed dependence of the United States on commerce provided the hope that a Chinese appeal for support, properly baited, would succeed. Though some nationalists, fearful of foreign economic domination and resentful over the exclusion policy, challenged the assumption of American benevolence, policy makers kept coming back to the pro-American strategic policy.

This Chinese policy was strong in its potential for advancing common Sino-American interests, but in practice it proved disappointing. Chinese policy makers who believed that Washington shaped policy to the requirements of merchants lured to China by profit were perhaps closest to the mark in the 1840s. Increasingly thereafter, however, policy based on the assumption that Americans could be manipulated by the prospect of trade preference, and later railway concessions and loans, to play the balance of power game in China's favor proved fatally flawed. Chinese policy makers encountered difficulties, first of all, in simply arousing the interest of American capitalists.[3] But even when Americans swallowed the economic bait, they remained unresponsive to Chinese direction. The open door policy, as officials in Peking and the provinces repeatedly had to learn, most often took the form of joint action with other powers or toothless notewriting and steered well clear of intimate arrangements with Chinese leaders, in peculiarly bad odor in Washington fron the 1890s onward. Chinese policy had fallen victim, it seems, to lingering conceptions of barbarian greed, to misapprehensions about the dependence of United States industries on foreign markets, to American rhetoric about the importance of the China market, and perhaps to wishful thinking.

These patterns of perception and interaction in which the United States and China had become locked by the early twentieth century endowed the relationship with a remarkable instability. This instability was accentuated by a tension inherent in the Chinese approach to the United States. The central goal of Chinese nationalists, galled by a century of foreign humiliation, was to bring down the structure of unequal treaties and establish true independence from outside interference and control. This ideal of national autonomy, as applicable to relations with the United States as to those with any other power, was sharply at odds with China's dependence on the United States as a major or even sole potential source of support against blatant territorial aggression. The nature of the Sino-American relationship at any one moment thus depended on whether Chinese policy makers were giving priority to thoroughgoing independence or accepting temporary dependence in the name of national security. The choice in favor of independence, on the one hand, set Chinese demands for equality and reciprocity in foreign relations up against American dreams of influence and uplift and put Sino-American relations in an antagonistic phase. On the other hand, the appearance of some threatening third power would move China toward an accommodation of interests necessary to secure American support; it would evoke in Americans the urge to act as China's savior from foreign aggression; and thus it would suppress for the time being the tensions between American claims as a patron of a modernizing and liberal China and Chinese resistance to foreign interference however benevolent.

The 1920s found the United States and China locked in an antagonistic phase, as Chinese nationalists more sharply than ever attacked the unequal treaties and called for bringing foreign economic and cultural penetration under control. Though Japan and Britain were their main targets, the United States nonetheless attracted fire. T'ang Shao-i, by now an elder statesman, criticized the very open door policy which he had earlier promoted. Chiang Kai-shek, the rising star in Chinese politics, called the Americans a dangerously two-faced people. "The Americans come to us with smiling faces and friendly talk," Chiang observed in 1926, "but in the end your government acts just like the Japanese." Mao Tse-tung, yet an obscure radical, agreed. He had had a view of the United States right out of Hsü Chi-yü until exposure to nationalist literature and Wilson's betrayal at Versailles had shaken his illusion. By the 1920s he had come to count the Americans as one of the "foreign masters" who used "a pretense of 'amity' in order that they may squeeze out more of the fat and blood of the Chinese people," and he castigated those who still entertained, as he once had, "a superstitious faith in America [as] a good friend who helped China."[4]

A group of increasingly influential career American diplomats schooled in the principles of pre-1914 China policy reacted with predictable dismay, even hostility, to this rising tide of Chinese nationalism and the disorder that accompanied its challenge to foreign interests. John V. A. MacMurray, the most outspoken of that group, regarded the intensifying Chinese protest as a kind of "folly," a venting of spleen by humiliated Chinese whipped into a frenzy of "hysterical self-assertion" under the influence of "Bolshevik and juvenile nationalistic influences." Even appearing to make concessions on the major issue of treaty revision was dangerous because, as he kept explaining through the 1920s and into the next decade, the Chinese "normally attach to force a degree of reverence that most Western peoples do not, and they are far quicker to sense a weakness and not merely to exploit it but to let themselves be tempted into reckless and unscrupulous browbeating of anyone who has thus lost their respect and esteem." Nelson T. Johnson and Stanley K. Hornbeck, other China experts in the foreign service, followed the general drift of MacMurray's analysis though they expressed their views in less strident terms.[5]

American policy makers, influenced in part by their advisers and in part by their own inherent assumptions about the proper basis for Sino-American relations, sought to ride out the Chinese challenge to the status quo. Already at the Washington Conference in 1921–1922 Secretary of State Charles Evans Hughes had made a reformed China the prerequisite for putting an end to the unequal treaties. Hughes' successor, Frank B. Kellogg, took the position that the United States would not in any case formalize such concessions until China had committed the other powers to a similar retreat. He suggested that in the meantime discontented Chinese regard the treaty system not as a sinister arrangement selfishly imposed by the powers "but merely as a *modus operandi* intended to remedy a vexatious condition which had for many years proved what seemed an almost insurmountable obstacle to the maintenance of friendly relations." The Hoover administration followed the policy of foot-dragging, insisting that China first become unified, ordered, and modernized before changing the essential terms of Sino-foreign relations.[6]

The revived Japanese menace in the late 1920s began to alter the dynamics of the Sino-American relationship. Japan's aggressive continental policy diverted nationalist hostility from the rest of the powers to herself and launched the Nationalist government of Chiang Kai-shek on the predictable search for American support. No less predictably, American policy makers offered their sympathy for Chinese resistance to foreign aggression. Secretary of State Henry L. Stimson laid down the basis for China policy in the 1930s when he announced after the seizure of Manchuria that the United States would not countenance such an outrage against international order and the open door. By late in the decade the American public overwhelmingly supported China against Japan.[7]

But Chiang Kai-shek was to discover, just as his predecessors had, that sympathy did not necessarily translate into meaningful support or close collaboration. The penchant of the Americans for providing help in their own way and at their own time prevailed in the years before Pearl Harbor and even after. For a time Roosevelt, like Stimson, tried to steer clear of a costly war against Japan. Though eventual American entry into the Pacific war made the United States and China de facto allies, Roosevelt rejected Chiang's proposal for formal alliance and kept him at arm's length throughout the conflict. Chiang himself attended but one of the wartime summits, and his military aides were excluded from secret, high-level interallied planning. China remained low on the list of strategic priorities, and what aid Roosevelt promised arrived little and late or not at all. Roosevelt tried to compensate for his neglect by cultivating Chinese national pride and publicly playing up the "special relationship." He formally recognized China as one of the great powers, promised the return of territory seized by Japan since 1895, and acceded to Chinese requests for a renegotiation of the unequal treaties. But Americans in China retained their privileged status and their predilection for reforming China's government, her army, and even her basic cultural values. General Joseph Stilwell, Roosevelt's representative and one of the military's "China hands" barely hid his contempt for Chiang ("the peanut") and Chiang's regime, while Roosevelt himself tried in 1944 to put Stilwell in control of the troubled Nationalist army. Early the next year at Yalta Roosevelt went over Chiang's head to strike a deal with Stalin over East Asia. Even in the most favorable circumstances a relationship of respect and equality with the United States remained for the Chinese an elusive ideal.[8]

The receding Japanese threat in 1944–1945 did not at once restore to the forefront of Sino-American relations the old unresolved contradiction between nationalism and the open door ideology. For Chiang dependence on the United States, and the humiliations that went with it, continued beyond the war because of his need for American support in his showdown with the Communists. For the Communists as well, the struggle for domestic power was the prime order of business. Mao found himself in the position, like Sun Yat-sen forty years earlier, of trying to assure the most powerful of the

foreigners that the revolution posed no threat, indeed
support as a progressive, popular, democratic movement
very least, Mao could hope, such an appeal would convince the
United States, the unrivaled power in the Pacific and the chief
source of international support for the Nationalists, to stand aside
from the internal struggle ahead.⁹

The American decision to side with the Nationalists followed
logically from the conclusion reached by late 1945 that the Soviet
Union posed a serious threat to China and the open door ideology.
Fundamental to American policy was the fear that an indigenous
communist party sponsored by the USSR would extinguish the
hallowed special relationship, deny China's progressive impulses,
and deal a serious blow to the American geopolitical position. Between 1945 and 1949 Washington provided money, weapons, and
logistical support (valued conservatively at $3 billion) to maintain
the Nationalists against the Communists, while George C. Marshall
sought in 1946 to defuse the political crisis by some negotiated
settlement that would bring the Communists into a Nationalist-
dominated government.¹⁰

Alongside the American preoccupation with foreign aggression in China in the late 1940s ran that other strain in the open
door ideology, the belief in American-directed reform. No less secure than Burlingame, Hay, or Reinsch in their sense of superiority,
policy makers in the Truman administration sought to show China
the way to political salvation. Thus with aid to the Nationalists
went proposals to stamp out corruption, improve efficiency in the
state apparatus, and increase democratic participation. Sustaining
this reformist impulse was the old, well entrenched conviction that
"native" leaders were not equal to the challenges of the modern
world. Just as Li Hung-chang and Yüan Shih-k'ai had failed in the
end to measure up to American standards for a Chinese strongman,
so now Chiang also failed. The image of the Nationalists as corrupt
and ineffectual, deeply planted by the end of World War II, grew if
anything stronger as a result of the civil war collaboration. But if
the Nationalists evoked images once associated with Ch'ing officialdom, the Communists called to mind early twentieth-century
nationalist agitators and revolutionaries, who had failed to appreciate American good will and heedlessly created national divisions
that made China prey to foreign aggression. Like their antecedents,
Communist leaders (virtually unknown as individuals to American
policy makers) were viewed as ideological fanatics, men without
roots in Chinese soil. Their totalitarian goals were, policy makers

decided hopefully, opposed to "the basic Chinese way of life," especially the "democratic individualism" of the people. The similarity of the old and new State Department views of Chinese revolutionaries became apparent when a dusty MacMurray critique of the American failure to deal forcefully with what he saw as the irresponsible, violently antiforeign, and Bolshevik-inspired Chinese agitation of the 1920s came to light and won the undiluted praise of Acheson's chief China advisers as a "basic and timeless" analysis.[11]

The Communist victory in 1949 set the stage for the culminating collision between Chinese nationalism and the open door ideology. Tensions were predictable as Chinese policy, no longer distracted by direct foreign aggression or internal upheaval, resumed the long postponed attack on imperialist influence. With Japan eliminated and Britain in retreat, the United States was now the dominant power. Moreover, continuing U.S. support for the Nationalist government perpetuated the old and now deeply offensive pattern of foreign interference in Chinese internal affairs. Even so, Mao moved cautiously down to mid-1949 to see if he could reach a diplomatic accommodation with the United States and thereby avoid exclusive dependence on the USSR, maintain trade ties through the period of reconstruction, and facilitate the liberation of Taiwan. No accommodation was possible, however, unless the United States accepted relations based on the principle of formal equality and full respect for China's integrity. That meant, explicitly, an end to support for the Nationalists regrouping on Taiwan and, implicitly, acquiescence in the Communist policy of following a revolutionary rather than a reformist path toward a socialist rather than a liberal order at home.[12]

The tensions that were bound to arise between the renewed Chinese search for autonomy and the fixed sense of American paternalism were exacerbated by the fear of U.S. policy makers that the Communist victory would give a major boost to Soviet imperialism and stimulate latent Chinese xenophobia. Accepting Mao's essential terms was thus out of the question. By mid-1949 all that was left of a now tattered open door policy was wavering support for the discredited and retreating Nationalists and a determination to isolate the Communist-controlled mainland both economically and diplomatically. With hope for accommodation gone, Mao now set out on a policy line that was to traumatize already shaken Americans and confirm their worst fears. The new policy began in the fall of 1949 with a propaganda blitz against the United States and lingering American influence in China, proceeded to the conclusion of an alliance with the USSR the following February, and ended with the complete elimination of the American diplomatic establishment by the spring of 1950. The Soviet Union now enjoyed the special relationship with China, and Peking branded the United States the successor to Britain, Tsarist Russia, and Japan as the principal foreign threat.

Korea, where Li Hung-chang had once tried to maneuver Americans to his advantage, provided the stage for the tragic denouement. The United States entered the conflict there in June 1950. By autumn General Douglas MacArthur was advancing his forces up the peninsula to the Chinese border, while Washington was offering assurances of benign intentions. More struck by MacArthur's advance than Washington's invocation of historic friendship, Peking entered the war in November 1950 after having tried and failed to communicate its alarm and determination to fight if need be. The ensuing conflict summoned up antagonisms lurking beneath the surface of Sino-American relations for fifty years. In China the anti-American campaign was rich in historical resonance. Like the 1905 boycott, it depended heavily on popular participation. Its much-trumpeted charges of germ warfare echoed earlier anxieties over physical debilitation caused by missionary potions and spells, while the hunt for lackeys of imperialism at home called to mind the plight of Christian converts of another day. The burden of the campaign's message—that the American ruling class was reckless, ruthless, and driven by imminent economic crisis—had been part of the nationalist repertoire for nearly half a century. For Americans the war revived the equally timeworn images of the Chinese as a faceless, insensible mass. The "human waves" that overwhelmed Americans in battle were the same ones that nativists of another century had feared would pour into the United States. Communist leaders emerged in the American popular imagination as the ruthless red mandarins—estranged from the people as much as any of the Chinese autocrats of the past.

The collision of 1949–1950 had an ironic impact on Americans in China and Chinese in the United States, the two groups which had set Sino-American relations in motion a century earlier. The collision loosened the grip of exclusion and saved Chinatown, once thought doomed to extinction. Newcomers had been completely cut off by the 1924 immigration act, and an assimilating native-born group had constituted a steadily increasing portion of the Chinese-American community. But already in 1943, just as the native-born were achieving majority status, Congress in token recognition to a wartime ally made Chinese eligible for immigration and naturalization. Though that measure provided for only 105 Chinese to enter the United States each year, the figure was substantially raised by later special congressional and executive acts intended as succor to the victims of the Communist takeover, including students and professionals, former government officials, displaced persons, and relatives of Chinese already settled in the United States. A general revision of the immigration law in 1965 expanded the formal Chinese quota to 20,000 and for the first time defined Chinese in national not racial terms. Renewed postwar immigration revived Chinatowns across the United States. By 1970 the Chinese population had reached 435,000, four times its nineteenth-century peak.[13]

At the same time that immigration revived, an American presence in China, once thought secure, disappeared with a rapidity and thoroughness that would have cheered the earlier, frustrated opponents of foreign penetration. The last of the American diplomats left under pressure in the spring of 1950, and soon after the outbreak of the Korean War Peking ousted the remaining businessmen and missionaries. Many, perhaps most, economic and philanthropic enterprises had hoped to keep operating under the new government. But Washington's adamant refusal to take a flexible approach had from the start hindered their efforts to preserve their place in China. From the perspective of policy makers, the humiliating loss of political prestige and influence at the hands of the Communists overshadowed the value of the remaining American cultural and economic activity in China and ruled out any attempt to salvage it. Washington's decision to impose trade controls and block currency exchange in December 1950 and Peking's resort to an official anti-American campaign, all prompted by China's entry into the Korean War, proved fatal to remanent American interests.[14]

Though the open door constituency was gone, the myths that it had played a pivotal role in generating about China and Chinese-American relations, continued to be influential in the United States. These myths had obscured Washington's understanding of the significance of the Communist revolution, and they had obstructed the formulation of a new China policy that might have saved the open door constituency. Now despite, or perhaps because of, the traumatic events that had obliterated the American presence, those myths prospered under continued official patronage. Dean Acheson eloquently articulated the Truman administration's confidence in the old approach to China. He identified the Chinese Communist party as "the spearhead of Russian imperialism," which even as he spoke was allegedly consolidating its control over North China. In time-honored fashion Acheson laid responsibility for this disaster that had overtaken China and the open door squarely on China's corrupt, self-serving, shortsighted rulers. He indicted them for failing to adopt the program of reform Americans had urged. But even in that dark hour Acheson affirmed hope for the eventual restoration of the old special relationship so that American skill and capital could once more go to work for China's betterment.[15]

Acheson spoke against a background of mounting domestic criticism of the Truman administration's "loss of China." The intensity of the attack, which took on the qualities of a fundamentalist crusade, should not obscure the agreement between the Truman administration and its critics over the basic assumptions of China policy. They differed over whether more should have been done but not over what had been lost—the special relationship itself. The Eisenhower, Kennedy, and Johnson administrations held to Acheson's China policy of isolating the People's Republic and supporting Taiwan, while invoking frequently and with appropriate solemnity the old vision of China transformed and made secure under American aegis. The leaders in "Peiping" were, John Foster Dulles and Dean Rusk earnestly and dutifully assured Americans, the bad boys of Asia who did not live up to "the practices of civilized nations" and whose "aggressive arrogance and obsessions of [their] own making" proved they were out of touch with international realities, not to mention the true spirit of their own country. Some day, American officials still promised, the Chinese people would regain their freedom and help rebuild the old ties of friendship.[16]

Into the Sino-American deadlock the Soviet Union at last intruded, providing a new common enemy and once more driving the United States and China together in a new phase of the old cycle. The Sino-Soviet ideological dispute, which had broken out in the late 1950s, and the tensions along a shared frontier in the following decade began, in the classic fashion, to divert Chinese hostility from the United States. Incontestably, the American role on Taiwan violated China's sovereignty and impeded the full realization of the goal of national reunification, while American involvement along China's periphery from Vietnam to Korea constituted from Peking's perspective a strategy of encirclement and counterrevolution. But whether the United States was a greater threat than the USSR was the issue leaders in Peking now had to debate. While the balance of judgment moved against Moscow,

Chinese leaders remained divided over whether strategic self-interest justified compromise over Taiwan and the risks of restored foreign dependence. Only with the overthrow of the "gang of four" and the accession to power of Teng Hsiao-p'ing was that question settled.[17]

On the American side, the Soviet achievement of parity in strategic nuclear weapons provided the impetus to seek an offset in a rapprochement with China. In 1969 Richard Nixon and Henry Kissinger set aside open door qualms that had obstructed a change in policy for two decades and began the pursuit of the long postponed accommodation. That venture took a major step forward in 1972 when they and their Chinese counterpart, Chou En-lai, indicated in the Shanghai communiqué a willingness to suspend, though not abandon, differences of opinion over Taiwan in the name of mutual self-interest. Finally in 1978 the effort Nixon, Kissinger, and Chou had begun was crowned with success as Teng and Jimmy Carter announced a restoration of full diplomatic relations.

What has happened to the old dream of a special relationship between the two countries? While some scholars in the past decade have held up China as a model from which Americans might learn and many others have warned against the dangers of paternalism in American dealings with China, they are pitted against a far stronger popular impulse that looks for all the world like variants on the old fantasies of guiding and sustaining a struggling China. Nixon's well-publicized visit to China in 1972 reawakened in a public long hostile to Communist China as "ignorant," "warlike," "sly," and "treacherous" a nostalgic vision of Chinese as a "hard-working," "intelligent," "artistic," "progressive," and "practical" people deserving of support and encouragement. Otherwise sober pundits began to speculate on the rich possibilities of a market made up of no less than a quarter of mankind. Recognition of the People's Republic, announced by Carter as a step toward restoring "a long history of friendship" between the American and Chinese peoples, put back in place part of the diplomatic structure dismantled in 1949 and 1950, set long suppressed dreams of a return to China swirling in the minds of educators (the legatees of the missionary impulse) and businessmen, and gave a major fillip to technological and scientific exchange.[18]

The history of U.S. dealings with China neatly illustrates how likely Americans with their unique historical experience and out-

The Special Relationship in Perspective

look are to ignore diversity in the world and instead reduce cultures radically different from our own to familiar, easily manageable terms. There is a danger in putting great national power at the service of such a flawed and essentially ethnocentric vision, especially in a world inexorably reduced in scale by modern technology and possessed of the secret of the atom, yet still far from homogeneous in its essential outlook and values. If more harmonious cross-cultural and interstate relations is an ideal worth pursuing, then Americans must rein in the fatal tendency to project our fantasies beyond our borders and recognize the limits—defined in considerable measure by the aspirations and conditions of others—to the changes that the United States can hope to effect. The need for restraint and sensitivity is particularly acute in approaching a China whose American policy still hinges unsteadily on the unresolved tension between the ideal of national autonomy and the requirements of national security.

In short, the time has come to abandon hopes of resurrecting the special relationship and accept as natural rather than aberrational the problems our largely divergent interests and experiences are bound to create. By a more enlightened attitude toward these problems, we may at least hope to minimize them, no small achievement in an already trouble-prone world. In that sense, historical perspective is, as John Fairbank pointed out some years back, "not a luxury but a necessity."[19]

Notes

Abbreviations

ABFM Letters and Papers of the American Board of Commissioners for Foreign Missions, North China Missions, Houghton Library, Harvard University.
CJCC *Chung-Jih chan-chang* [The Sino-Japanese War] (Peking, 1956), compiled by Shao Hsün-cheng et al.
CKCC *Ch'ing Kuang-hsü ch'ao Chung-Jih chiao-she shih-liao* [Historical materials on Sino-Japanese negotiations during the Kuang-hsü reign] (Taipei reprint, 1963), compiled by Ku-kung po-wu yüan.
CKS *Chung-Mei kuan-hsi shih-liao* [Historical materials on Chinese-American relations] (Taipei, 1968–), compiled by Chung-yang yen-chiu yüan, chin-tai shih yen-chiu so.
CR Consular Reports (various posts, to 1906), records of the Department of State, Record Group 59, National Archives, Washington, D.C.
CWC *Chang Wen-hsiang-kung ch'üan-chi* [Collected works of Chang Chih-tung] (Taipei reprint, 1963), compiled by Wang Shu-t'ung.
CWS *Ch'ing-chi wai-chiao shih-liao* [Historical materials on late Ch'ing diplomacy] (Peking, 1935), compiled by Wang Yen-wei and Wang Liang.
CYCT *Chung-Fa Yüeh-nan chiao-she tang* [Records of the Sino-French negotiations over Vietnam] (Taipei, 1962), compiled by Chung-yang yen-chiu yüan, chin-tai shih yen-shiu so.
DF Decimal File (from 1910), records of the Department of State, Record Group 59, National Archives, Washington, D.C.
DIC Diplomatic Instructions: China (1843–1906), records of the Department of State, Record Group 59, National Archives, Washington, D.C.
FRUS *Papers Relating to the Foreign Relations of the United States* (Washington, D.C., yearly series), compiled by the Department of State.
HT *Hsüan-t'ung ch'ao* [Hsüan-t'ung reign (1909–1911)].
ISHK *I-shu han-kao* [Letters to the foreign office] in *Li Wen-chung-kung ch'üan-shu* [Collected works of Li Hung-chang] (Nanking, 1905), compiled by Wu Ju-lun.
ITS *I-ho-t'uan tang-an shih-liao* [Historical materials on the Boxers] (Peking, 1959), compiled by Kuo-chia tang-an chü, Ming-Ch'ing tang-an kuan.

Abbreviations

IWSM   Ch'ing-tai ch'ou-pan i-wu shih-mo [Complete account of the management of barbarian affairs] (Peking, 1930), compiled by Ku-kung po-wu yüan.
KH   Kuang-hsü ch'ao[Kuang-hsü reign (1875–1908)].
LC   Manuscript Division, Library of Congress, Washington, D.C.
MCD   Minister to China: Dispatches (1843–1906), records of the Department of State, Record Group 59, National Archives, Washington, D.C.
MPHS   Mei-kuo p'o-hai Hua-kung shih-liao [Historical materials on American oppression of Chinese laborers] (Peking, 1959), compiled by Chu Shih-chia.
NF   Numerical File (1906–1910), records of the Department of State, Record Group 59, National Archives, Washington, D.C.
RG   Record Group in the National Archives, Washington, D.C.
SL   Ta-Ch'ing li-ch'ao shih-lu [Veritable records of successive reigns of the Ch'ing dynasty] (Manchuria, 1937).
SSM   Shih-chiu shih-chi Mei-kuo ch'in-Hua tang-an shih-liao hsüan-chi [Selected archival materials on American aggression against China during the nineteenth century] (Peking, 1959), compiled by Chu Shih-chia.
TC   T'ung-chih ch'ao [T'ung-chih reign (1862–1874)].
WWP   Wai-wu pu tang-an [Records of the Chinese foreign office], Diplomatic Archives, Chung-yang yen-chiu yüan, chin-tai shih yen-chiu so, Nankang, Taipei, Taiwan.

Notes to Pages ix–1

Preface

1. That other classic, A. Whitney Griswold's fluent interpretive tour de force, *The Far Eastern Policy of the United States* (New Haven, Conn., 1938), which carries forward Dennett's cyclical theory of American policy, suffers from these same weaknesses. Dorothy Borg, "Two Histories of the Far Eastern Policy of the United States: Tyler Dennett and A. Whitney Griswold," in Borg and Shumpei Okamoto, eds., *Pearl Harbor as History: Japanese-American Relations, 1931–1941* (New York, 1973), pp. 561–71; Robert Ferrell, "The Griswold Theory of Our Far Eastern Policy," in Borg, comp., *Historians and American Far Eastern Policy* (New York, 1966), pp. 14–21; and Ernest May, "Factors Influencing Historians' Attitudes: Tyler Dennett," also in Borg, pp. 34–36.

Like Dennett and Griswold, surveys of Sino-American relations published in the People's Republic of China are in general curiously American-centered. They too concentrate on diplomacy in developing their key contention that the open door was a policy of economic exploitation. And their interpretive emphasis on the dangerously aggressive nature of American economic imperialism—a theme to which even the cultural imperialism of the missionary and the travail of the Chinese immigrant are reduced—was no less shaped by contemporary developments (in particular the collision with the United States in the late forties and early fifties over recognition, Taiwan, and Korea). T'ao Chü-yin's *Mei-kuo ch'in-Hua shih-liao* [Historical materials on American aggression against China] (Shanghai, 1951) and Liu Ta-nien's *Mei-kuo ch'in-Hua shih* [A history of American aggression against China] (Peking, 1951), are two of the more interesting of these surveys.

A revived interest in the United States in American–East Asian relations over the last decade or so has produced four general works worth noting. Akira Iriye's *Across the Pacific: An Inner History of American–East Asian Relations* (New York, 1967) is an ambitious attempt to juxtapose national images and policies on a trans-Pacific stage. Warren I. Cohen's *America's Response to China* (New York, 1971) offers more careful but narrowly drawn coverage. The essays in Ernest R. May and James C. Thomson, Jr., eds., *American–East Asian Relations: A Survey* (Cambridge, Mass., 1972) describe the literature and in some cases suggest neglected themes and topics. In *Americans and Chinese: A Historical Essay and a Bibliography* (Cambridge, Mass., 1963), Kwang-Ching Liu has provided a full listing of English-language sources and research aids prefaced by a plea for a more broadly conceived approach to Sino-American relations. Despite the appearance of these newer works, Harold Isaacs' older and highly influential *Scratches on Our Minds: American Images of China and India* (New York, 1958) still deserves attention for its many shrewd insights.

2. Garrett Mattingly, *The Armada* (Boston, 1959), p. 375.

3. These are Howard L. Boorman, ed., *Biographical Dictionary of Republican China* (New York, 1967–1979); Arthur W. Hummel, ed., *Eminent Chinese of the Ch'ing Period (1644–1912)* (Washington, D.C., 1943–1944); Allen Johnson et al., eds., *Dictionary of American Biography* (New York, 1946–1974).

Part One: First Contacts

1. This account of the Terranova incident is based on the documents in U.S. House of Representatives, 26th Cong., 2d Sess., *Executive Documents* (no. 71; "Po-

litical Relations between the United States and China"), pp. 9–52; and governor general Juan Yüan's memorial of October 28, 1821 in Ku-kung po-wu yüan, comp., Ch'ing-tai wai-chiao shih-liao: Tao-kuang ch'ao [Historical materials on Ch'ing foreign relations: the Tao-kuang reign] (Taipei reprint, 1968), 1:7–9. The translation of this memorial found in U.S. House, Executive Documents, pp. 46–50, differs from the original only in minor details. The East India Company Factory Records (India Office, London), Consultations (1821), 224:30–31, 68, 75, 80–82, offer some supplementary insights into the case.

    2. U.S. House, Executive Documents, p. 39.

    3. Clarence L. Ver Steeg, "Financing and Outfitting of the First U.S. Ship to China," Pacific Historical Review (1953), vol. 22.

    4. The Chinese legal code provided, in the case of foreigners, for the swift and condign punishment best calculated to keep them in check. That code also stipulated strangulation as the punishment appropriate for a homicide resulting from a quarrel (more serious than a purely accidental homicide in line with the premium Chinese law put on the maintenance of social harmony. Strangulation constituted in fact a lesser form of capital punishment in the eyes of the Chinese since it preserved the body intact, unlike execution by beheading or slicing (either of which allowed the spirit to escape the mutilated body and become a wanderer). Strangulation was also relatively humane, bringing death (according to some authorities) about as swiftly as hanging in the West. Fu Lo-shu, A Documentary History of Sino-Western Relations (1644–1820) (Tucson, Ariz., 1966), 1:186, 187, 297, 319; Derk Bodde and Clarence Morris, Law in Imperial China (Cambridge, Mass., 1967), pp. 91–93; Sir George Thomas Staunton, Ta Tsing Leu Lee (London, 1810), pp. 36, 311, 314, 523; Ernest Alabaster, Notes and Commentaries on Chinese Criminal Law (London, 1899), pp. 260–62, 288.

Chapter 1: The Rise of the Open Door Constituency, 1784–1860

There is only one work that examines both sides of Sino-American relations in the years treated here—Li Ting-i's Chung-Mei wai-chiao shih, 1784–1860 [A history of Sino-American relations, 1784–1860], published in Taipei in 1960 as a projected (but apparently aborted) multivolume history. The appearance in the interim of new specialized studies has, however, reduced its usefulness as an interpretative synthesis. This observation applies with even more force to Kenneth Scott Latourette's survey of the pretreaty days, The History of Early Relations between the United States and China, 1784–1844 (New Haven, Conn., 1917). Five articles, all in Lieh Tao, ed., Ya-p'ien chan-cheng shih lun-wen chuan-chi [A collection of essays on the history of the Opium War] (Peking, 1958), offer an overview from a Chinese Communist perspective. Their predictable indictment of American activities is built almost exclusively from American sources.

    The place to begin on the American China trade is Jacques M. Downs, "American Merchants and the China Opium Trade, 1800–1840," Business History Review (Winter 1968), vol. 42. It is notable for its mining of business records, for delineating the workings of the American opium trade, and for underlining the importance of the Boston merchant-capitalists to the conduct of that trade. Charles Stelle's articles on "American Trade in Opium to China" down to 1839 which appeared in the Pacific Historical Review in 1940 and 1941 have been largely though not entirely superseded by Downs. Louis Dermigny, La Chine et l'Occident: Le Commerce à Canton au XVIIIe Siècle, 1719–1833 (Paris, 1964), has the considerable virtue of setting the Americans (treated in vol. III, ch. 3) in the context of the global system of trade of which Canton was only a part. A most helpful work on the later phase of American mercantile activity is Stephen C. Lockwood, Augustine Heard and Company, 1858–1862: American Merchants in China (Cambridge, Mass., 1971). Like Downs, it uses business records to chart commercial patterns and personal attitudes. A similar account is badly needed for Russell and Company. Memoirs by China traders offer a valuable window on their long-lost world. Samuel Shaw and William Hunter have given us two of the best. Yuan Chung Tung, "American China-Trade, American-Chinese Relations and the Taiping Rebellion, 1853–1858," Journal of Asian History (1969), vol. 3, offers some scattered insights.

The most valuable work on American missionaries in China for this period, as in general, has come out of Harvard. Clifton J. Phillips' Protestant America and the Pagan World: The First Half Century of the American Board of Commissioners for Foreign Missions, 1810–1860 was a doctoral dissertation neglected for fifteen years until brought out as a Harvard East Asian monograph in 1969. Even more important is Ellsworth Carlson's The Foochow Missionaries, 1847–1880, published as part of that same series in 1974. By drawing on Chinese sources as well as mission records, Carlson provides a sense of the interaction between one missionary community and the Chinese world it sought to penetrate. Carlson, himself once a participant in the missionary enterprise in China, is more defensive about mission work than authors of other recent works but his historical candor and ample documentation allow readers to draw their own conclusions. The lives of two active missionaries who also played an important role as diplomats are accessible through accounts by Edward V. Gulick, Peter Parker and the Opening of China (Cambridge, Mass., 1973), and Frederick Wells Williams, The Life and Letters of Samuel Wells Williams (New York, 1889). More work is needed on individual missionaries, including Bridgman and Williams (whose biography is unabashedly filiopietistic, as one might expect of a son writing about his father whose chair of Chinese studies at Yale he had succeeded to). More is also needed on individual mission stations operating in a Chinese context (along the lines provided by Carlson) and on the relations of missionaries in China to the boards and public at home in these years.

The development of American diplomatic interest in China has received extraordinarily full but uneven treatment. Te-kong Tong, United States Diplomacy in China, 1844–60 (Seattle, Wash., 1964), is marred by the author's intrusive Anglophobia and his overriding preoccupation with formal diplomacy, and consequently with formal diplomatic documentation. His interpretation of American policy in this period is straitjacketed by his questionable determination to show "the gradual triumph of the open door principle" (p. 285). A good corrective to Tong's American-centered approach is Earl H. Pritchard, "The Origins of the Most-Favored-Nation and the Open Door Policies," Far Eastern Quarterly (February 1942), vol. 1. Estimates of the watershed Cushing mission have gone in cycles. Thomas Hart Benton's Thirty Years' View (New York, 1854–1856), ch. 122, was a politically partisan attack that made some telling points against Cushing. The case for the defense appeared in 1923 in Claude M. Fuess' The Life of Caleb Cushing, ch. 10. From Fuess emerges an admiring view of Cushing as an exemplary scholar-statesman. A decade later P. C. Kuo in "Caleb Cushing and the Treaty of Wanghia, 1844," Journal of Modern History (1933), vol. 5, responded to Fuess by reviving the substance of the charges made by Benton. The most recent reappraisal, Richard E. Welch, Jr., "Caleb Cushing's Chinese Mission and the Treaty of Wanghia: A Review," Oregon Historical Quarterly (December 1957), vol. 58, marks a turn back toward a sympathetic view. American policy in the 1850s is treated thematically in Ssu-yu Teng, The Taiping Rebellion and the Western Powers (Oxford, 1971), pp. 220–27, 250–63. Although it adds little to the coverage in Tong, Teng's work is helpful in putting American diplomacy in context. So too is John K. Fairbank's dense, magisterial Trade and Diplomacy on the China Coast: The Opening of the Treaty Ports, 1842–1854 (Cambridge, Mass., 1953); a readable new synthesis with few surprises by Peter W. Fay, The Opium War, 1840–1842 (Chapel Hill, N.C., 1975); Tan Chung's provocative "Interpretations of The Opium War (1840–1842): A Critical Appraisal," Ch'ing-shih wen-t'i (December 1977), vol. 3, supp. 1; and by way of background Earl Pritchard's The Crucial Years of Early Anglo-Chinese Relations, 1750–1800 (Pullman, Wash., 1936).

Interest in the nature of American attitudes toward China has been stirred by Stuart C. Miller's controversial revisionist study, The Unwelcome Immigrant: The American Image of the Chinese, 1785–1882 (Berkeley, Calif., 1969). That part of his interpretation most germane to this chapter is his argument (worked out in chs. 1–6) that the "negative" attitude of Americans in China, once communicated home, gave rise to a "negative" national point of view even before the arrival of Chinese immigrants in sizable numbers. The older Enlightenment view, Miller contends, had little enduring influence. Miller's conclusions are at odds with earlier works by Latourette, Tyler Dennett, and Harold Isaacs, all of which emphasize good will and admiration as the salient features of early American attitudes toward China. Jonathan Goldstein's Philadelphia and the China Trade, 1682–1846: Commercial, Cultural, and Attitudinal Effects (University Park, Pa., 1978), is a case study (of which more are needed) suggesting that favorable attitudes persisted among some Americans into the early nineteenth century and that Sinophobia was neither as strong nor as pervasive as Miller contends.

1. Samuel Eliot Morison, *The Maritime History of Massachusetts* (Boston, 1921), p. 22; Jonathan Goldstein, *Philadelphia and the China Trade, 1682–1846: Commercial, Cultural, and Attitudinal Effects* (University Park, Pa., 1978), p. 13.

2. Arthur M. Johnson and Barry E. Supple, *Boston Capitalists and Western Railroads: A Study in Nineteenth-Century Investment Process* (Cambridge, Mass., 1967), pp. 21–25; Carl Seaburg and Stanley Paterson, *Merchant Prince of Boston: Colonel T. H. Perkins, 1764–1854* (Cambridge, Mass., 1971), pp. 155–57; Jacques M. Downs, "American Merchants and the China Opium Trade, 1800–1840," *Business History Review* (Winter 1968), 42:429–30.

3. Goldstein, *Philadelphia*, p. 17; James Kirker, *Adventures to China: Americans in the Southern Oceans, 1792–1812* (New York, 1970).

4. This and the paragraph that follows is based for the most part on Downs, "American Merchants"; and Peter Temin, *The Jacksonian Economy* (New York, 1969), pp. 80–82, 87, 173.

5. Downs, "American Merchants," pp. 435–39; Robert B. Forbes, *Personal Reminiscences*, 2d ed. (Boston, 1882), pp. [418–19].

6. John K. Fairbank, *Trade and Diplomacy on the China Coast: The Opening of the Treaty Ports, 1842–1854* (Cambridge, Mass., 1953), 1:46–47, 52–53; Josiah Quincy, ed., *The Journals of Major Samuel Shaw* (Boston, 1847), pp. 173–80.

7. Michael Greenberg, *British Trade and the Opening of China, 1800–42* (New York, 1951), pp. 59, 68; William Milburn, *Oriental Commerce* (London, 1825 ed.), pp. 456–57; Downs, "American Merchants," pp. 426, 430; Johnson and Supple, *Boston Capitalists*, pp. 23–24.

8. Greenberg, *British Trade*, pp. 72–73.

9. *Ibid.*, pp. 54–57.

10. In 1806 Americans had claimed 6 percent of the imports into Canton. After the disruption caused by Jefferson's trade restrictions and war with Britain, they resumed the upward movement, beginning at 13 percent in 1817–1819 and climbing to 19 percent for 1830–1833. The share of Canton exports carried by Americans amounted to 30 percent by 1805–1806, and reached a record high of 41 percent just after the Treaty of Ghent (1815) before subsiding to 25 percent by 1830–1833. Total American trade had hit its peak in 1817–1819 (an annual average of $15 million), and despite the decline thereafter in the share and value of American exports from Canton the total still held at $12 million in 1830–1833. For American shippers this substantial trade proved a major boon. Between 1784 and 1804 as many as thirty-one of their vessels went to China a year; thereafter well into the 1840s the annual figures ranged between thirty and forty. Yen Chung-p'ing et al., *Chung-kuo chin-tai ching-chi shih t'ung-chi tzu-liao hsüan-chi* [Selected statistical materials on modern Chinese economic history] (Peking, 1955), pp. 4–5; Timothy Pitkin, *Statistical View of the Commerce of the United States of America* (New Haven, Conn., 1835), p. 252; Earl Pritchard, "The Struggle for Control of the China Trade in the Eighteenth Century," *Pacific Historical Review* (September 1934), 3:292–95; Goldstein, *Philadelphia*, p. 34.

11. Julian P. Boyd, ed., *The Papers of Thomas Jefferson* (Princeton, N.J., 1950–), 13:3–4; Thomas R. Trowbridge, Jr., "The Diary of Mr. Ebenezer Townsend, Jr., the Supercargo of the Sealing Ship 'Neptune' on Her Voyage to the South Pacific and Canton," *Papers of the New Haven Colony Historical Society* (1888), 4:1–2; Downs, "American Merchants," p. 422n; Robert Ernest, *Rufus King: American Federalist* (Chapel Hill, N.C., 1968), p. 353n; Kenneth W. Porter, *John Jacob Astor, Business Man* (Cambridge, Mass., 1931), 2:916–17; James B. Hedges, *The Browns of*

Providence: The Nineteenth Century (Providence, R.I., 1968), pp. 18, 24, 146; Johnson and Supple, Boston Capitalists, p. 28.

12. Goldstein, Philadelphia, p. 36; Boyd, The Papers of Thomas Jefferson, 7:353–55, 638, 13:3–4, 14:599–602; Harold C. Syrett, ed., The Papers of Alexander Hamilton (New York, 1961–1979), 1:384, 387, 388, 8:20–21, 9:53–54, 19:234; Charles R. King, ed., The Life and Correspondence of Rufus King (New York, 1894–1900), 1:155; Charles Francis Adams, ed., Memoirs of John Quincy Adams (Philadelphia, 1874–1877), 4:181.

13. Albert E. Bergh, ed., The Writings of Thomas Jefferson (Washington, D.C., 1907), 5:183.

14. Quincy, The Journals of Major Samuel Shaw, p. 388; Lyman H. Butterfield, ed., Diary and Autobiography of John Adams (Cambridge, Mass., 1961), 3:140–41; Charles Francis Adams, ed., The Works of John Adams (Boston, 1851–1865), pp. 343–44; Boyd, The Papers of Thomas Jefferson, 8:154, 10:135; Bergh, The Writings of Thomas Jefferson, 12:134; H. A. Washington, ed., The Writings of Thomas Jefferson (Washington, D.C., 1853–1854), 5:325–26. Jefferson's dignitary seems, in fact, to have been a petty Canton shopkeeper abetting a scheme by Astor and Perkins to get a ship carrying $45,000 in goods off to Canton and bring it home filled with items sure to command a good price as long as the embargo remained in effect. Jefferson eventually learned that he had been duped but it was too late to block the departure. Porter, Astor, 1:144–50, 424–28.

15. Gaillard Hunt, ed., Journals of the Continental Congress, 1774–1789 (Washington, D.C., 1904–1937), 26:58–59; Charles O. Paullin, Diplomatic Negotiations of American Naval Officers, 1778–1883 (Baltimore, 1912), pp. 154–82; E. Mowbray Tate, "American Merchant and Naval Contacts with China, 1784–1850," American Neptune (July 1971), 31:180–81.

16. Paullin, Diplomatic Negotiations, pp. 183–84; Tate, "American Merchant and Naval Contacts," p. 184; Claude M. Feuss, The Life of Caleb Cushing (New York, 1923), 1:403–4.

17. Tate, "American Merchant and Naval Contacts," pp. 181–83.

18. Fletcher Webster, ed., The Writings and Speeches of Daniel Webster (Boston, 1903), 12:139–40, 142–43.

19. James M. Merrill, "The Asiatic Squadron: 1835–1907," American Neptune (April 1969), 29:107–8; Tate, "American Merchant and Naval Contacts," pp. 183–89; Paullin, Diplomatic Negotiations, pp. 189–212.

20. Cushing to John Tyler, December 27, 1842, Cushing Papers (LC); Feuss, The Life of Caleb Cushing, 1:414–15; Richard E. Welch, Jr., "Caleb Cushing's Chinese Mission and the Treaty of Wanghia: A Review," Oregon Historical Quarterly (December 1957), 58:330n; Fletcher Webster, The Writings and Speeches of Daniel Webster, 12:145–46.

21. Fletcher Webster, The Writings and Speeches of Daniel Webster, 12:143–46.

22. Feuss, The Life of Caleb Cushing, 1:430–31; Tate, "American Merchant and Naval Contacts," p. 186.

23. The account of diplomacy to 1860 that follows is, unless otherwise indicated, based on Te-kong Tong, United States Diplomacy in China, 1844–60 (Seattle, Wash., 1964); Ssu-yu Teng, The Taiping Rebellion and the Western Powers (Oxford, 1971), pp. 220–27, 250–63.

24. John Bassett Moore, ed., The Works of James Buchanan (Philadelphia, 1909), 6:143–44, 7:202.

25. Specifically, Marshall wanted to offer to sustain the imperial side on the

condition that the Emperor grant a general political amnesty to pacify the countryside, while to conciliate foreigners he would have to promise complete freedom of movement, open the Yangtze to steam navigation, agree to conduct foreign affairs along Western lines, and grant general religious freedom throughout the empire. Fairbank, *Trade and Diplomacy*, 1:415; MCD (July 10, 1853); Chester A. Bain, "Commodore Matthew Perry, Humphrey Marshall, and the Taiping Rebellion," *Far Eastern Quarterly* (May 1951), vol. 10; Laurence A. Schneider, "Humphrey Marshall, Commissioner to China, 1853–1854," *Register of the Kentucky Historical Society* (April 1965), vol. 63.

26. Yuan Chung Teng, "American China-Trade, American-Chinese Relations and the Taiping Rebellion, 1853–1858," *Journal of Asian History* (1969), 3:101–2.

27. Edward V. Gulick, *Peter Parker and the Opening of China* (Cambridge, Mass., 1973), p. 190; MCD (December 12, 1856); Huang Chia-mo, *Mei-kuo yü T'ai-wan, 1784–1895* [The United States and Taiwan, 1784–1895] (Taipei, 1966). ch. 4; Thomas R. Cox, "Harbingers of Change: American Merchants and the Formosa Annexation Scheme," *Pacific Historical Review* (May 1973), vol. 42.

Naval commanders in this period were no less than diplomats impatient with inactivity—and with the Chinese. When, in the midst of Anglo-Chinese conflict in Canton harbor in the fall of 1856, American sailors came under fire while trying to sound waters near Chinese forts guarding the approach to the city, Commander James Armstrong at once seized and destroyed the forts to preserve national honor. Peter Parker, smarting from repeated rebuffs by the Chinese governor general in Canton, praised Armstrong for having given that official "a lesson that he will not soon forget." Two years later as the Chinese coastal defenses mauled British and French forces trying to advance on Peking, the ostensibly neutral American commander on the scene, Josiah Tattnall, intervened to shield disabled British vessels and haul them to safety. Harold Colvocoresses, "The Capture and Destruction of the Barrier Forts," *United States Naval Institute Proceedings* (May 1938), vol. 64; Julius W. Pratt, "Our First 'War' in China: The Diary of William Henry Powell, 1856," *American Historical Review* (July 1948), vol. 53; Edith R. Curtis, "Blood is Thicker than Water," *American Neptune* (July 1967), vol. 27; Y. C. Teng, "American China-Trade," p. 111.

28. Moore, *The Works of James Buchanan*, 10:141–42, 347.

29. As foreign enterprise moved north, so did these Chinese merchants who came from the three well-to-do counties adjoining Canton (Hsiang-shan, Nan-hai and P'an-yü). They and new men from other regions (especially Chekiang and Kiangsi provinces) filled the same functions the cohong had. These merchant intermediaries, generally known as compradors, still depended on their reputation for honesty and dependability, on their ability to collect market information and to supply exports (principally teas and silks) and sell off imports (chiefly cotton cloth and opium), on the convenience of their bonded warehouses, on the availability of their capital to finance trade, and finally on their ability to supply to foreign firms and guarantee the performance of an array of native workers ranging from linguists and shroffs to boatmen and warehousemen. Compradors, like cohong merchants before them, benefited by the opportunity to trade under the name of foreign merchants and ship under a foreign flag, thereby eluding official oversight and "squeeze." Yen-p'ing Hao, *The Comprador in Nineteenth Century China: Bridge between East and West* (Cambridge, Mass., 1970), esp. chs. 1–3; Fairbank, *Trade and Diplomacy*, 1:248–51.

30. In 1863 it came under the efficient direction of a shrewd, hard-working, ambitious and, in time, politically influential Ulsterman, Robert Hart, and there con-

trol remained for half a century. Stanley F. Wright, *Hart and the Chinese Customs* (Belfast, 1950).

31. For example, five American ships carried 3,000 coolies out of Swatow in 1855, roughly one-half of that port's human "export" for that year. Between 1847 and 1859 nearly 51,000 Chinese coolies set off to Cuba in American holds. The coolie trade had the great economic virtue of keeping employed those ships that could no longer compete for other goods, whether because of age or disrepair. John K. Fairbank, "Legalization of Opium Trade before the Treaties of 1858," *Chinese Social and Political Science Review* (July 1933), vol. 17; Y. C. Teng, "American China-Trade," pp. 94, 109; P'eng Tse-i, comp., *Chung-kuo chin-tai shou-kung-yeh shih tzu-liao (1840–1949)* [Historical materials on modern Chinese handicraft industries, 1840–1949] (Peking, 1957), 1:473–74; Shih-shan Tsai, "Reaction to Exclusion: Ch'ing Attitudes Toward Overseas Chinese in the United States, 1848–1906" (Ph.D. diss., University of Oregon, 1970), p. 67.

32. With the rise of commerce on the West Coast and the 1849 repeal of the British navigation acts opening London to American China traders, clippers traced a new pattern of voyages beginning in New York, thence to California and Canton, and back to New York via the Cape of Good Hope and often London. By sacrificing bulk for speed, the clippers cut travel time between China and New York to eighty or fewer days, thus getting their cargoes, usually teas, opium, and other high value, lightweight goods, to market before later arrivals drove prices down. By the 1850s, however, overbuilding had created an oversupply, driving profits down and forcing the handsome vessels into less glamorous work such as the coolie trade. Robert G. Albion, "New York Port and Its Disappointed Rivals, 1815–1860", *Journal of Economic and Business History* (August 1931), vol. 3; Robert G. Albion, *The Rise of New York Port [1815–1860]* (New York, 1939), pp. 201, 203, 399; Goldstein, *Philadelphia*, p. 66; Stephen C. Lockwood, *Augustine Heard and Company, 1858–1862: American Merchants in China* (Cambridge, Mass., 1971); Edward K. Haviland, "American Steam Navigation in China, 1845–1878," part 1, *American Neptune* (July 1956), vol. 16.

33. Y. C. Teng, "American China-Trade," pp. 96–97, 104–7; Tong, *United States Diplomacy in China*, pp. 196–97; Greenberg, *British Trade*, pp. 205, 209–10, 212; Lockwood, *Augustine Heard*, pp. 79–90, 94.

34. Sarah Forbes Hughes, ed., *Letters and Recollections of John Murray Forbes* (Boston, 1899), chs. 3 and 4; Henrietta M. Larson, "A China Trader Turns Investor—A Biographical Chapter in American Business History," *Harvard Business Review* (April 1934), 12:351; Johnson and Supple, *Boston Capitalists*, pp. 28, 31, 358n; Fairbank, *Trade and Diplomacy*, 2:25n.

35. Peter W. Fay, "The Protestant Mission and the Opium War," *Pacific Historical Review* (May 1971), 40:145.

36. John A. Andrew III, *Rebuilding the Christian Commonwealth: New England Congregationalists and Foreign Missions, 1800–1830* (Lexington, Ky., 1976); Clifton J. Phillips, *Protestant America and the Pagan World: The First Half Century of the American Board of Commissioners for Foreign Missions, 1810–1860* (Cambridge, Mass., 1969), pp. 30–31.

37. Phillips, *Protestant America*, pp. 195–96. Catholic missionaries, whose operations in China dating back to the late sixteenth century were already at this time widespread, did not count Americans among their numbers until the early twentieth century.

38. This and the paragraph that follows draw on Ellsworth C. Carlson, *The Foo-*

chow Missionaries, 1847–1880 (Cambridge, Mass., 1974), pp. 69–75; George H. Danton, The Cultural Contacts of the United States and China: The Earliest Sino-American Culture Contacts, 1784–1844 (New York, 1931), p. 62; Paul A. Varg, Missionaries, Chinese and Diplomats: The American Protestant Missionary Movement in China, 1890–1952 (Princeton, N.J., 1958), pp. 6–7; Frederick Wells Williams, The Life and Letters of Samuel Wells Williams (New York, 1889), pp. 173, 180.

39. Elijah C. Bridgman in Missionary Herald (March 1846), 42:94; Elizabeth L. Malcolm, "The Chinese Repository and Western Literature on China, 1800 to 1850," Modern Asian Studies (April 1973), vol. 7.

40. Danton, Cultural Contacts, pp. 52–71; George B. Stevens and W. Fisher Marwick, The Life, Letters, and Journals of the Rev. and Hon. Peter Parker, M.D. (Boston, 1896), pp. 122, 133.

41. Rufus Anderson, "The Theory of Missions to the Heathen . . ." (pamphlet; Boston, 1845), reproduced in full in Philip D. Curtin, ed., Imperialism (New York, 1971), pp. 212, 214, 219, 221, 223; F. W. Williams, The Life and Letters, p. 180; Gulick, Peter Parker, pp. 20–21, 35, 133–37.

42. The following two paragraphs are based on Carlson, The Foochow Missionaries, pp. 6, 48–65 passim, 86–87, 171–73; Bridgman in Missionary Herald (March 1846), 42:94–95; Sun, comp., Chung-kuo chin-tai kung-yeh shih tzu-liao, ti-i chi, 1840–1895 nien [Historical materials on modern Chinese industry, first series, 1840–1895] (Peking, 1957), pp. 113–14.

43. Carlson, The Foochow Missionaries, pp. 15–17, 83.

44. Ibid., pp. 64–65, 72; Stuart C. Miller, "Ends and Means: Missionary Justification of Force in Nineteenth Century China," in John K. Fairbank, ed., The Missionary Enterprise in China and America (Cambridge, Mass., 1974), pp. 249–64.

45. Carlson, The Foochow Missionaries, pp. 18–33, 36–44.

46. Edmund Roberts, Embassy to the Eastern Courts of Cochin-China, Siam, and Muscat (New York, 1837), pp. 5–6, 130; W. S. W. Ruschenberger, A Voyage Round the World (Philadelphia, 1838), pp. 389, 420–34 passim; Feuss, The Life of Caleb Cushing, 1:405–6, 416n, 430–31; Charles Francis Adams, Memoirs of John Quincy Adams, 10:44–45, 11:166–67; Phillips, Protestant America, p. 198; Gulick, Peter Parker, pp. 98–100, 107; FRUS 1866, part 1, pp. 476–77. The hostile Humphrey Marshall and Chinese observers both contended that Parker's command of the language was fairly rudimentary. Tong, United States Diplomacy in China, pp. 32–33; Earl Swisher, China's Management of the American Barbarians: A Study of Sino-American Relations, 1841–1861, with Documents (New York, 1951), pp. 153, 155.

47. F. W. Williams, The Life and Letters, pp. 234–35, 242–43, 274; W. A. P. Martin, A Cycle of Cathay (New York, 1900), pp. 147, 181–84, 190.

48. Leonard W. Labaree and William B. Willcox, eds., The Papers of Benjamin Franklin (New Haven, Conn., 1959–), 9:82, 10:182–88, 389, 11:230, 12:11–12, 17:107, 18:188, 19:69, 136, 138, 268, 20:442–43; Albert H. Smyth, ed., The Writings of Benjamin Franklin (New York, 1905–1907), 8:24; John Bigelow, ed., The Works of Benjamin Franklin (New York, 1904), 11:177–78; Goldstein, Philadelphia, pp. 7, 15; George W. Corner, ed., The Autobiography of Benjamin Rush (Princeton, N.J., 1948), pp. 175–76.

49. The James Madison Letters (New York, 1884), 3:80, 90, 209; Boyd, The Papers of Thomas Jefferson, 8:216, 633.

50. Labaree and Willcox, The Papers of Benjamin Franklin, 16:201, 20:442–43; Smyth, The Writings of Benjamin Franklin, 9:200–8; Butterfield, Diary and Autobiography of John Adams, 2:247; Corner, The Autobiography of Benjamin Rush,

pp. 175–76, 222, 245–46; John C. Fitzpatrick, ed., *The Writings of George Washington* (Washington, D.C., 1931–1944), 28:223, 239, 30:179, 354n.

51. Ping-chia Kuo, "Canton and Salem: The Impact of Chinese Culture upon New England Life During the Post-Revolutionary Era," *New England Quarterly* (1930), vol. 3; Goldstein, *Philadelphia*, 40:77–79; *A Small Book About China and the Chinese* (Auburn, [N.Y. ?], 1837); *"Ten Thousand Chinese Things": A Descriptive Catalogue of the Chinese Collection in Philadelphia* (Philadelphia, 1839), esp. pp. 94–120; E. C. Wines, *A Peep at China in Mr. Dunn's Chinese Collection* (Philadelphia, 1839). Bridgman, after examining one of the catalogs, predictably pronounced it "a little too favorable." The catalog prepared for another Chinese exhibit that toured Boston and Philadelphia closely follows the tone of the earlier works. John R. Peters, Jr., *Miscellaneous Remarks upon the . . . Chinese* (various editions; Philadelphia and Boston, 1845–1847). Two travelogues from these years offer mixed evidence. Comments on Canton by Silas Holbrook, *Sketches by a Traveler* (Boston, 1830), pp. 252–89, are in the main in a light critical vein, while Osmond Tiffany, Jrs.'s *Canton Chinese* (Boston, 1849), pp. viii–ix, 266, finds much to praise about the Chinese people (though not their civilization).

52. Ruth Miller Elson, *Guardians of Tradition: American Schoolbooks of the Nineteenth Century* (Lincoln, Nebr., 1964), p. 162; Labaree and Willcox, *The Papers of Benjamin Franklin*, 15:175, 218, 220; Bergh, *The Writings of Thomas Jefferson*, 19:262–63; "Adams' Lecture on the War with China," *Chinese Repository* (1842), vol. 11, reprinted in *Proceedings of the Massachusetts Historical Society* (October 1909–June 1910), 43:310, 313.

53. Gulick, *Peter Parker*, pp. 32–33; F. W. Williams, *The Life and Letters*, pp. 64, 80, 174, 237, 344; Martin R. Ring, "Anson Burlingame, S. Wells Williams and China, 1861–1870" (Ph.D. diss., Tulane University, 1972), pp. 43–44; David Abeel, *Journal of a Residence in China and the Neighboring Countries*, 2d ed. (New York, 1836), p. 141; Roberts, *Embassy*, p. 159.

54. Lockwood, *Augustine Heard and Company*, p. 64; Feuss, *The Life of Caleb Cushing*, 1:417–18, 430–31; F. W. Williams, *The Life and Letters*, pp. 197, 223, 248, 257, 263, 268, 299, 325, 342.

55. The cooperative (or symbiotic) side of their relationship is reflected in their agreement in broad terms on the need to convince Washington of the importance of the American stake in China and in their support of one another's efforts. Some merchants, such as David Olyphant and John M. Forbes, patronized the missionaries by providing free transport, underwriting some of the costs of mission medical and printing projects, and handling personal investments (much to the benefit of Parker and Williams). The missionaries, through their pioneering Sinological work, opened a window on China for diplomats and merchants, who in turn commended this expertise to Washington.

56. The Taiping leader was Hung Hsiu-ch'üan, a charismatic and visionary figure, known to have not only digested mission literature picked up in Canton in 1836, but also studied with Reverend Issachar Roberts in that city for three months in 1847 (though neither Hung nor Roberts spoke the same Chinese dialect). Yuan Chung Teng, "Reverend Issachar Jacox Roberts and the Taiping Rebellion," *Journal of Asian Studies* (November 1963), vol. 23; Eugene Boardman, *Christian Influence upon the Ideology of the Taiping Rebellion, 1851–1864* (Madison, Wis., 1952), chs. 1–4; Jen Yu-wen, *The Taiping Revolutionary Movement* (New Haven, Conn., 1973), pp. 14–15, 18, 20–22.

57. S. Y. Teng, *The Taiping Rebellion*, pp. 188–95, 226, 255–56; F. W. Wil-

Chapter 2: Bibliographical Essay

liams, *The Life and Letters*, pp. 201, 336; Martin, *A Cycle of Cathay*, pp. 133, 142; John B. Littell, "Missionaries and Politics in China," *Political Science Quarterly* (December 1928), 43:566–88; Y. C. Teng, "Reverend Issachar Jacox Roberts," pp. 59, 61–66; Eliza J. G. Bridgman, ed., *The Pioneer of American Missions in China: The Life and Labors of Elijah Coleman Bridgman* (New York, 1864), p. 200.

58. Y. C. Teng, "American China-Trade," pp. 99–101, 103, 110, 114–15; Lockwood, *Augustine Heard and Company*, p. 69.

59. Bain, "Commodore Matthew Perry," pp. 265, 268; Y. C. Teng, "American China-Trade," pp. 101–3, 110, 116; S. Y. Teng, *The Taiping Rebellion*, pp. 220–27, 251–63.

60. Feuss, *The Life of Caleb Cushing*, 1:417–18; Carlson, *The Foochow Missionaries*, pp. 5–7, 35; Ring, "Anson Burlingame, S. Wells Williams and China," p. 40; Tong, *United States Diplomacy in China*, pp. 101–2.

61. Cushing suffered brickbats for "his important airs" and for the pointlessness of his proposed trip to Peking. Merchants disliked Marshall's senseless Anglophobia. Missionaries were unhappy with his agnosticism and aversion to the Taiping, and with McLane's Roman Catholicism. W. A. P. Martin thought Reed a man with "no fixed principles," while the Heard partners pronounced him "a humbug, plausible, neat and shallow," who was only interested in using his service in China to build up his "political capital at home." Reed's treaty of 1858 was by their judgment "a perfectly useless abortion." They judged his successor Ward no better. Tong, *United States Diplomacy in China*, pp. 23n, 140, 142; Martin, *A Cycle of Cathay*, p. 184; Lockwood, *Augustine Heard and Company*, pp. 64–65, 133n.

62. Trowbridge, "The Diary of Mr. Ebenezer Townsend, Jr.," pp. 89–101; Hughes, *Letters and Recollections*, 1:86; William C. Hunter, *The "Fan Kwae" at Canton Before Treaty Days, 1825–1844* (London, 1882), esp. p. 26; Hunter, *Bits of Old China* (London, 1885), passim; Robert Bennett Forbes, *Personal Reminiscences*, 2d ed. (Boston, 1882), p. 374; John P. Cushing, "Memo for Mr. Forbes Respecting Canton Affairs . . . ," *Business History Review* (Spring 1966), 40:99–100; Goldstein, *Philadelphia*, pp. 69–70, 72–73; Quincy, *The Journals*, pp. 183–84, 195, 249, 354–55; Syrett, *The Papers of Alexander Hamilton*, 9:50, 53–54.

63. E. C. [Bridgman], "Negotiations with China," *Chinese Repository* (January 1835), 3:419–21, 426–28; Abeel, *Journal of a Residence*, p. 143.

64. Roberts, *Embassy*, pp. 130, 133–34, 151–52; Ruschenberger, *A Voyage*, pp. 389, 394–402, 420–34.

65. Jacques M. Downs, "Fair Game: Exploitive Role-Myths and the American Opium Trade," *Pacific Historical Review* (May 1972), 41:145; Roberts, *Embassy*, p. 151; Ruschenberger, *A Voyage*, p. 389; F. W. Williams, *The Life and Letters*, p. 122; Ring, "Anson Burlingame, S. Wells Williams and China," pp. 10–11; Miller, "Ends and Means," pp. 25, 54, 256; G. R. Williamson, ed., *Memoir of the Rev. David Abeel, D.D.* (New York, 1848), p. 180.

66. F. W. Williams, *The Life and Letters*, pp. 147–49, 155.

Chapter 2: The Chinese Discover America, 1784–1879

The record of early Chinese attitudes and policy toward the United States is remarkably full. Though no general account based on that record has yet to appear, the initial reaction of Chinese intellectuals to the West-

ern presence has been explored in some detail. Perhaps the single most valuable piece in this regard is Ch'en Sheng-lin, "Ya-p'ien chan-cheng ch'ien-hou Chung-kuo jen tui Mei-kuo ti liao-chieh ho chieh-shao" [The Chinese discovery of the United States around the time of the Opium War], Chung-shan ta-hsüeh hsüeh-pao (1980), nos. 1 and 2. This carefully researched and argued essay, reflecting new trends in historical writing in China, emphasizes the struggle of "enlightened, progressive patriots," Lin Tse-hsü, Wei Yüan, and Liang T'ing-nan, against the obscurantist "closed door policy" sustained by the "feudal" ruling class.

Wang Chia-chien's Wei Yüan tui hsi-fang ti jen-shih chi ch'i hai-fang ssu-hsiang [Wei Yüan's knowledge of the West and his ideas on maritime defense] (Taipei, 1964), and Jane Kate Leonard's 1971 Cornell University doctoral dissertation, "Wei Yüan and the Hai-kuo t'u-chih: A Geopolitical Analysis of Western Expansion in Maritime Asia," explore the views of that key figure in greater detail. Fred W. Drake is the authority on Hsü Chi-yü. His work, which first appeared in 1965 in Harvard's Papers on China, culminated in a 1975 monograph, China Charts the World: Hsu Chi-yü and His Geography of 1848 (Cambridge, Mass.). The latter largely supersedes Dorothy Ann Rockwell's early (1957) account of Hsü's "Brief study" in Papers on China. Suzanne Wilson Barnett has underlined the important role of missionaries as conveyors of secular knowledge to the Chinese in her "Protestant Expansion and Chinese Views of the West." It, along with essays by Drake and Leonard, is to be found in a special issue of Modern Asian Studies (April 1972), vol. 6, devoted to China's discovery of the West.

Tracing the impact on policy of early perceptions of the United States is a task considerably facilitated by Earl Swisher's China's Management of the American Barbarians: A Study of Sino-American Relations, 1841–1861 (New York reprint, 1972). It contains extensive translations from official documents in the Ch'ou-pan i-wu shih-mo [Complete record of the management of barbarian affairs] collection. Swisher has to be supplemented by other printed primary sources (identified in the notes for this chapter). The topical and chronological indexes in the first two volumes of Chung-yang yen-chiu yüan, chin-tai shih yen-chiu so, comp., Chung-Mei kuan-hsi shih-liao [Historical materials on Sino-American relations] (Taipei, 1968–), are helpful in winnowing through this mass of often routine and frequently overlapping diplomatic material. The studies by Fu Lo-shu, Chang Hsin-pao, John Fairbank, J. Y. Wong, and Masataka Banno (all cited below) are useful in putting interest in the United States in the contemporary policy context. The official projects of the 1860s and early 1870s involving the United States are best approached through Knight Biggerstaff, "The Official Chinese Attitude Toward the Burlingame Mission," American Historical Review (July 1936), vol. 41; Richard J. Smith, Mercenaries and Mandarins: The Ever-Victorious Army in Nineteenth Century China (Millwood, N.Y., 1978), which scales down to size the long inflated reputation of Ward and Burgevine; Thomas E. LaFargue, "Chinese Educational Commission to the United States: A Government Experiment in

Chapter 2: Bibliographical Essay

Western Education," Far Eastern Quarterly (November 1941), vol. 1; Thomas E. LaFargue, China's First Hundred (Pullman, Wash., 1942); and William Hung's contribution as translator and annotator of "Huang Tsun-hsien's Poem 'The Closure of the Educational Mission in America,' " Harvard Journal of Asiatic Studies (June 1955), vol. 18. Yung Wing's own autobiography, My Life in China and America (New York, 1909), opens but a small window on his bicultural experience. Edmund Worthy, "Yung Wing in America," Pacific Historical Review (August 1965), vol. 34, squeezes a few new insights from a thin collection of personal papers at Yale's Sterling Library.

A substantial body of historical studies offers easy entrée to the American side of the immigration question in its early stages. My account is particularly in debt to Elmer C. Sandmeyer, The Anti-Chinese Movement in California (Urbana, Ill., 1939); Gunther Barth, Bitter Strength: A History of the Chinese in the United States, 1850–1870 (Cambridge, Mass., 1964); and Alexander Saxton, The Indispensable Enemy: Labor and the Anti-Chinese Movement in California (Berkeley, Calif., 1971). Although in many respects supplanted by these accounts, Mary Coolidge's pioneering Chinese Immigration (New York, 1909) remains a helpful overview. The best general account in Chinese for the nineteenth and early twentieth century is Liu Pei-chi's Mei-kuo Hua-ch'iao shih [A history of the Chinese in the U.S.] (Taipei, 1976). Roger Daniels has recently brought together ten previously published journal articles that constitute a good sampling of a large literature dealing with localized instances of Anti-Chinese Violence in North America (New York, 1978).

The Chinese side poses difficulties for the inquiring historian. We lack and badly need a close examination of those social and economic conditions in the Canton delta which conspired to drive Chinese to the United States and elsewhere abroad. Robert G. Lee, "The Origins of Chinese Immigration to the United States, 1848–1882," in Chinese Historical Society of America, comp., The Life, Influence and the Role of Chinese in the United States, 1776–1960 (San Francisco, 1976), provides a helpful agenda for such research. The most recent attempt at plugging this gap, June Mei's "Socioeconomic Origins of Emigration: Guangdong to California, 1850–1882," Modern China (October 1979), vol. 5, emphasizes the disruptive impact of Western imperialism on the region. Unfortunately, the supporting evidence falls considerably short of sustaining the author's sweeping generalizations. Older works that are still helpful are Frederic Wakeman, Jr., Strangers at the Gate: Social Disorder in South China, 1839–1861 (Berkeley, Calif., 1966); Wan Lo, "Communal Strife in Mid-Nineteenth Century Kwangtung: The Establishment of the Ch'ih-ch'i," Papers on China (1965), vol. 19; and Myron L. Cohen, "The Hakka or 'Guest People': Dialect as a Sociocultural Variable in Southeastern China," Ethnohistory (Summer 1968), vol. 15.

Chapter 2: Bibliographical Essay

The literature does not do justice to the largely poor, illiterate, and transient members of the Chinese community during their first decades in the United States. Much of what has been written violates the Zen Buddhist injunction against mistaking the finger pointing at the moon for the moon. It confuses the lives of those sojourners with the white fantasies and misconceptions which pervaded the contemporary polemical literature. This tendency is evident in the earliest serious histories of the "Chinese question" (for example, Hubert H. Bancroft's History of California, 1860–1890 [San Francisco, 1890], pp. 335–62), and has persisted even in recent scholarship. Barth's pathbreaking Bitter Strength is the most obvious and frequently criticized example of this tendency to take sensationalistic nativist indictments of the Chinese at face value. The same penchant toward coaxing the story of the Chinese from contemporary English-language sources is also evident in newer works by sociologist Stanford Lyman, "Conflict and the Web of Group Affiliation in San Francisco's Chinatown, 1850–1910," Pacific Historical Review (November 1974), vol. 43; and by Ivan Light, "From Vice District to Tourist Attraction: The Moral Career of American Chinatowns, 1880–1940," Pacific Historical Review (August 1974), vol. 43. Their emphasis on conflict within the Chinese community has suggested new lines of investigation; however, their particular formulations of the nature of the conflict (e.g., "mercantile class" versus "gangster-politician establishment" in Light [p. 379] or Chinatown as "something of a Hobbesian cockpit" in Lyman [p. 499]) seem both crude and unsupported by their evidence. Ping Chiu, Chinese Labor in California, 1850–1880: An Economic Study (Madison, Wis., 1963), stands almost alone as an effort to understand at least one aspect of the life of the Chinese at mid-century without falling prey to prevailing contemporary stereotypes.

There is much to be said in favor of approaching Chinese emigration and settlement in the United States tangentially, through works on Chinese social history. A good place to begin in understanding the values and institutions the Chinese brought with them to the United States is with Hu Hsien-chin, The Common Descent Group in China and Its Function (New York, 1948); Hui-chen Wang Liu, The Traditional Chinese Clan Rules (Locust Valley, N.Y., 1959); Ho Ping-ti, Chung-kuo hui-kuan shih-lun [An Historical survey of Landsmannschaften in China] (Taipei, 1966); Jean Chesneaux, Secret Societies in China in the Nineteenth and Twentieth Centuries, trans. by Gillian Nettle (Ann Arbor, Mich., 1971); and Frederic Wakeman, Jr., "The Secret Societies of Kwangtung, 1800–1856," in Jean Chesneaux, ed., Popular Movements and Secret Societies in China, 1840–1950 (Stanford, Calif., 1972).

Studies of the Chinese in Southeast Asia, who emigrated for many of the same reasons Chinese in the United States did, offer another set of valuable clues to the nature of the Chinese community in the United States. These studies also underline the importance of the host culture's response to the Chinese presence in inclining them toward assimilation, the creation of a hybrid culture, or the formation of a protective cultural enclave.

The best accounts are G. William Skinner, Chinese Society in Thailand: An Analytical History (Ithaca, N.Y., 1957); and Edgar Wickberg, The Chinese in Philippine Life, 1850–1898 (New Haven, Conn., 1965). They should be supplemented by T'ien Ju-k'ang, The Chinese of Sarawak: A Study of Social Structure (London, 1955); Antonio S. Tan, The Chinese in the Philippines, 1898–1935: A Study of Their National Awakening (Quezon City, Philippines, 1972); and T. G. McGee, "Peasants in the City: A Paradox, A Paradox, A Most Ingenious Paradox," Human Organization (Summer 1973), vol. 32, a stimulating examination of how peasants fit into a capitalist, urban economy. Mary F. Somers Heidhues, Southeast Asia's Chinese Minorities (New York, 1975), is a skillful, up-to-date synthesis of the extensive literature on the subject. Clarence E. Glick, Sojourners and Settlers: Chinese Migrants in Hawaii (Honolulu, 1980), adds yet another comparative dimension.

Armed with a clearer sense of Chinese society and the experience of overseas Chinese generally, it is possible to return to the reams of contemporary literature in English on the Chinese in the United States better able to separate the historical wheat from the polemical chaff. Among the more nourishing grains are Stewart Culin, "The I Hing or 'Patriotic Rising,' A Secret Society among the Chinese in America," Report of the Proceedings of the Numismatic and Antiquarian Society of Philadelphia for the Years 1887–89 (1891); Walter N. Fong, "The Chinese Six Companies," Overland Monthly, 2d series (May 1894), vol. 23; and his "Chinese Labor Unions in America," Chautauquan (July 1896), vol. 23.

1. Lien-sheng Yang, "Historical Notes on the Chinese World Order," in John K. Fairbank, ed., The Chinese World Order: Traditional China's Foreign Relations (Cambridge, Mass., 1968), p. 27.
2. J. Y. Wong, Yeh Ming-ch'en: Viceroy of Liang Kuang, 1852–8 (New York, 1976), p. 158.
3. L. Carrington Goodrich, "China's First Knowledge of the Americas," Geographical Review (July 1938), 28:404–5, 407; Fu Lo-shu, A Documentary Chronicle of Sino-Western Relations (1644–1820) (Tucson, Ariz., 1966), pp. 302–3, 363, 393–94, 408–13, 600n; Yang Ping-nan, Hai-lu [A maritime record] (Taipei reprint, 1962), 2:75; Chung-yang yen-chiu yüan, chin-tai shih yen-chiu yüan so, comp., Chin-tai Chung-kuo tui hsi-fang chi ch'i lieh-ch'iang jen-shih tzu-liao hui-pien, 1st series: Tao-kuang yüan-nien chih Hsien-feng shih-i nien [Collected materials on modern China's knowledge of the West and the powers, 1st series: 1821–1861] (Taipei, 1972–), pp. 61–62; Ch'i Ssu-ho et al., comps., Ya-p'ien chan-cheng [The Opium War] (Shanghai, 1954), 1:75, 89, 91.
4. IWSM: Tao-kuang ch'ao [The Tao-kuang reign], 47:23–24; Ya-p'ien chan-cheng, 4:531.
5. Revised editions appeared in 1846 and again in 1867, each carrying a title different from the original.
6. Chin-tai Chung-kuo tui hsi-fang, pp. 197–208, 848, 852–53.
7. Chin-tai Chung-kuo tui hsi-fang, p. 908. After gaining the highest degree in the state exam system in 1844, Wei spent a decade in official harness, but even so

he still found time for scholarship, including expanded editions of Hai-kuo t'u-chih in 1847 and 1852.

8. Chin-tai Chung-kuo tui hsi-fang, pp. 925–26, 949. One portion of Hai-kuo t'u-chih dealing with the United States appears in translation in Ssu-yü Teng and John K. Fairbank, China's Response to the West: A Documentary Survey, 1839–1923 (Cambridge, Mass., 1954), pp. 32–33. See Chin-tai Chung-kuo tui hsi-fang, p. 1003; and Ya-p'ien chan-cheng, 6:165–66, for extracts from Wei's Sheng-wu chi concerning the United States. A partial translation is available in Edward Parker, Chinese Account of the Opium War (Shanghai, 1888), pp. 72–74.

9. Fred W. Drake, China Charts the World: Hsu Chi-yü and His Geography of 1848 (Cambridge, Mass., 1975), p. 1.

10. Ibid., p. 36.

11. The following paragraphs on Hsü and the United States are based on the relevant section of Ying-huan chih-lüeh reproduced in Chin-tai Chung-kuo tui hsi-fang, pp. 526–39; the extended summary and translation in Drake, China Charts the World, pp. 154–67; and the brief extract in Teng and Fairbank, China's Response, pp. 44–46.

12. Chung-kuo shih-hsüeh hui, comp., T'ai-p'ing t'ien-kuo [The Taiping kingdom] (Shanghai, 1957), 2:529; and the translation in Franz Michael, The Taiping Rebellion: Documents and Comments (Seattle, Wash., 1966–1971), 3:758–60. This favorable estimate of the United States formed part of Hung's rationale for an attempt in 1860 (unsuccessful, as it turned out) to cultivate the powers. See Yuan Chung Teng, "The Failure of Hung Jen-k'an's Foreign Policy," Journal of Asian Studies (November 1968), vol. 28.

13. Chin-tai Chung-kuo tui hsi-fang, pp. 460–61; Chang Tze-mu, Ying-hai lun [A discussion of the circuit of the sea] (circa 1876), 1:2–3; Paul A. Cohen, Between Tradition and Modernity: Wang T'ao and Reform in Late Ch'ing China (Cambridge, Mass., 1974), pp. 98–99; Noriko Kamachi; "American Influences on Chinese Reform Thought: Huang Tsun-hsien in California, 1882–1885," Pacific Historical Review (May 1978), 47:242; Wang Chia-chien, Wei Yüan nien-p'u [A chronological biography of Wei Yüan] (Taipei, 1967), pp. 87–97.

14. Jane Kate Leonard, "Wei Yüan and the Hai-kuo t'u-chih: A Geopolitical Analysis of Western Expansion in Maritime Asia" (Ph.D. diss., Cornell University, 1971), ch. 6; Chin-tai Chung-kuo tui hsi-fang, pp. 925–26, 1003; Drake, China Charts the World, chs. 7–8, 10.

15. Josiah Quincy, ed., The Journals of Major Samuel Shaw (Boston, 1847), pp. 199–200.

16. Ya-p'ien chan-cheng, 2:201; Chang Hsin-pao, Commissioner Lin and the Opium War (Cambridge, Mass., 1964).

17. Pan arranged the purchase of an American merchantman to augment China's outclassed naval forces and hired an American skilled in the construction of torpedoes and explosives to work for the Chinese. Earl Swisher, China's Management of the American Barbarians: A Study of Sino-American Relations, 1841–1861 (New York reprint, 1972), pp. 57, 60, 98–99.

18. Ibid., pp. 136–37, 142–43, 154, 156–58, 160, 162–64, 171, 182–84, 190; Tekong Tong, United States Diplomacy in China, 1844–60 (Seattle, Wash., 1964), pp. 12, 87.

19. After Wu had moved from Canton to Shanghai, his business (conducted in close association with Russell and Company) had prospered. He had bought his way into officialdom and established a reputation for being (according to one of his su-

periors) "thoroughly conversant with barbarian psychology." Chung-yang yen-chiu yüan, chin-tai shih yen-chiu so, comp., *Ssu-kuo hsin-tang* [New files concerning the four powers] (Taipei, 1966), "Ying-kuo tang," p. 149; CKS: *Chia-ch'ing Tao-kuang Hsien-feng ch'ao* [The Chia-ch'ing, Tao-kuang, and Hsien-feng reigns], pp. 132–33, 165–67, 169; SSM, pp. 179–80; Swisher, *China's Management*, pp. 191, 203, 206–8, 217–19, 223–25, 228.

20. In 1858 one official, Wang Mao-yin, did urge the Emperor and those around him to take Wei's "Treatise" as a guide to handling the English. It is not apparent that any took his advice. IWSM: *Hsien-feng ch'ao* [The Hsien-feng reign], 28:45–46.

21. A. Barrister (pseudonym?), ed., *Journals Kept by Mr. Gully and Capt. Denham during a Captivity in China in the Year 1842* (London, 1844), pp. 46, 134–35; Swisher, *China's Management*, pp. 78, 97, 175, 177, 182–83; Drake, *China Charts the World*, pp. 45–46, 48.

22. Swisher, *China's Management*, pp. 191, 203, 266.

23. Yeh was not entirely consistent since in 1854 he hired two Americans to organize a foreign force to attack nearby rebel positions only to have McLane put a stop to this "unneutral" behavior. Wong, *Yeh*, pp. 158–60, 182; Swisher, *China's Management*, pp. 198–99, 208, 213–14, 266, 291, 293, 294, 299–303, 314; SSM, p. 180; John K. Fairbank, *Trade and Diplomacy on the China Coast: The Opening of the Treaty Ports, 1842–1854* (Cambridge, Mass., 1964), pp. 408–9, 431–32; William C. Hunter, *Bits of Old China* (Shanghai, 1911), pp. 92–93.

24. Swisher, *China's Management*, pp. 403, 408, 409–10, 413; *Ssu-kuo hsin-tang*, "Ying-kuo tang," p. 428. One informed critic of barbarian management thought the problem was deeper than even Ho realized. He questioned whether trade and other concessions would create lasting gratitude among the mercantile community. Might not the grant of one set of demands simply open the way for new ones in an unending stream? How could a weak power control stronger ones and prevent their squeezing more concessions than China would be willing to make? Wang Chia-chien, *Wei Yüan tui hsi-fang ti jen-shih chi ch'i hai-fang ssu-hsiang* [Wei Yüan's knowledge of the West and his ideas on maritime defense] (Taipei, 1964), pp. 149–53.

25. During the battle over Canton in the fall of 1856 American naval vessels had destroyed outlying forts while the American consul had been seen, flag in hand, joining in the British assault on the city walls. Swisher, *China's Management*, pp. 314–15, 324, 327–28, 331, 335, 343–44, 347; Tong, *United States Diplomacy*, pp. 185–87; CKS: *Chia-ch'ing Tao-kuang Hsien-feng ch'ao*, pp. 235–48, 274–78.

26. Swisher, *China's Management*, pp. 347, 349–50, 383, 388, 406, 413, 427, 430–32, 441, 447, 454–55; Tong, *United States Diplomacy*, pp. 220, 227; IWSM: *Hsien-feng ch'ao*, 14:21; *Ssu-kuo hsin-tang*, "Ying-kuo tang," pp. 381, 387, 429–30.

27. Swisher, *China's Management*, pp. 441–42, 446, 448, 455, 459–60, 462–63, 466–68, 471–73, 477–82, 484–88, 504–5; W. A. P. Martin, *A Cycle of Cathay* (New York, 1900), pp. 183, 186.

28. Arriving in Peking after an arduous overland journey, Ward was physically restricted, closely watched, and at loggerheads with his hosts over performing the kowtow before the Emperor. Ward finally presented his credentials to Kuei-liang and left the capital to exchange the ratified treaty at another site. Afterwards Ho Kuei-ch'ing, who continued to insist that the Americans were "entirely subservient to the English," moved slowly in meeting Ward's request to begin talks on the detailed trade regulations called for by the new treaty. Ward, he reported back to the capital, was resentful because he had been "treated like a captive chieftain in Peking" and

humiliated in the eyes of the foreign community. Swisher, *China's Management,* pp. 522–24, 527, 529–30, 534, 555–58, 560–61, 565–69, 571, 576, 579, 581–84, 587–89, 591–98, 602–4, 607–9, 612–14, 625–27, 631; CKS: *Chia-ch'ing Tao-kuang Hsien-feng ch'ao,* pp. 319–22, 325; SSM, p. 88; Tong, *United States Diplomacy,* pp. 266–67; Masataka Banno, *China and the West, 1858–1861: The Origins of the Tsungli Yamen* (Cambridge, Mass., 1964), pp. 110–26.

29. CKS: *Chia-ch'ing Tao-kuang Hsien-feng ch'ao,* p. 361; Tong, *United States Diplomacy,* p. 281; SSM, 156, 160; Swisher, *China's Management,* pp. 650, 652–53, 657, 661–64, 667–68, 670, 672–77, 679.

30. Fittingly, Hsü Chi-yü now came back into official favor. He was made a member of the newly instituted foreign office and director of the government's college of foreign languages (T'ung-wen kuan). His "Brief survey" was reprinted in 1866 under government auspices. Swisher, *China's Management,* pp. 690–91.

31. *FRUS 1897,* part 1, p. 513; IWSM:TC, 51:27–32; Knight Biggerstaff, "The Official Chinese Attitude toward the Burlingame Mission," *American Historical Review* (July 1936), 41:684–86.

32. ISHK, 2:24–26, 43–44, 51–57.

33. Richard J. Smith, *Mercenaries and Mandarins: The Ever-Victorious Army in Nineteenth Century China* (Millwood, N.Y., 1978), pp. 56–57, 109.

34. Yung Wing (Jung Hung) was a pivotal figure in the educational mission to the United States. Born in 1828 near Macao of a peasant family, he had attended a mission school in Hong Kong and at eighteen went to the United States for further schooling. Though a convert to Christianity and a naturalized U.S. citizen with an American wife, he had nursed a strong and ultimately unfulfilled ambition to play an influential political role in China. Sending students to the United States, an idea that originated with him, was but one of a long string of proposals for strengthening China that he was to make to Ch'ing authorities. Unable to secure a satisfactory position in the imperial power structure, Yung threw his support to the cause of revolution after 1898. He died in the United States in 1912. The substantial secondary literature on the educational mission and Yung Wing, discussed at the outset of the notes for this chapter, is the basis for the generalizations on the mission here and below.

35. Burlingame's instructions of December 20, 1868 in Wang I et al., comps., *T'ung-chih t'iao-yüeh* [Treaties of the T'ung-chih reign] (Taipei reprint, 1963), p. 176; H. D. Gordon, "Japan's Abortive Colonial Venture in Taiwan, 1874," *Journal of Modern History* (June 1965), vol. 37.

36. Smith, *Mercenaries and Mandarins,* pp. 111–14; Martin R. Ring, "The Burgevine Case and Extrality in China, 1863–1866," *Papers on China* (1969), vol. 22A; IWSM:TC, 33:1–3; Chung-yang yen-chiu yüan, chin-tai shih yen-chiu so, comp., *Hai-fang tang* [Records on maritime defense] (Taipei, 1957) "Kou-mai chuan-p'ao," part 1, p. 358; George Seward to Burlingame, June 5, 1865, Burlingame Family Papers (LC).

37. Not until 1905 were Chinese allowed into West Point. Annapolis, even then, remained closed. CR: Shanghai (January 4 and February 17, 1872); *FRUS 1875,* pp. 227–28; *FRUS 1883,* p. 126; RG 94, file 978425; NF 7421.

38. Mary Coolidge, *Chinese Immigration* (New York, 1909), pp. 498–500; Mary F. Somers Heidhues, *Southeast Asia's Chinese Minorities* (New York, 1975), p. 16.

39. Sam Yup consisted of the districts of Nan-hai, P'an-yü, and Shun-te. Sze Yup divided into Hsin-ning (later Toi-shan), Hsin-hui, En-p'ing and K'ai-p'ing. It was sometimes referred to as the five districts (Ng Yup) when listed along with

adjoining Ho-shan district, which shared the general physical and cultural characteristics of the other four. The Sam Yup and Sze Yup dialects were mutually unintelligible. Further diversity of dialect existed within the Sze Yup group. The speech common to Hsin-ning tended to prevail because of the large number of immigrants from that area but other Sze Yup dialects could at least be understood by anyone from that general area.

40. Frederic Wakeman, Jr., *Strangers at the Gate: Social Disorder in South China, 1839–1861* (Berkeley, Calif., 1966), esp. pp. 179–80 for population estimates; Wan Lo, "Communal Strife in Mid-Nineteenth Century Kwangtung: The Establishment of the Ch'ih-ch'i," *Papers on China* (1965), vol. 19; Yuk Ow et al., eds., *Lu-Mei san-i tsung-hui-kuan chien-shih* [A concise history of the main branch of the Sam Yup Association in the United States] (San Francisco, 1975), pp. 56–57; Robert G. Lee, "The Origins of Chinese Immigration to the United States, 1848–1882," in Chinese Historical Society of America, comp., *The Life, Influence and the Role of Chinese in the United States, 1776–1960* (San Francisco, 1976), pp. 185–88. Dwight H. Perkins, *Agricultural Development in China, 1368–1968* (Chicago, 1969), has challenged the contention that Kwangtung was suffering an acute demographic crisis in this period. His argument for scaling down the estimated population figure for 1851 is not entirely compelling, and of course even if it were, it would still leave the Canton delta a possible exception to general trends in the province. See pp. 19, 45, 206–7, 212, 214, 234, and 236. The statistical basis for any generalization remains highly problematic.

41. Somers Heidhues, *Southeast Asia's Chinese Minorities*, p. 16.

42. Some Chinese in California had brought in contract laborers but by the early 1850s had given up the practice as unprofitable and ill-suited to the American environment. Kil Young Zo, "Chinese Emigration into the United States, 1850–1880" (Ph.D. diss., Columbia University, 1971), pp. 88–98; M. Foster Farley, "The Chinese Coolie Trade, 1845–1875," *Journal of Asian and African Studies* (July–October 1968), vol. 3; Robert L. Irick, "Ch'ing Policy Toward the Coolie Trade, 1847–1878" (Ph.D. diss., Harvard University, 1971).

43. The details on methods of emigration covered in this and the following paragraph are conveniently summarized in Zo, "Chinese Emigration," pp. 95–104; Sing-wu Wang, *The Organization of Chinese Emigration, 1848–1888* (San Francisco, 1978), pp. 89–112, 197.

44. Zo, "Chinese Emigration," p. 124; Elmer C. Sandmeyer, *The Anti-Chinese Movement in California* (Urbana, Ill., 1939), p. 59; CWS:KH, 79:27–32; Walter N. Fong, "The Chinese Six Companies," *Overland Monthly* (May 1894), 2d series, 23:524; Li Kuei, "Tung-hsing jih-chi" [Diary of a trip to the east] in Wang Hsi-ch'i, comp., *Hsiao-fang-hu-chai yü-ti ts'ung-ch'ao* [A collection of geographical works from the Hsiao-fang-hu studio] (Taipei reprint, 1962), 12:93, 95.

45. Hui-chen Wang Liu, *The Traditional Chinese Clan Rules* (Locust Valley, N.Y., 1959); Hu Hsien-chin, *The Common Descent Group in China and Its Function* (New York, 1948); W. E. Willmott, "Chinese Clan Associations in Vancouver," *Man* (March–April 1964), vol. 64.

46. Ho Ping-ti, *Chung-kuo hui-kuan shih-lun* [An historical survey of Landsmannschaften in China] (Taipei, 1966); William Speer, "Democracy of the Chinese," *Harper's Magazine* (November 1868), 37:845–47.

47. Hosea B. Morse, *The Gilds of China* (London, 1909); C. F. R. Allen, trans., "Regulations of the Canton Guild at Foochow," in Justus Doolittle, *A Vocabulary and Handbook of the Chinese Language* (Foochow, 1872), 2:399–402; Alexander

Saxton, *The Indispensable Enemy: Labor and the Anti-Chinese Movement in California* (Berkeley, Calif., 1971), pp. 215, 217; Walter N. Fong, "Chinese Labor Unions in America," *Chautauquan* (July 1896), vol. 23; Thomas W. Chinn et al., *A History of the Chinese in California* (San Francisco, 1969), pp. 50–52, 54, 61.

48. Stewart Culin, "The I Hing or 'Patriotic Rising,' A Secret Society among the Chinese in America," *Report of the Proceedings of the Numismatic and Antiquarian Society of Philadelphia for the Years 1887–89* (1891); G. William Skinner, *Chinese Society in Thailand: An Analytical History* (Ithaca, N.Y., 1957), p. 140; Stanford Lyman et al., "Rules of a Chinese Secret Society in British Columbia," in Lyman, *Asians in the West* (Reno, Nev., 1970); Feng Tzu-yu, *Ko-ming i-shih* [Reminiscences of the revolution] (Ch'ang-sha, Chungking, and Shanghai, 1939–1947), 2:122; Chinn, *A History of the Chinese in California*, p. 68; Li Ch'ang-fu, *Chung-kuo chih-min shih* [A history of Chinese migration] (Shanghai, 1937), pp. 238–39.

49. Thanks to arrangements with steamship companies, the Six Companies could prevent the flight of debtors and collect a departure tax, which was applied to local relief with any surplus remitted to China in times of disaster. Fong, "The Chinese Six Companies"; Li Kuei, "Tung-hsing jih-chi," 12:95.

50. Fong, "The Chinese Six Companies," p. 524; and Alexander Saxton, "The Army of Canton in the High Sierras," *Pacific Historical Review* (May 1966), 35:149–50.

51. Chinese on the Central Pacific Railway earned the equivalent of about $30 per month or roughly two-thirds the wages of whites, who got room and board in addition. Chinese employed in industry or agriculture also earned about $1.00 a day. Chinese hired to work in the mines initially received $1.00 or $1.25 a day but time and experience pushed that up to $1.75 or $2.00. Fragmentary evidence suggests Chinese living expenses were about the same as those of whites—about $16 per month. Ow, *Lu-Mei san-i*, p. 57; Saxton, "The Army of Canton," p. 149; Ping Chiu, *Chinese Labor in California, 1850–1880: An Economic Study* (Madison, Wis., 1963), pp. 35, 39, 72, 126–27; Stewart Culin, "Customs of the Chinese in America," *Journal of American Folklore* (July–September 1890), 3:192–93, and his "The I Hing," pp. 54, 57; Chinn, *A History of the Chinese in California*, p. 20; Rodman W. Paul, *California Gold: The Beginning of Mining in the Frontier West* (Lincoln, Nebr., 1947), pp. 351–52.

52. Quoted in Huang Fu-luan, *Hua-ch'iao yü Chung-kuo ko-ming* [Overseas Chinese and China's revolution] (Hong Kong, 1954), preface, p. 7.

53. Betty Lee Sung, *The Story of the Chinese in America* (New York, 1971), p. 320. If the census figures used by Lucie Cheng Hirata, "Chinese Immigrant Women in Nineteenth-Century California," in Carol R. Burkin and Mary Beth Norton, eds., *Women of America: A History* (Boston, 1979), are accurate, then by 1880 prostitution declined to a point that it employed only a quarter of all Chinese women in the United States, while the numbers of housewives and female wage-earners rose to the point that they accounted for fully 70 percent. It is, however, conceivable that the figures are not accurate. Chinese respondents to the 1880 census may have found it prudent in the wake of nativist attacks on Chinese immorality to hide prostitutes in another occupational category.

54. Culin, "Customs of the Chinese," p. 199; "Life Story of a Chinaman" in Hamilton Holt, ed., *The Life Stories of Undistinguished Americans* (New York, 1906).

55. Nancy Farrar, *The Chinese in El Paso* (El Paso, Tex., 1972); Stewart Culin, "Social Organization of the Chinese in America," *American Anthropologist* (October 1891), vol. 4.

56. Over one-half of the Sze Yup people came from Hsin-ning (Toi-shan). Aside from Sze Yup and Sam Yup, the two other major groups were the people from Hsiang-shan district and the Hakkas. Hsiang-shan accounted for a little more than a third (14,000) of the population in the mid-1850s, but fell to numerical parity with Sam Yup by the mid-1870s. The Hakka population never exceeded 7 percent or 5,000 people. Chinn, *A History of the Chinese in California*, p. 20.

57. California's 1877 anti-Chinese memorial to Congress, included in Wu Cheng-Tsu, ed., *"Chink!": A Documentary History of Anti-Chinese Prejudice in America* (New York, 1972), pp. 113–23, is a cogent and coherent statement. See also the testimony in California Senate Special Committee on Chinese Immigration, *Chinese Immigration* (Sacramento, Calif., 1878), pp. 46–51, 59, 80–81, 122–23, 128–33; *Valedictory Address of Gen. A. M. Winn, President, to the Mechanics State Council of California* (pamphlet; San Francisco, 1871), pp. 4–5; Henry George, "The Chinese on the Pacific Coast" appended in *ibid*, pp. 18–19; M. J. Dee, "Chinese Immigration," *North American Review* (1878), 126:524. Luther Spoehr's "Sambo and the Heathen Chinee: Californians' Racial Stereotypes in the Late 1870s," *Pacific Historical Review* (May 1973), vol. 42, helps underline the point that the Chinese fell short of American cultural standards, unlike blacks whose inferiority was thought to be both biological and cultural.

58. David B. Davis, "Some Themes of Counter Subversion: An Analysis of Anti-Masonic, Anti-Catholic, and Anti-Mormon Literature," *Mississippi Valley Historical Review* (September 1960), vol. 47, contains some stimulating observations on the nature of nativism in preceding decades.

59. Moses Rischin, "Immigration, Migration and Minorities in California: A Reassessment," *Pacific Historical Review* (February 1972), 41:74. See also John Higham, *Strangers in the Land: Patterns of American Nativism, 1860–1925* (New York, 1963), on the uses of nativism during periods of stress.

60. Saxton, *The Indispensable Enemy*, chs. 1–6.

61. Chiu, *Chinese Labor*, develops this point in detail. See also R. Hal Williams, *The Democratic Party and California Politics, 1880–1896* (Stanford, Calif., 1973), p. 15; Coolidge, *Chinese Immigration*, p. 498.

62. The constitution, the pet project of the Workingmen's party, won overwhelming popular approval. But the party itself was already on the decline. Kearney with his inflammatory rhetoric had won the party its mass following and helped divert the urban violence of 1877 into political channels. He could not, however, hold the party together, and by 1882 it had disappeared as a political force. Ralph Kauer, "The Workingmen's Party of California," *Pacific Historical Review* (September 1944), vol. 13; Saxton, *The Indispensable Enemy*, pp. 116–32, 138–48, 152–56.

63. Sandmeyer, *The Anti-Chinese Movement*, pp. 41–45, 51–55, 62–63, 74–76; and Sandmeyer, "California Anti-Chinese Legislation and the Federal Courts," *Pacific Historical Review* (1936), vol. 5. A miscegenation law applicable to the Chinese did not appear on the books until 1901. Other western states copied California's pioneering legislation and sometimes added to it (e.g., prohibitions against land ownership).

64. *Congressional Record*, 44th Cong., 1st Sess., pp. 2850–58, 3099–102, 4418–21, 4678; 45th Cong., 2d Sess., pp. 1544–53, 2439–40, 4328–32; U.S. Congress, *Report of the Joint Special Committee to Investigate Chinese Immigration* (Senate report 689; Washington, D.C., 1877), pp. iii–viii.

65. Robert Seager, "Some Denominational Reactions to Chinese Immigration to California, 1856–1892," *Pacific Historical Review* (February 1959), vol. 28; Michael

L. Stahler, "William Speer: Champion of California's Chinese, 1852–1857," *Journal of Presbyterian History* (Summer 1970), vol. 48; Speer, "Democracy of the Chinese," pp. 839, 848; A. W. Loomis, "The Six Chinese Companies," *Overland Monthly* (September 1868), 1:227; Loomis, "How Our Chinamen are Employed," *Overland Monthly* (March 1869), vol. 2; Otis Gibson testimony in California Senate, *Chinese Immigration*, pp. 33–34; Gibson, *The Chinese in America* (Cincinnati, 1877), pp. 49–58, 111–14, 124–26.

66. *Congressional Record*, 44th Cong., 1st Sess., pp. 4418–21; Sidney Andrews, "Wo Lee and His Kinfolk," *Atlantic Monthly* (February 1870), vol. 25.

Part Two: Emerging Patterns of Interaction

1. *FRUS 1886*, pp. 101–33; J. H. Goodnough, "David G. Thomas' Memories of the Chinese Riot," *Annals of Wyoming* (July 1947), vol. 19; and *MPHS*, pp. 78–79, describe the massacre from a variety of contemporary perspectives.

2. *FRUS 1886*, p. 129.

3. Paul Crane and Alfred Larson, "The Chinese Massacre," *Annals of Wyoming*, (January and April 1940), vol. 12; Aaron A. Sargent, "The Wyoming Anti-Chinese Riot," *Overland Monthly* (November 1885), 2d series, 6:508; the reply to Sargent by "J," "The Wyoming Anti-Chinese Riot—Another View," ibid. (December 1885); Alexander Saxton, *The Indispensable Enemy: Labor and the Anti-Chinese Movement in California* (Berkeley, Calif. 1971), pp. 202–5.

Chapter 3: The Politics and Diplomacy of Exclusion, 1879–1895

The initial section of this chapter owes much to the intellectual provocation supplied by Stuart C. Miller's Unwelcome Immigrant. Miller directly challenges the "California thesis" laid out most clearly in Elmer Sandmeyer, The Anti-Chinese Movement in California (Urbana, Ill., 1939), but also evident in an earlier work by Mary Coolidge, Chinese Immigration (New York, 1909), and a later work by Gunther Barth, Bitter Strength: A History of the Chinese in the United States, 1850–1870 (Berkeley, Calif., 1964). Miller argues that a racist view of the Chinese as outlandish and unassimilable heathens was not the exclusive possession of California or the West (as Sandmeyer, Coolidge, and Barth contend), but pervaded the entire United States in the latter half of the nineteenth century. Exclusion thus emerges in a new light in Miller's account—as a popular nationwide concern and not an aberration resulting from the intense agitation of one region. Stanford Lyman, Chinese Americans (New York, 1974), pp. 62–65; Roger Daniels' introduction, "The Anti-Chinese Movement in Historical Perspective," to Elmer C. Sandmeyer, The Anti-Chinese Movement in California (Urbana, Ill., reprint, 1973), p. 5; and Jack L. Hammersmith, "West Virginia, the 'Heathen Chinee,' and the 'California Conspiracy'," West Virginia History (April 1973), vol. 34, all follow Miller in their emphasis on nationwide racism.

There is no single account that gives adequate general coverage of the late Ch'ing defense of overseas Chinese. Robert L. Irick, "Ch'ing Policy Toward the Coolie Trade, 1847–1878" (Ph.D. diss., Harvard University, 1971); Yen Ching-hwang, "Ch'ing Changing Images of the Overseas Chinese (1644–1912)," Modern Asian Studies (April 1981), vol. 15; and Michael R. Godley, "The Late Ch'ing Courtship of the Chinese in Southeast Asia," Journal of Asian Studies (February 1975), vol. 34, offer insight on the evolution of Chinese policy. The literature on the effort to safeguard the Chinese in the United States is equally fragmentary. The watershed treaty of 1880 is treated by Li Ting-i in "Tsao-ch'i Hua-jen i-Mei chi 'An-chi-li' t'iao-yüeh chih ch'ien-ting' [Early Chinese emigration to the United States and the signing of the Angell treaty], Lien-ho shu-yüan hsüeh-pao (Hong Kong, 1964), no. 3, and Li Tsung-t'ung and Liu Feng-han, comps., Li Hung-tsao hsien-sheng nien-p'u [Chronological biography of Li Hung-tsao] (Taipei, 1969), pp. 258–300. The latter vividly establishes the Chinese context of the treaty talks, while Coolidge, Chinese Immigration, ch. 10, provides a still helpful overview. The pressures that led to the treaties of 1888 and 1894 have been treated only from the American perspective—in John A. S. Grenville and George B. Young, Politics, Strategy, and American Diplomacy: Studies in Foreign Policy, 1873–1917 (New Haven, Conn., 1966), pp. 51–63, and in George E. Paulsen, "The Gresham-Yang Treaty," Pacific Historical Review (August 1968), vol. 37. The reaction in China, above all in the Canton region, to the American exclusionist onslaught remains a potentially fascinating yet woefully neglected topic. The views that Chinese ministers to the United States Ch'en Lan-pin, Chang Yin-huan, and Tsui Kuo-yin expressed in their diaries (now available in print) might also repay further study.

    1. Two Presidents gained office without a majority of the popular vote, and two by majorities of less than 25,000.
    2. Donald B. Johnson and Kirk H. Porter, comps., National Party Platforms, 1840–1972, 5th ed. (Urbana, Ill., 1975), pp. 50, 57.
    3. Olive R. Seward, William H. Seward's Travels Around the World (New York, 1873), p. 200; W. A. P. Martin, A Cycle of Cathay (New York, 1900), p. 376.
    4. Grant made passing reference in his first annual address in December 1869 to putting an end to enslaved coolie labor, but at the same time he endorsed the recently concluded Burlingame treaty which provided for free immigration. James B. Richardson, comp., A Compilation of the Messages and Papers of the Presidents (Washington, D.C., 1896–1900; supplement, 1903), 7:37, 8:288; Johnson and Porter, National Party Platforms, p. 54.
    5. Richardson, A Compilation, 7:516–19, 569; Charles C. Tansill, The Foreign Policy of Thomas F. Bayard (Bronx, N.Y., 1940), pp. 128–30; Charles R. Williams, The Life of Rutherford Birchard Hayes (Boston, 1914), 2:213–18; Charles R. Williams, ed., Diary and Letters of Rutherford Birchard Hayes (Columbus, Ohio, 1922–1926), 3:522–26; George Sinkler, The Racial Attitudes of American Presidents from Abraham Lincoln to Theodore Roosevelt (Garden City, N.Y., 1971), pp. 193–96, 400–1; Gary Pennanen, "Public Opinion and the Chinese Question, 1876–1879,"

Ohio History (1968), 77:143–45, 148; and Johnson and Porter, National Party Platforms, p. 62.

6. In 1879 and 1880 opponents of exclusion came up with an ingenious variety of proposals intended to extricate the Hayes administration from its bind. Some, for example, searched for precedents in China's laws and treaties which could be invoked to justify immigration controls, while others suggested cutting off the queue would scare away some Chinese, render the remainder less conspicuous, hasten assimilation, and thereby dampen the cry for exclusion. Seward memo, March 25, 1879, quoted in extenso in Tansill, The Foreign Policy of Thomas F. Bayard, p. 131; Evarts in FRUS 1880, pp. 301–2; Trescott to Evarts, August 15, 1880, Evarts Papers (LC); Otis Gibson in Angell to Evarts, May 10, 1880, Evarts Papers (Sterling Library, Yale University); and Angell to Evarts, March 11, 1880, quoted in extenso in Shirley W. Smith, James Burrill Angell (Ann Arbor, Mich., 1954), p. 120.

7. Tansill, The Foreign Policy of Thomas F. Bayard, p. 131; Seward memo (n.d.; circa 1908?), misc. file in Seward Papers (New-York Historical Society); Seward's "The United States and China" (pamphlet dated October 28, 1902) in Seward Papers; and David L. Anderson, "The Diplomacy of Discrimination: Chinese Exclusion, 1876–1882," California History (Spring 1978), 57:34–38.

8. The details of the Li-Grant deal are covered in chapter 4. Esson M. Gale, "President James Burrill Angell's Diary as United States Commissioner and Minister to China, 1880–1881," Michigan Alumnus Quarterly Review (May 1943), 49:195, 198–99, 203; Angell to Evarts, May 11, 1880, Evarts Papers (LC); Smith, James Burrill Angell, pp. 116–29; Brainerd Dyer, The Public Career of William M. Evarts (Berkeley, Calif., 1933), pp. 221–22; Anderson, "The Diplomacy of Discrimination," pp. 40–41; DIC (June 7, 1880); MCD (October 26, 1880); Coolidge, Chinese Immigration, pp. 152–53.

9. Harry J. Brown and Frederick D. Williams, eds., The Diary of James A. Garfield (East Lansing, Mich., 1967–), 3:87; Theodore C. Smith, The Life and Letters of James Abram Garfield (New Haven, Conn., 1925), 2:677; Burke A. Hinsdale, The Republican Text-book for the Campaign of 1880 (New York, 1880), pp. 151–52; Sinkler, Racial Attitudes, p. 211.

10. Richardson, A Compilation, 8:112–18; George F. Howe, Chester A. Arthur: A Quarter Century of Machine Politics (New York, 1935), pp. 116, 168–69; and Thomas C. Reeves, Gentleman Boss: The Life of Chester Alan Arthur (New York, 1975), pp. 278–79. The final version of this and subsequent legislation appears in FRUS 1892, pp. 107–13.

11. Charles Denby, China and Her People (Boston, 1906), 1:15–16; Richardson, A Compilation, 8:329; and George F. Parker, ed., The Writings and Speeches of Grover Cleveland (New York, 1892), pp. 11, 36.

12. Cleveland to Bayard, December 18, 1887, Bayard Papers (LC); and the political correspondence in Cleveland Papers (LC), especially from English, Irish, Tarkey, and Kearney under September 2, 8, 15, 16, 18, and October 1, 3, 6, 1888.

13. Richardson, A Compilation, 8:631, 634.

14. Tansill, The Foreign Policy of Thomas F. Bayard, pp. 154, 160, 162–71, 174–77; Sinkler, Racial Attitudes, pp. 237–38; R. Hal Williams, The Democratic Party and California Politics, 1880–1896 (Stanford, Calif., 1973), pp. 124, 129–30.

15. Harry J. Sievers, Benjamin Harrison: Hoosier Statesman, (New York, 1952–1968), 2:223–24, 322–23, 409; Lew Wallace, Life of Gen. Ben Harrison (Philadelphia, 1888), pp. 342–48; Charles Hedges, comp., Speeches of Benjamin Harrison (New York, 1892), pp. 111–12; Sinkler, Racial Attitudes, pp. 284–87; Williams, The Democratic Party, p. 122.

Notes to Pages 94–99

16. James G. Blaine, *Political Discussions: Legislative, Diplomatic, and Popular, 1856–1886* (Norwich, Conn., 1887), pp. 216–45 (esp. pp. 242–44 on negligible missionary and commercial interests); Blaine, *Twenty Years of Congress* (Norwich, Conn., 1884–1886), 2:651–56; Mary Abigail Dodge, *Biography of James G. Blaine* (Norwich, Conn., 1895), pp. 486–87; Alice F. Tyler, *The Foreign Policy of James G. Blaine* (Minneapolis, Minn., 1927), pp. 18, 256, 261. In a mid-1887 colloquy with Cleveland, thirteen western congressmen had strongly and unanimously expressed their willingness to see trade interests subordinated to exclusion. John A. S. Grenville and George B. Young, *Politics, Strategy, and American Diplomacy: Studies in Foreign Policy, 1873–1917* (New Haven, Conn., 1966), pp. 55–56.

17. Coolidge, *Chinese Immigration*, pp. 209–18.

18. *FRUS 1893*, pp. 234–64 passim; George E. Paulsen, "The Gresham-Yang Treaty," *Pacific Historical Review* (August 1968), vol. 37.

19. I refer here primarily to Stuart C. Miller's *Unwelcome Immigrant: The American Image of the Chinese, 1785–1882* (Berkeley, Calif., 1969).

20. *Congressional Record*, 45th Cong., 3d Sess., pp. 800–1; 47th Cong., 1st Sess., p. 3412; 50th Cong., 1st Sess., p. 8369.

21. Kil Young Zo, "Chinese Emigration into the United States, 1850–1880" (Ph.D. diss., Columbia University, 1971), p. 22; W. A. P. Martin, *A Cycle of Cathay*, p. 160; *IWSM:TC*, 50:32, Edgar Wickberg, *The Chinese in Philippine Life, 1850–1898* (New Haven, Conn., 1965), p. 211.

22. Wickberg, *The Chinese in Philippine Life*, p. 227; Li Wen-chih et al., comps., *Chung-kuo chin-tai nung-yeh shih tzu-liao, ti-i chi* [Historical materials on modern Chinese agriculture, first series] (Peking, 1957), 1:941–42.

23. Earl Swisher, *China's Management of the American Barbarians: A Study of Sino-American Relations, 1841–1861* (New York reprint, 1972), pp. 199–203; *CKS: Hsien-feng ch'ao* [Hsien-feng reign], pp. 203, 330–31, 342–53; Chung-yang yen-chiu yüan, chin-tai shih yen-chiu so, comps., *Ssu-kuo hsin-tang* [New files concerning the four powers] (Taipei, 1966), "Ying-kuo tang," pp. 1028–29, and "Mei-kuo tang," p. 184; *CKS:TC*, pp. 311–16, 613–15, 782, 999–1000, 1005, 1029–30, 1039, 1040–41, 1052–56; Watt Stewart, *Chinese Bondage in Peru: A History of the Chinese Coolie in Peru, 1849–1874* (Durham, N.C., 1951), pp. 176–77; Sing-wu Wang, "The Attitude of the Ch'ing Court Toward Chinese Emigration," *Chinese Culture*, (December 1968), 9:65–75.

24. *IWSM:TC*, 55:12–13; Chih Chün, comp., *Ch'i-pu-fu-chai cheng-shu* [Official papers of Chou Chia-mei] (Taipei reprint, 1973), 1:4, 27, 2:26–27; Tseng Chi-tse, *Tseng Hui-min-kung i-chi* [Collected works of Tseng Chi-tse] (Taipei reprint, 1968), "Wen-chi," 3:6; *CWC*, 15:7ff. and 23:8–15; *CWS:KH*, 68:5–6, 74:22–27, 75:18–21, 87:14–18; Knight Biggerstaff, "The Establishment of Permanent Chinese Diplomatic Missions Abroad," *Chinese Social and Political Science Review* (April 1936), vol. 20; Wickberg, *The Chinese in Philippine Life*, pp. 217, 226.

25. *SL:KH*, 327:1, 450:9; Ch'en Pi, *Wang-yen-t'ang tsou-kao* [Collected memorials of Ch'en Pi] (Taipei reprint, 1968), 1:4; *CWS:KH*, 119:21; Hsüeh Fu-ch'eng, *Yung-an wen-pien* [Papers of Hsüeh Fu-ch'eng] (Taipei reprint, 1973), "Hai-wai wen-pien," 1:12–15; Wang Ching-yü, comp., *Chung-kuo chin-tai kung-yeh shih tzu-liao, ti-erh chi, 1895–1914 nien* [Historical materials on modern Chinese industry, second series, 1895–1914] (Peking, 1957), 2:982–83; Wickberg, *The Chinese in Philippine Life*, pp. 226–28, 233–34; *IWSM:TC*, 55:19–20; Immanuel C. Y. Hsü, *China's Entrance into the Family of Nations: The Diplomatic Phase, 1858–1880* (Cambridge, Mass., 1960), pp. 161, 171–72.

26. Biggerstaff, "The Establishment of Permanent Chinese Diplomatic Missions

Abroad," p. 13; Li Kuei, "Tung-hsing jih-chi" [Diary of a trip to the east] in Wang Hsi-ch'i, comp., *Hsiao-fang-hu-chai yü-ti ts'ung-ch'ao* [A collection of geographical works from the Hsiao-fang-hu studio] (Taipei reprint, 1962), 12:95; *CWS:KH*, 4:17–19, 5:17, 9:19–20, 14:31–32, 25:22–23, 31:22–23; Lu Feng-shih, comp., *Hsin-tsuan yüeh-chang ta-ch'üan* [A new compilation of treaty provisions] (Shanghai, 1909), 13:8.

27. *CYCT*, 4:1919; Chung-kuo k'o-hsüeh yüan, li-shih yen-chiu so, ti-san so, comp., *Liu K'un-i i-chi* [Collected works of Liu K'un-i] (Peking, 1959), p. 730; *CWC*, 15:8, 24:25–28; *CJCC*, 5:262–64; Chih Chün, *Ch'i-pu-fu-chai cheng-shu*, 5:50–52; Noriko Kamachi, "American Influences on Chinese Reform Thought: Huang Tsun-hsien in California, 1882–1885," *Pacific Historical Review* (May 1978), 47:256.

28. A number of other ranking foreign office officials including Shen Kuei-fen and Wang Wen-shao joined Li and Pao-yün to form an informal negotiating team. All were members of the powerful Grand Council.

29. Li Tsung-t'ung and Liu Feng-han, comps., *Li Hung-tsao hsien-sheng nien-p'u* [Chronological biography of Li Hung-tsao] (Taipei, 1969), pp. 258, 272–80; Smith, *James Burrill Angell*, pp. 132–35; *CWS:KH*, 22:17.

30. Gale, "Angell Diary," p. 203; Smith, *James Burrill Angell*, pp. 135–36; *CWS:KH*, 24:8–10, 12–14; Li and Liu, *Li Hung-tsao*, pp. 234, 280–90, 294–300.

31. *CWS:KH*, 63:18; MCD (Denby to Bayard, February 9, March 10, July 31, 1886); Miriam Levering, "The Chungking Riots of 1886," *Papers on China* (May 1969), 22A:160.

32. *CWS:KH*, 63:26, 64:9, 11–12, 67:4–10, 78:13, 17; *FRUS 1886*, p. 82; Tansill, *The Foreign Policy of Thomas F. Bayard*, pp. 143–44.

33. *CWS:KH*, 61:18, 37, 63:26, 66:12, 68:8–9, 12, 70:12; *MPHS*, pp. 102–3, 110, 112–15, 123; Theodore M. Liu, "Chang Yin-huan, Minister Plenipotentiary to the United States, 1886–1889" (M.A. thesis, Columbia University, 1956), pp. 45–57; John W. Foster, *Diplomatic Memoirs* (Boston, 1909), 2:285–87.

34. This set of views persisted in the legation into the 1890s: Jung Hung to Li Hung-chang (n.d., circa 1876), summarized in *FRUS 1876*, p. 58; Li Kuei, "Tung-hsing jih-chi," 12:95; Chang Chih-tung and Chang Yin-huan, joint memorial of March 30, 1886, in *CWC*, 15:8; Chang Chih-tung, memorial of June 13, 1886, in *CWS:KH*, 67:8; Chang Yin-huan, supplementary memorial of August 13, 1886, in *CWS:KH*, 68:8; Prince Ch'ing to Denby, August 3, 1886, in *MPHS*, p. 121; Cheng Tsao-ju to the foreign office, January 20, 1886, in *CWS:KH*, 79:32; Hsü Chüeh, *Fu-an i-chi* [Works of Hsü Chüeh] (Taipei reprint, 1970), 2:701–9; Yang-ju in *CWS:KH*, 97:5–7; Tsui Kuo-yin, *CWS:KH*, 97:5.

35. *CWS:KH*, 63:17–18, 79:32–36.

36. Chang was at the beginning of an active but ultimately tragic diplomatic career. He had briefly served in the foreign office during the Sino-French War before taking up the immigration problem in the United States. He later became involved in negotiating an end to China's losing war with Japan and handing over the Liao-tung leasehold to Russia. With his reputation already tainted by these foreign policy reverses, Chang made the mistake of supporting the reform movement in 1898. Its collapse was followed by his exile and finally his execution at the height of the Boxer movement. Ho Ping-ti, "Chang Yin-huan shih-chi" [Notes on Chang Yin-huan], in Pao Tsun-p'eng et al., comps., *Wei-hsin yü pao-shou* [Reform and conservation] (Taipei, 1959).

37. *CWS:KH*, 68:9, 70:12, 75:38, 76:1–2; Tansill, *The Foreign Policy of Thomas F. Bayard*, pp. 145–46, 154, 157, 160–65; Liu, "Chang Yin-huan," pp. 62–67; Coolidge, *Chinese Immigration*, pp. 191–200.

38. CWS:KH, 76:14–15, 23, 77:3, 10, 17, 25, 26, 78:17, 79:27–32; ISHK, 19:18–20; CWC, 24:25–28; Denby to Bayard, September 6, 1888, Bayard Papers (LC); Tansill, The Foreign Policy of Thomas F. Bayard, pp. 170–72, 174, 177–80; FRUS 1888, part 1, pp. 354–55.

39. Examples of this line of argument appear in Jung Hung to the Secretary of State, March 9, 1880 (in MPHS, pp. 60–61); Li Hung-chang in the fall of 1885 (quoted in Tansill, The Foreign Policy of Thomas F. Bayard, pp. 138–39n); Chang Yin-huan, diary entry of June 26, 1886 (in A Ying [pseudonym for Ch'ien Hsing-ts'un], comp., Fan Mei Hua-kung chin-yüeh wen-hsüeh chi [Collected literature on opposition to the American treaty excluding Chinese laborers] (Peking, 1960), pp. 581–82; Prince Ch'ing to Denby, n.d. [circa 1890?] (in MPHS, p. 137); Li Hung-chang in conversations with Americans on June 29 and October 25, 1892 (CR:Tientsin and FRUS 1892, p. 134); and Yang-ju, memorial of early May 1894 (in MPHS, p. 141).

The only effective form of retaliation the Chinese actually resorted to was to refuse to receive American ministers with an exclusionist record. The practical result was virtually to close the legation doors beginning in the mid-1880s to Californians who had earlier enjoyed an inside track in the appointment process—thanks as much to the importance of their state's swing votes to Republican presidential aspirants as to their state's vista on the Pacific. (J. Ross Browne had been the first of the Californians in Peking, followed by Frederick Low, Benjamin Avery, and John F. Swift [a member of the commission that Hayes sent out in 1880]). Cleveland had to abandon one serious California candidate for the Peking post due to Chinese objections, and President Harrison had to give up on two of his candidates before Washington began to take the Chinese veto of exclusionists as a given in making appointments. Lina F. Browne, ed., J. Ross Browne: His Letters, Journals and Writings (Albuquerque, N.M., 1969), pp. 343–44; Tansill, The Foreign Policy of Thomas F. Bayard, p. 136n; A. T. Volwiler, ed., The Correspondence between Benjamin Harrison and James G. Blaine, 1882–1893 (Philadelphia, 1940), pp. 145, 191, 220, 229; CWS:KH, 80:10, 15, 19, 20, 27; ISHK, 20:13–14.

40. FRUS 1893, pp. 234–35, 244–53, 255, 257–59; Paulsen, "The Gresham-Yang Treaty," pp. 281–90.

41. FRUS 1893, pp. 263–64; Paulsen, "The Gresham-Yang Treaty," pp. 291–97; CWS:KH, 88:23–25, 89:30–34, 94:22–24, 107:32–33; Hsü, Fu-an i-chi, 1:207–33, 237, 241–42, 2:734–38.

42. Coolidge, Chinese Immigration, pp. 498–501; Betty Lee Sung, The Story of the Chinese in America (New York, 1971), p. 320.

43. Ping Chiu, Chinese Labor in California, 1850–1880: An Economic Study (Madison, Wis., 1963), pp. 115, 125–27; CWS:KH, 63:17–18, 89:30–34; FRUS 1879, pp. 237–41. See also the recent reexamination by Paul M. Ong, "Chinese Labor in Early San Francisco: Racial Segmentation and Industrial Expansion," Amerasia Journal (Spring/Summer 1981), vol. 8.

44. Thomas W. Chinn et al., A History of the Chinese in California (San Francisco, 1969), p. 10; Roger W. Lotchin, San Francisco, 1846–1856: From Hamlet to City (New York, 1974), p. 120; Edward C. Lydon, "The Anti-Chinese Movement in Santa Cruz, California: 1859–1900," in Chinese Historical Society of America, comp., The Life, Influence and the Role of Chinese in the United States, 1776–1900 (San Francisco, 1976), pp. 231–38; Gregg L. Carter, "Social Demography of the Chinese in Nevada: 1870–1880," Nevada Historical Society Quarterly (Summer 1975), 18:74.

45. Sung, The Story of the Chinese in America, p. 320; Ivan Light, "From Vice District to Tourist Attraction: The Moral Career of American Chinatowns, 1880–1940," Pacific Historical Review (August 1974), vol. 43; CWS:KH, 15:7, 94:23–24.

46. In the first case (Chae Chan Ping v. the United States), the Chinese failed to overturn the Scott Act though they spent $100,000 trying. The second case (Fong Yue Ting v. the United States) saw a legal team headed by Joseph Choate mount an unsuccessful attack on the registration provisions of the Geary Act. Lu Feng-shih, *Hsin-tsuan yüeh-chang*, 60:36–37; *Remarks of the Chinese Merchants of San Francisco upon Governor Bigler's Message* . . . (pamphlet; San Francisco, 1855); William Speer, *The Oldest and Newest Empire* (Hartford, Conn., 1870), pp. 588–603; *Facts upon the Other Side of the Chinese Question* . . . (pamphlet; 1876); *Memorial of the Six Chinese Companies* (pamphlet; San Francisco, 1877); Otis Gibson, *The Chinese in America* (Cincinnati, 1877), pp. 315–23; Fred A. Bee, *Memorial: The Other Side of the Chinese Question* (pamphlet; San Francisco, 1886).

47. Yuk Ow et al., eds., *Lu-Mei san-i tsung-hui-kuan chien-shih* [A concise history of the main branch of the Sam Yup Association in the United States] (San Francisco, 1975), p. 60; Gong Eng Ying and Bruce Grant, *Tong War!* (New York, 1930), pp. 25–37, 55–98; John E. Bennett, "The Chinese Tong Wars in San Francisco," *Harper's Weekly* (August 11, 1900), vol. 44; Pardee Lowe, *Father and Glorious Descendant* (Boston, 1943), p. 82.

48. This and the following paragraphs on the legation's strained relation to the Chinese community derive from CWS:KH, 68:8, 79:27–36, 88:1–5, 89:30–34, 97:5–7, 107:32–33; Hsü, *Fu-an i-chi*, 1:175–80, 211–16, 221–31, 237, 241–42, 250–52, 255, 2:701–9, 734–38; Liu, "Chang Yin-huan," pp. 40–45; Ow, *Lu-Mei san-i tsung-hui-kuan chien-shih*, p. 60; Bennett, "The Chinese Tong Wars in San Francisco," p. 747; Stewart Culin, "Customs of the Chinese in America," *Journal of American Folklore*, (July–September 1890), 3:194.

49. Both Chang and Yang passed over in silence the prostitution in Chinatown about which Sinophobes made much. Tolerated in China, it was for the two diplomats an unexceptional feature of this isolated society dominated by single males.

50. Chang and Yang had wanted to secure direct control over wrongdoers by concluding an extradition treaty with the United States, but officials in the State Department and congressmen thought the Chinese community already too autonomous and the Chinese legation likely to wield any extradition powers in a despotic fashion. The State Department avoided a flat refusal for fear that the Chinese, with their preoccupation with reciprocal rights, might cite American extraterritoriality in China as a more sweeping example of precisely what the Chinese wanted for their own subjects in the United States. The Chinese legation finally abandoned the idea.

Chapter 4: The United States in Li Hung-chang's Foreign Policy, 1879–1895

There is no work that offers a full and adequate treatment of Li Hung-chang's approaches to the United States. Of accounts in English, Dennett's Americans in Eastern Asia, part 5, comes closest with its emphasis on the diplomacy of good offices, but it is understandably thin on the Chinese side. Chinese historians writing in the 1950s have provided the interpretive antidote to Dennett's picture of an American policy devoted to the independence of Asian states but unwary in matters of power politics. They find in late nineteenth-century American interest in China's border states the beginnings of an aggressive policy, often conducted in collaboration with Japan, that was to lead to the mid-twentieth-century encirclement of

Chapter 4: Bibliographical Essay

China. Li Hung-chang, denigrated in Dennett's account, fares even less well in these works, which brand him a traitor. The orthodox Chinese Communist view is best articulated in brief compass by Ch'ing Ju-chi, Mei-kuo ch'in-Hua shih [A history of American aggression against China] (Peking, 1956), vol. 2, part 5, ch. 2; Liu Ta-nien, Mei-kuo ch'in-Hua shih [A history of American aggression against China] (Peking, 1951), chs. 4 and 5; and by Ssu Shou-yen's and Shang Yüeh's essays in Chung-Jih chia-wu chan-cheng lun-chi [Essays on the Sino-Japanese War of 1894] (Peking, 1954).

In reconstructing the rise and decline of Li's policy I have relied heavily on the Chinese documentary collections and the testimony of the Americans Li dealt with. But a substantial body of secondary works, generally more sympathetic to Li than the above, proved indispensable in establishing background and context. For bringing together the pieces of Li's long and eventful career, Tou Tsung-i's Li Hung-chang nien (jih) p'u [A (daily) chronology of Li Hung-chang's life] (Hong Kong, 1968) is essential. Articles by Kwang-Ching Liu—"The Confucian as Patriot and Pragmatist: Li Hung-chang's Formative Years, 1823–1866," Harvard Journal of Asiatic Studies (1970), vol. 30, and "Li Hung-chang in Chihli: The Emergence of a Policy, 1870–1875," in Albert Feuerwerker et al., eds., Approaches to Modern Chinese History (Berkeley, Calif., 1967)—and Kenneth E. Folsom's monograph, Friends, Guests, and Colleagues: The Mu-Fu System in the Late Ch'ing Period (Berkeley, Calif., 1968), chs. 5–7, have helped to fill out the picture at important points.

The Liuchiu and Vietnam crises, the subjects of scant historical attention, are best pursued through Hyman Kublin, "The Attitude of China during the Liu-ch'iu Controversy, 1871–1881," Pacific Historical Review (May 1949), vol. 18, and Lloyd Eastman, Throne and Mandarins: China's Search for a Policy during the Sino-French Controversy, 1880–1885 (Cambridge, Mass., 1967). William H. Morken, "America Looks West: The Search for a China Policy, 1876–1885" (Ph.D. diss., Claremont Graduate School, 1974), ch. 2, is the first American diplomatic history to examine Grant's involvement in the Liuchiu dispute although extracts from Li Hung-chang's correspondence on the subject have long been available in English—in an appendix to Charles S. Leavenworth's The Loochow Islands (Shanghai, 1905). American interest in the Vietnam crisis gets a brief nod in David Pletcher, The Awkward Years: American Foreign Policy under Garfield and Arthur (Columbia, Mo., 1962).

The prolonged and complicated Korean problem has been well studied. Key-Hiuk Kim, The Last Phase of the East Asian World Order: Korea, Japan, and the Chinese Empire, 1860–1882 (Berkeley, Calif., 1980), and C. I. Eugene Kim and Han-Kyo Kim, Korea and the Politics of Imperialism, 1876–1910 (Berkeley, Calif., 1967), are the best general works on the subject, but they should be supplemented by Frederick Foo Chien, The Opening of Korea: A Study of Chinese Diplomacy, 1876–1885 (Hamden, Conn., 1967), with its detailed treatment of Sino-American contacts, and Martina Deuchler, Confucian Gentlemen and Barbarian Envoys: The Opening of

Korea, 1875–1885 (Seattle, Wash., 1977). Yur-Bok Lee, Diplomatic Relations Between the United States and Korea, 1866–1887 (New York, 1970), written exclusively from American sources, is Sinophobic in tone and full of praise for Americans who supported Korean independence against Chinese interference. Mary C. Wright, "The Adaptability of Ch'ing Diplomacy: The Case of Korea," Journal of Asian Studies (1957–1958), vol. 17, and Wang Hsin-chung, Chung-Jih chia-wu chan-cheng chih wai-chiao pei-ching [Diplomatic background of the Sino-Japanese War] (Taipei reprint, 1964), follow the evolution of Chinese policy. On the other side, Charles O. Paullin, Diplomatic Negotiations of American Naval Officers, 1778–1883 (Baltimore, 1912), ch. 10, covers the Shufeldt Mission.

The outbreak and course of the Sino-Japanese War has yet to receive adequate, comprehensive treatment. For the moment one must look for instruction on Chinese policy to Edmund S. K. Fung, "The Peace Efforts of Li Hung-chang on the Eve of the Sino-Japanese War (June–July 1894)," Papers on Far Eastern History (March 1971), vol. 3, and his "Ch'ing Policy in the Sino-Japanese War," Journal of Asian History (1973), vol. 7. Both pieces are critical of Li's policy, in part reflecting the biases of the foreign diplomatic sources Fung used to fill gaps in the Chinese documentation. The American response to the war is treated by Marilyn B. Young, The Rhetoric of Empire: American China Policy, 1895–1901 (Cambridge, Mass., 1968), ch. 1; and Jeffrey M. Dorwart, The Pigtail War: American Involvement in the Sino-Japanese War of 1894–1895 (Amherst, Mass., 1975).

    1. Weng T'ung-ho quoted in Hsiao Kung-ch'üan, "Weng T'ung-ho and the Reform Movement of 1898," Tsing Hua Journal of Chinese Studies (April 1957), 1:116.
    2. Dealings with the foreign office were regarded with distaste by most diplomats in Peking. A visit began with the minister and his aides traveling down the streets of what was "by all odds the dustiest and dirtiest city in the world," crowded with people given to insulting foreigners. The diplomats would enter into the mean courtyard of a public building as much in disrepair as any other in the capital and then proceed into a large, drafty reception room containing no more than a long table and some dingy chairs. The attending officials, listlessly gossiping or dozing, would utter a few polite words and then, confronted with the business at hand, begin to draw on their infinite capacity for delay and circumlocution. At least so it all seemed to a generation of Westerners who wanted crisp acquiescence and efficiency. Clive Bingham, A Year in China, 1898–1900 (London, 1901), pp. 41–44.
    3. Americans tended to categorize Li as a prudent "progressive"—and to exaggerate his political power (e.g., by seeing him as the "Prime Minister" of China).
    4. John Russell Young, Men and Memories: Personal Reminiscences (New York, 1901), pp. 311–12.
    5. In the case of Liuchiu as well as Korea and Vietnam (both of which are dealt with below) the formal tribute relationship rested on recognition by the vassal state ruler of China's cultural and political overlordship and in return legitimation of that ruler by the Chinese Emperor. The relationship was sustained by periodic tribute missions which confirmed Chinese suzerainty and exchanged gifts. These missions

also engaged in trade. The Board of Rites was the body within the Chinese government traditionally responsible for managing relations with these tribute states. John K. Fairbank, ed., *The Chinese World Order: Traditional China's Foreign Relations* (Cambridge, Mass., 1968).

6. CKCC, 1:21–22, 24–25, 30; CWS:KH, 15:11–13.

7. CWS:KH, 15:11–13; FRUS 1879, pp. 606–8; James H. Wilson, *China*, 3d ed. (New York, 1901), pp. 90–91; Kenneth E. Folsom, *Friends, Guests and Colleagues: The Mu-Fu System in the Late Ch'ing Period* (Berkeley, Calif., 1968).

8. ISHK, 8:36, 39–41; John Russell Young, *Around the World with General Grant* (New York, 1879), pp. 411–12.

9. Young, *Around the World*, pp. 314, 319; ISHK, 8:36–39.

10. William B. Hesseltine, *Ulysses S. Grant: Politician* (New York, 1935), pp. 427, 431–39; Jesse R. Grant, *In the Days of My Father General Grant* (New York, 1929), p. 319; Hamlin Garland, *Ulysses S. Grant* (New York, 1920), pp. 464, 467. Young, *Around the World*, and John Y. Simon, ed., *The Personal Memoirs of Julia Dent Grant* (New York, 1975), chs. 7–9, deal in detail with the trip.

11. Grant's presidential position—a rejection of exclusion coupled with a condemnation of unfree Chinese labor and the polluting presence of Chinese prostitution—is dealt with in chapter 3. For references to immigration during his visit to Asia, including representations by Hokkien and Cantonese merchants on April 5, 1879 and Grant's reply (reaffirming his opposition to contract labor), see the press clipping in Grant Papers (LC); Simon, *Personal Memoirs*, pp. 280, 291; Young, *Around the World*, pp. 343, 432–33, 601.

12. ISHK, 8:40–44; Young, *Around the World*, pp. 411–12, 415–16, 432–33; Young, *Men and Memories*, pp. 320–22. Grant's bargain over immigration appears only in ISHK, 8:40–41. No one in Grant's immediate party refers to the deal. The Tientsin consul, who was on the periphery of the talks and a confidant neither of Li nor Grant, attributes the bargain to the Chinese and claims Grant declined it. CR:Tientsin (August 12, 1879).

13. ISHK, 8:40–41, 9:10; Simon, *Personal Memoirs*, p. 291; Young, *Around the World*, pp. 432–33; Young to Evarts, March 26, 1880, Evarts Papers (LC). Elsewhere, in *Men and Memories*, pp. 304–5, Young incorrectly claimed Li was indifferent to the plight of immigrants.

14. Young, *Around the World*, pp. 545–46, 558–60, 581, offers a sanitized account of the stay in Japan. Japan's case is laid out in detail in Young to Li, July 5, 1879, ISHK, 9:10–14. Grant's formal written proposals, addressed to Prince Kung and Prince Iwakura Tomomi, August 13, 1879, are in Grant Papers. Grant also communicated these proposals informally to Li, August 1 and 20, 1879, ISHK, 9:32–33, and CWS:KH, 16:24. For further evidence, see Richard T. Chang, "General Grant's 1879 Visit to Japan," *Monumenta Nipponica* (1969), 24:381 (summarizing the Japanese record of Grant's audience with the Emperor); D. B. W. O'Carkee [?] to Young, July 14 [?], 1879, Young Papers (LC); Young to Li, July 25, 1879 and n.d. [received September 7, 1879], in ISHK, 9:30–32, 41–44; and foreign office memorial, September 7, 1879, CWS:KH, 16:19–21.

15. The best expressions of Grant's general views at this time are in his letters to General E. F. Beale, June 7, 1879, Young Papers; to Beale, August 10, 1879, Grant Papers; and to Adam Badeau, June 22, July 16, August 25, 1879, in Badeau, *Grant in Peace* (Hartford, Conn., 1887), pp. 515, 517, 519. Grant expressed himself in nearly the same terms to Prince Kung and Prince Iwakura (August 13, 1879, Grant Papers) and to Li (August 1, 1879, ISHK, 9:32–33). Grant's views are to a degree foreshad-

owed in a passage in his 1874 address to Congress (in James B. Richardson, comp., *A Compilation of the Messages and Papers of the Presidents* [Washington, D.C., 1896–1900; supplement, 1903], 7:288) and echoed in Young's correspondence with Li (July 5 and 25, 1879, *ISHK*, 9:10–14, 30–32).

16. Some among Grant's advisers preferred to tarry longer abroad so that the triumphant homecoming might coincide more closely with the opening of the Republican convention. However, his wife, tired of travel, carried the day.

17. Young served as advance man, visiting both President Hayes and Secretary of State Evarts the day before Grant's arrival. Once in town, Grant appears to have stayed away from Hayes, identified with the reform faction of the Republican party hostile to the Grant stalwarts. However, he did stop by the State Department. Nothing in the Hayes, Young, Evarts, or Grant Papers reveals the substance of Grant's remarks though it was probably on this occasion that he revealed to Evarts that the Chinese would now cooperate on immigration. Evarts to Young, November 20, 1879, Evarts Papers (LC); John Hay to Young, December 30, 1879, Young Papers; Hesseltine, *Ulysses S. Grant*, pp. 431–39; Jesse R. Grant, *In the Days*, pp. 318–19, 326; Badeau, *Grant*, pp. 313, 320; Garland, *Ulysses S. Grant*, pp. 469–85; William H. Morken, "America Looks West: The Search for a China Policy, 1876–1885" (Ph.D. diss., Claremont Graduate School, 1974), p. 69.

18. Esson M. Gale, "President James Burrill Angell's Diary as United States Commissioner to China, 1880–1881," *Michigan Alumnus Quarterly Review* (May 1943), 49:205.

19. *CKCC*, 2:1–40 passim; Wang Yün-sheng, *Liu-shih nien lai Chung-kuo yü Jih-pen* [China and Japan in the last sixty years] (Tientsin, 1932–1934), 1:139–43.

20. Li to Grant, August 23, 1879, *ISHK*, 9:35–37; Li to foreign office, September 8 and 24, 1879, *ISHK*, 9:38–39, 44–45, 10:2; Li to Grant and Young, September 24, 1879, *ISHK*, 10:3–4, and Young Papers. Grant's efforts had disappointed the foreign office as much as they had Li. It had sidestepped his partition proposal as doing an injustice to the Liuchiu dependency, and it had given only lukewarm endorsement to Grant's idea of an alliance. In the estimate of the foreign office, however, Grant's shortcomings were less to blame for the lack of progress toward a settlement than the craft, covetousness, and unreasonableness of the Japanese. *CWS:KH*, 16:19–21.

21. Li's views on immigration are recorded in Young to Evarts, March 26, 1880, Evarts Papers (LC); MCD (August 8, 1883); Chung-yang yen-chiu yüan, chin-tai shih yen-chiu so, comp., *Hai-fang tang* [Records on maritime defense] (Taipei, 1957), "Tien-hsien," part 1, p. 384; Pethick to James H. Wilson, December 1, 1888, Wilson Papers (LC); *FRUS 1892*, pp. 134–35; W. A. P. Martin, *A Cycle of Cathay* (New York, 1900), pp. 352–53; CR: Tientsin (June 29, 1893).

22. This and the paragraph that follows are based on *IWSM:TC*, 45:10–13, 47:20–22, 57:22–24, 94:37, 97:15–17, 22–28; *CKS:TC*, pp. 388–94, 509–14, 829–30; *CKCC*, 1:1, 31–32.

23. *ISHK*, 9:34; *CWS:KH*, 16:11–17; *CKCC*, 2:6–8; *CJCC*, 2:338–41. Not until February 1881 did the foreign office secure imperial approval for shifting formal responsibility for Korean affairs from the Board of Rites, which was slow and unable to maintain confidentiality, to Li and the minister to Japan. *CKCC*, 2:31.

24. Mary C. Wright, "The Adaptability of Ch'ing Diplomacy: The Case of Korea," *Journal of Asian Studies* (1957–1958), 17:370–73; Ernest N. Paolino, *The Foundations of the American Empire: William Henry Seward and U.S. Foreign Policy* (Ithaca, N.Y., 1973), pp. 196–203.

25. *Congressional Record*, 45th Cong., 2d Sess., pp. 2324, 2600–1.

26. Shufeldt quoted in Frederick Foo Chien, *The Opening of Korea: A Study of Chinese Diplomacy, 1876–1885* (Hamden, Conn., 1967), pp. 77–78, 260; and in Kenneth J. Hagan, *American Gunboat Diplomacy and the Old Navy, 1877–1889* (Westport, Conn., 1973), p. 37.

27. CKS:TC, pp. 745–48, 752, 770–72, 783, 786–88, 790–800, 806, 808, 814–17, 830–31, 834–35; IWSM:TC, 80:19–20.

28. CJCC, 2:347–50.

29. CKS:TC, pp. 751, 757, 762–64; IWSM:TC, 94:37.

30. CJCC, 2:6–8, 17; CWS:KH, 16:11–13.

31. This and the following paragraph are based on CKCC, 2:6–8, 31–33, 3:8–16, 5:5–6; CWS:KH, 23:25–36, 26:13–24, 27:11–18; CJCC, 2:8–12, 15–16, 148–60; Chou Ch'üeh-shen-kung ch'üan-chi [Complete works of Chou Fu] (Taipei reprint, 1966), pp. 971–73; Charles O. Paullin, *Diplomatic Negotiations of American Naval Officers, 1778–1883* (Baltimore, 1912), p. 302.

32. When Shufeldt's extraordinary poison-pen letter found its way into print in China in May, even before the conclusion of the treaty, his candidacy for the appointment as naval adviser died a quiet but definitive death. After his mission to China Shufeldt returned briefly to the routine of the peacetime navy before retiring in 1884. Holcombe to Shufeldt, November 30, December 19 and 27, 1881, Shufeldt Papers (LC); Paullin, *Diplomatic Negotiations*, pp. 301, 313; John K. Fairbank et al., eds., *The I.G. in Peking: The Letters of Robert Hart, Chinese Maritime Customs, 1868–1907* (Cambridge, Mass., 1975), 1:380, 387–88, 412–13.

33. CJCC, 2:11–16; ISHK, 13:23–24, 15:19–20; DIC (August 4, 1882 and November 16, 1885); Adee to Young, October 12, 1883, Young Papers; George M. McCune and John A. Harrison, eds., *Korean-American Relations: Documents Pertaining to Far Eastern Diplomacy of the United States*, vol. 1: *The Initial Period, 1883–1886* (Berkeley, 1951), pp. 25–28; MCD (August 8, 1883).

34. Chien, *The Opening of Korea*, pp. 181–84; McCune and Harrison, *Korean-American Relations*, 1:31, 53–65, 139–40; Charles C. Tansill, *The Foreign Policy of Thomas F. Bayard* (Bronx, N.Y., 1940), pp. 426–42.

35. ISHK, 15:13–14; MCD (October 12, 1885); Yur-Bok Lee, *Diplomatic Relations Between the United States and Korea, 1866–1887* (New York, 1970), p. 93; Chien, *The Opening of Korea*, pp. 135–36.

36. Allen subsequently took charge of the Seoul legation where he labored mightily for the Korean cause. Denny, a former consul in Tientsin, had assisted Grant and Young during their stay in China and won their praise as an able official on whom Li could rely. Li learned otherwise and in 1889 instigated an official protest against Denny's rumored appointment as minister to Korea. CKCC, 10:29–42, 11:11; CJCC, 2:82–85, 92–97, 125–33, 136–41; CWS:KH, 73:24, 74:18–19, 80:10, 15, 26; Fred H. Harrington, *God, Mammon, and the Japanese: Dr. Horace N. Allen and Korean-American Relations, 1884–1905* (Madison, Wis., 1944), part 4; Robert R. Swartout, Jr., *Mandarins, Gunboats, and Power Politics: Owen Nickerson Denny and International Rivalries in Korea* (Honolulu, 1980); McCune and Harrison, *Korean-American Relations*, 1:147–56; Pethick to James H. Wilson, December 1, 1888, Wilson Papers.

37. Hsü Chüeh, *Fu-an i-chi* [Works of Hsü Chüeh] (Taipei reprint, 1970), 2:703.

38. MCD (August 8, 1883); Shao Hsün-cheng et al., comps., *Chung-Fa chan-cheng* [The Sino-French War] (Shanghai, 1955), 4:52, 60–62, 132–34, 136; CYCT, pp. 1250–53, 1301–3. Young's interest in peace derived in part from a fear that the Sino-French crisis would intensify popular antiforeign feeling in China and endanger

even innocent bystanders. The destruction of American property in the course of rioting in Canton in 1883 made him sensitive to the problem, and an accidental Chinese attack on an American ship in September 1884 was a forceful reminder. *Chung-Fa chang-cheng*, 5:2–5, 12–13; SSM, p. 218. Young's appointment and general views are covered in chapter 5.

39. That irrepressible advocate of a negotiated settlement, Li Hung-chang, nearly succeeded in removing the indemnity demand in talks with the French. But within a month an accidental frontier clash prompted Ferry to revive the indemnity demand and threaten war unless China paid. CWS:KH, 40:27; *Chung-Fa chan-cheng*, 4:166.

40. CYCT, pp. 1778, 1788, 1790–91, 1863, 1873, 2091–93, 2121; *Chung-Fa chan-cheng*, 4:173–76, 5:66, 448–49, 451–60, 463–66, 470–72, 475–77, 482, 486–87, 494–95; CWS:KH, 42:25, 32–33, 43:1, 13–14.

41. Li and his chief aide in this transfer, Ma Chien-chung, suffered considerable official criticism and even an imperial reprimand, but the ships were spared destruction and kept profitably employed through the wartime, and in the summer of 1885 were restored to China. Chang Jo-ku, *Ma Hsiang-po hsien-sheng nien-p'u* [A chronological biography of Ma Liang] (Ch'ang-sha, 1939), pp. 152–53, 162; *Chung-Fa chan-cheng*, 4:167–73, 5:458, 460; CWS:KH, 44:20, 45:14–16, 49:22–23, 59:14–15; Chung-kuo shih-hsüeh hui, comp., *Yang-wu yün-tung* [The westernization movement] (Shanghai, 1959), 4:75, 103.

42. CYCT, pp. 1876–77, 1888, 2081, 2237; *Chung-Fa chan-cheng*, 4:28, 34, 103–7, 184, 198, 200, 5:501–5, 542–45, 549–52, 558; CWS:KH, 45:27, 46:10.

43. David M. Pletcher, *The Awkward Years: American Foreign Policy under Garfield and Arthur* (Columbia, Mo., 1962), p. 217.

44. CWS:KH, 99:3; CR: Tientsin (July 7, 1894); FRUS 1894, Appendix I, pp. 37–38.

45. Gresham privately warned the Chinese to beware of the Europeans, assured them of American good will, and helpfully suggested holding out for some battlefield successes before attempting to bargain with Japan. FRUS 1894, Appendix I, pp. 70–84; CWS:KH, 99:3, 9, 100:4, 101:5, 106:19; CJCC, 5:263, 266.

46. Denby had accompanied his father, the minister, to China to join the customs service in 1887 at age seventeen and stayed on, working in the legation as secretary. In March 1894 he became chargé when his father inopportunely decided to go home for medical treatment, confidently predicting a quiet summer ahead. Denby, Jr., later worked as agent for the American China Development Company, as head of the foreign-controlled Tientsin administration after the Boxer summer, and finally back in the foreign service as the State Department China expert and as the Shanghai consul general.

47. After the execution the legation came to the conclusion that the ostensible schoolboys were indeed part of a large and well-organized spy network. FRUS 1894, pp. 101, 109, 116; CWS:KH, 92:18, 93:14–16, 21, 94:2, 15, 25; SSM, pp. 219–22, 224–27; CJCC, 5:3, 6–7, 261, 265–66; George E. Paulsen, "Secretary Gresham, Senator Lodge, and American Good Offices in China, 1894," *Pacific Historical Review* (May 1967), vol. 36.

Denby, Jr., had also irritated the Chinese by pressing Japanese proposals for maintaining Sino-Japanese trade that corresponded with American interest in limiting commercial disruption along the China coast during the war. Li hoped, however, that a trade cut-off might create a debilitating grain shortage in Japan. After consulting with Li, the foreign office refused to discuss these unwelcome Japanese proposals further. FRUS 1894, pp. 169–75; CWS:KH, 94:14–15; SSM, pp. 222–24.

48. Denby's role, described below, can be followed in CWS:KH, 103:8, 107:11, 16–17, 110:26, 112:10, 113:7; Charles Denby, *China and Her People* (Boston, 1906), 2:130–32, 134–36, 138; MCD (November 16 and December 1 and 8, 1894); Denby letter quoted in John W. Foster, *Diplomatic Memoirs* (Boston, 1909), 2:106.

49. CWS:KH, 100:2.

50. Denby's views on reform in early 1895 are treated in detail in chapter 5.

51. On this last occasion Foster made some brief remarks on the peace treaty that struck his audience at the foreign office as lukewarm, and then delivered at length (by Foster's own recollection) "some pointed advice as to the reforms required in government" with the earnestness Chinese officials had come to expect of well-intentioned Americans. Foster, *Diplomatic Memoirs*, 2:102–18, 120–23, 127, 129–38, 150–60; CJCC, 4:553; CWS:KH, 111:13.

52. Protests over the terms of the peace began to pour in in late April (CWS:KH, 38:23) and continued into early May, when silence was imposed by imperial command (CWS:KH, 44:19).

53. "Li Fu-hsiang yu-li ko-kuo jih-chi" [A diary of Li Hung-chang's international tour] in Tso Shun-sheng, comp., *Chung-kuo chin-pai nien shih tzu-liao hsü-pien* [Chinese historical materials for the last hundred years—second collection] (Shanghai, 1933), pp. 413–15; Foster, *Diplomatic Memoirs*, 2:319; Young, *Men and Memories*, p. 325; Thomas J. McCormick, "The Wilson-McCook Scheme of 1896–1897," *Pacific Historical Review* (February 1967), 36:49–54; and Gerald G. Eggert, *Richard Olney: Evolution of a Statesman* (University Park, Pa., 1974), pp. 274–76.

Chapter 5: American Policy and Private Interests, 1860–1899

The concept of the open door can be taken in two ways—as a policy laid down in the 1890s or as a long gestating set of ideas. In the former sense it has received exhaustive treatment, so that it virtually dominates the voluminous literature on late nineteenth-century American China policy. The major sport of open door historians has been to trace the paternity of the policy. Hay's claim got its scholarly due first in Tyler Dennett's admiring John Hay: From Poetry to Politics *(New York, 1933), chs. 24–25, and recently with more restraint in Kenton J. Clymer,* John Hay: The Gentleman as Diplomat *(Ann Arbor: Mich., 1975), ch. 6. Paul A. Varg,* Open Door Diplomat: The Life of W. W. Rockhill *(Urbana, Ill., 1952), and A. Whitney Griswold,* The Far Eastern Policy of the United States *(New York, 1938), ch. 2, made the case respectively for Hay's China expert and for Alfred Hippisley, the English employee of the Imperial Maritime Customs visiting in the United States in 1899. An early effort to adjudicate these various claims, Harvey Pressman's historiographical essay—"Hay, Rockhill, and China's Integrity: A Reappraisal,"* Papers on China *(1959), vol. 13—has been improved on in Marilyn B. Young's broad and balanced account,* The Rhetoric of Empire: American China Policy, 1895–1901 *(Cambridge, Mass., 1968).*

Some, not content to limit themselves to the 1890s, have pushed the search for the father of the open door back to earlier American diplomats, for example, to Burlingame in Frederick Wells Williams' worshipful Anson

Burlingame and the First Chinese Mission to Foreign Powers (New York, 1912), and to Humphrey Marshall in Te-kong Tong, United States Diplomacy in China, 1844–60 (Seattle, Wash., 1964), ch. 9. Others, unwilling to leave the open door an American offspring, have given credit to British and Chinese policy makers. T. F. Tsiang, "The Extension of Equal Commercial Privileges to Other Nations than the British after the Treaty of Nanking," Chinese Social and Political Science Review (October 1931), vol. 15, launched the effort; Earl H. Pritchard, "The Origins of the Most-Favored-Nation and the Open Door Policies," Far Eastern Quarterly (February 1942), vol. 1, an interpretive classic, pulled the threads of the argument together; and John K. Fairbank, "'American China Policy' to 1898: A Misconception," Pacific Historical Review (November 1970), vol. 39, has pithily restated it.

The open door policy has received its most thorough reappraisal in the last several decades at the hands of "new left" historians, who see it as a response by an interlocking political and corporate elite to the structural needs of a maturing capitalist economy. They contend that overseas markets, especially in Asia as well as Latin America, were essential to absorb overproduction and thereby preserve order and prosperity at home. William Appleman Williams, The Tragedy of American Diplomacy (New York, 1959), is the seminal work. Walter LaFeber's The New Empire: An Interpretation of American Expansion, 1860–1898 (Ithaca, N.Y., 1963), is a sophisticated survey along similar interpretive lines. It hews closely in its treatment of China policy to Tyler Dennett's Americans in Eastern Asia for the pre-1890s and thereafter to Thomas J. McCormick's dissertation, since published as China Market: America's Quest for Informal Empire, 1893–1901 (Chicago, 1967). McCormick provides the most detailed case for this school's view of the turn of the century open door policy as a realistic strategy of economic expansion masterfully applied. This bold interpretation takes direct issue with a range of earlier and eminent interpreters of the open door as errant legalism (George Kennan in American Diplomacy, 1900–1950 [Chicago, 1951], ch. 2), as part of a great national aberration in the Pacific following victory over Spain (Samuel Flagg Bemis, A Diplomatic History of the United States, 5th ed. [New York, 1965], chs. 26–27), or as a policy foisted on the unwitting Hay by cunning Englishmen (Griswold's The Far Eastern Policy of the United States, noted above).

The new left view has in turn evoked some critical evaluations: Marilyn Young, "American Expansion, 1870–1900: The Far East," in Barton Bernstein, ed., Towards a New Past (New York, 1968); Paul A. Varg, "The Myth of the China Market, 1890–1914," American Historical Review (February 1968), vol. 73; William H. Becker, "American Manufacturers and Foreign Markets, 1870–1900: Business Historians and the 'New Economic Determinists,'" Business History Review (Winter 1973), vol. 47; and Michael H. Hunt, "Americans in the China Market: Economic Opportunities and Economic Nationalism, 1890s–1931," ibid. (Autumn 1977), vol. 51. Becker and Hunt both argue the need for greater use of the sources and

Chapter 5: Bibliographical Essay

methods of economic history alongside those of political and diplomatic history favored by the new left.

There are a sizable number of notable studies and collections (more fully identified in the notes below) directly relevant to Sino-American economic relations: Kwang-Ching Liu's meticulous study of Russell and Company's steamship operations; Ralph and Muriel Hidy's authorized history of Standard Oil; Yen Chung-p'ing's synthesis on textiles; Sun Yü-t'ang's statistical work; and the invaluable compilations on modern Chinese industry and handicrafts by Sun Yü-t'ang, Ch'en Chen, P'eng Tse-i, and Wang Ching-yü. Sun, "The Historical Development and Aggressive Nature of American Imperialist Investment in China (1784–1914)," trans. by Michael H. Hunt, Chinese Studies in History (Spring 1975), vol. 8, develops in brief compass the orthodox view in the People's Republic that the United States played an economically disruptive, oppressive, and exploitative role in China. Other noteworthy works, done in an eclectic mode, on the role of economic interest groups are James J. Lorence, "Coordinating Business Interests and the Open Door Policy: The American Asiatic Association, 1898–1904," in Jerry Israel, ed., Building the Organizational Society (New York, 1972); William R. Braisted, "The United States and the American China Development Company," Far Eastern Quarterly (February 1952), vol. 11; and above all Charles S. Campbell, Jr.'s slim classic, Special Business Interests and the Open Door Policy (New Haven, Conn., 1951).

To understand the open door in its other guise—as a set of ideas promoted throughout the nineteenth century by interest groups—it is necessary to trace the development of those interests as missionaries and diplomats as well as traders and investors developed and understood them. John K. Fairbank and his students have sustained scholarly interest in the mission movement. Samples of their work relevant to the late nineteenth century can be found in Papers on China, which makes available research done in Harvard seminars. Two of their monographs—Paul A. Cohen's China and Christianity: The Missionary Movement and the Growth of Chinese Antiforeignism, 1860–1870 (Cambridge, Mass., 1963), and Ellsworth C. Carlson, The Foochow Missionaries, 1847–1880 (Cambridge, Mass., 1974)—are both essential reading.

The "Harvard approach" tends to emphasize gentry leadership in the antimission movement and to describe mission activity in such neutral, if not somewhat apologetic, terms as "cultural interaction," "cultural transmission and adjustment," and "missionary contributions." Chinese historians have taken issue with both these points. Lü Shih-ch'iang working on Taiwan has made the best case for the popular and patriotic sources of antimissionary feeling in Chung-kuo kuan-shen fan-chiao ti yüan-yin (1860–1874) [The causes of the anti-Christian movement among Chinese officials and gentry (1860–1874)] (Taipei, 1966) and in a series of articles in Chung-yang yen-chiu yüan, chin-tai shih yen-chiu so chi-k'an (June 1971), vol. 2, (Dec. 1972), vol. 3, and (May 1973), vol. 4. Historians in the People's Republic, in essential agreement with Lü, have labeled missionaries as "cultural imperialists," whose work aroused mass opposition. (See for example Li Shih-yüeh, "Chia-wu chan-cheng ch'ien san-shih-nien chien fan-yang-chiao yün-tung" [The movement against Western religion during the thirty years prior to the war of 1894], Li-shih yen-chiu [1958], no. 6.) The ongoing series Chiao-wu chiao-an tang [Archives on mission affairs and mission incidents] (Taipei, 1973–), compiled by Chung-yang yen-chiu yüan, chin-tai shih yen-chiu so, contains materials for case studies that might test these divergent points of view.

The American policy response to mission imbroglios in the 1890s has received special attention in George E. Paulsen, "The Szechuan Riots of 1895 and American 'Missionary Diplomacy,'" Journal of Asian Studies (February 1969), vol. 28, and briefly but perceptively in Young, The Rhetoric of Empire, pp. 76–87. Missionary policy before the 1890s has yet to be carefully charted.

The views of leading American diplomats through the latter part of the nineteenth century have received only piecemeal attention. For Burlingame and Denby, unpublished Ph.D. dissertations offer the soundest and most detailed coverage: Martin R. Ring, "Anson Burlingame, S. Wells Williams and China, 1861–1870" (Tulane University, 1972), and John W. Cassey, "The Mission of Charles Denby and International Rivalries in the Far East, 1885–1899" (University of Southern California, 1959). Williams, Anson Burlingame (noted above), is still useful and little improved upon by David L. Anderson, "Anson Burlingame: American Architect of the Cooperative Policy in China, 1861–1871," Diplomatic History (Summer 1977), vol. 1, also intent on refurbishing the Burlingame legend. Paul H. Clyde has surveyed the ministers who served during the Grant administration in his summary of the dispatches of Browne, Low, and Seward in Pacific Historical Review (September 1932), vol. 1, and (March and December 1933), vol. 2. A sketch of Denby appears in David Healy, US Expansionism: The Imperialist Urge in the 1890s (Madison, Wis., 1970), ch. 10. Young, The Rhetoric of Empire, and McCormick, China Market, provide treatment inter alia of both Denby and Conger.

---

1. The result was an unfavorable trade balance with China, frequently by a three-to-one margin (thanks to an undiminished American taste for Chinese tea and silk). All in all, the American share of China's total foreign trade was minor through the late nineteenth century (6 percent in 1880 and 5 percent in 1890). And as a share of total U.S. exports, exports to China became ever less important—2.8 percent in 1860, .8 percent in 1870, and .1 percent in 1880. The precise value of China trade passing through Hong Kong is unknown, so that both import and export figures given here may understate the real levels by perhaps 10 to 20 percent. U.S. Bureau of the Census Historical Statistics of the United States from Colonial Times to 1957 (Washington, D.C., 1960), pp. 537, 550–53; C. Yang and H. B. Hau, Statistics of China's Foreign Trade During the Last Sixty-Five Years (Shanghai [?], 1931), pp. 143, 148; Hsiao Liang-lin, China's Foreign Trade Statistics (Cambridge, Mass., 1974), pp. 22–23, 162.

2. Olyphant remained active in the coolie trade and acquired an iron works in Hong Kong. Russell, with an even longer reach, extended its banking activities from financing the trade of others to making loans to provincial officials. It also joined with Chinese merchants in processing export goods. It established a silk filature in Shanghai in 1878. By the time of its sale in 1891 the factory had increased its capacity twentyfold (to 1,000 filatures) and employed over 1,100 workers. Russell also put $58,000 into a mechanical tea-firing plant on Taiwan that started up in 1888. Augustine Heard, Jr., to Anson Burlingame, July 13, 1865, Burlingame, Family Papers (LC); Edward K. Haviland, "American Steam Navigation in China, 1845–1878," parts 2–7, *American Neptune* (October 1956–January 1958), vols. 16–18; Wang Ching-yü, "Shih-chiu shih-chi wai-kuo ch'in-Hua ch'i-yeh chung ti Hua-shang fu-ku huo-tung" [Chinese merchant investment in foreign firms which encroached on China in the nineteenth century], *Li-shih yen-chiu* (1965), no. 4, pp. 42–47, 50; Sun Yü-t'ang, *Chung-Jih chia-wu chang-cheng ch'ien wai-kuo tzu-pen tsai Chung-kuo ching-ying ti chin-tai kung-yeh* [Modern enterprises in China financed by foreign capital before 1894] (Shanghai, 1955), pp. 23–25, 34, 80; Sun, comp., *Chung-kuo chin-tai kung-yeh shih tzu-liao, ti-i chi, 1840–1895 nien* [Historical materials on modern Chinese industry, first series, 1840–1895] (Peking, 1957), pp. 85, 245; Hsü I-sheng, *Chung-kuo chin-tai wai-chai shih t'ung-chi tzu-liao, 1853–1927* [Statistical materials on the history of foreign investments in China in modern times, 1853–1927] (Peking, 1962), pp. 4–5, 14; CWS:KH, 54:15–16.

3. Stephen C. Lockwood, *Augustine Heard and Company, 1858–1862: American Merchants in China* (Cambridge, Mass., 1971), pp. 103–14, 118–19, 128; Charles F. Remer, *The Foreign Trade of China* (Shanghai, 1926), pp. 40–41, 78–80.

4. The estimated $6 million that Americans had invested in China in 1875 (roughly 6 percent of total American overseas holdings) was tied up in real estate and business inventory as well as shipping. Charles F. Remer, *American Investment in China* (Honolulu, 1929), pp. 21–23; U.S. Bureau of the Census, *Historical Statistics*, p. 565; Hsiao, *China's Foreign Trade*, pp. 254–55, 259.

5. American shipping took exceptional jumps upward in 1884–1885 (when Russell and Company temporarily resumed operations of its former vessels to prevent French seizure) and 1905–1907 (when the Russo-Japanese War stimulated a brisk but brief trade boom). Kwang-Ching Liu, *Anglo-American Steamship Rivalry in China, 1862–1876* (Cambridge, Mass., 1962); Haviland, "American Steam Navigation," parts 2, 3, and 7; Wang, "Shih-chiu shih-chi wai-kuo," pp. 40–41, 45.

6. Aside from the Pacific Mail, the best known line with American ties was Robert Dollar's, which established a Shanghai office in 1906. However, Dollar shifted operations to Canada during World War I and then in 1925 sold out to the British. Thomas R. Cox, "The Passage to India Revisited: Asian Trade and the Development of the West, 1850–1900," in John A. Carroll, ed., *Reflections of Western Historians* (Tucson, Ariz., 1969); Thomas R. Cox, *Mills and Markets: A History of the Pacific Coast Lumber Industry to 1900* (Seattle, Wash., 1974), ch. 7; Cleona Lewis, *America's Stake in International Investments* (Washington, D.C., 1938), p. 323; John A. Kemble, "A Hundred Years of the Pacific Mail," *American Neptune* (1950), vol. 10; Yen Chung-p'ing et al., comps., *Chung-kuo chin-tai ching-chi shih t'ung-chi tzu-liao hsüan-chi* [Selected statistical materials on modern Chinese economic history] (Peking, 1955), pp. 241, 244, 246; John B. Hutchins, *The American Maritime Industry and Public Policy, 1789–1914: An Economic History* (Cambridge, Mass., 1941).

7. Some investors held on, such as in the case of the up-to-date American paper mill at Shanghai, established in 1881 with capital of $102,000 (an unknown but

probably substantial amount from Chinese sources). It prospered into the early twentieth century, increasing its capital sixfold and its capacity from two to fourteen tons per day. But probably more typical of the trends of the time were the American shipyards, one in Hong Kong (sold, probably in the early 1860s so that the proceeds could be invested in the American West) and the other in Shanghai (sold, again after only several years of operation, to Li Hung-chang to form part of a government arsenal). Sun, Chung-Jih chia-wu, pp. 11, 45–46, 74, 75, 79.

8. Later, when China granted a cable monopoly to the Danish Great Northern, American diplomats continued to press in favor of American interests. Martin R. Ring, "Anson Burlingame, S. Wells Williams and China, 1861–1870" (Ph.D. diss., Tulane University, 1972), pp. 210–12; Ernest N. Paolino, The Foundations of the American Empire: William Henry Seward and U.S. Foreign Policy (Ithaca, N.Y., 1973), pp. 58–65; Chung-yang yen-chiu yüan, chin-tai shih yen-chiu so, comp., Hai-fang tang [Records on maritime defense] (Taipei, 1957), "Tien-hsien," part 1, pp. 5, 63, 66–69, 73, 75–78, 127, 133–34, 249–51, 278, 282–92, 353, 358, 364, 379, 384.

9. Others were an American cotton mill that opened in Canton in 1871 using Chinese merchant capital and that lasted only six months, and a silk factory which some Americans in Shanghai never got beyond the planning stage. Wang, "Shih-chiu shih-chi wai-kou," pp. 56, 58; FRUS 1883, pp. 129–41, 152–68, 180, 187–97, 206–8; Sun, Chung-Jih chia-wu, pp. 24–25, 31; Sun, Chung-kuo chin-tai kung-yeh, pp. 162–65, 1037–39, 1050–53; George E. Paulsen, "Machinery for the Mills of China: 1882–1896," Monumenta Serica (1968), 27:320–29; Yen Chung-p'ing, Chung-kuo mien-fang-chih shih-kao, 1289–1937 [Draft history of cotton textiles in China, 1289–1937] (Peking, 1955), pp. 147–48; Wu T'ien-jen, Huang Kung-tu hsien-sheng chuan-kao [A draft biography of Huang Tsun-hsien] (Hong Kong, 1972), pp. 105–6.

10. MCD (August 8, 1883); Hsü, Chung-kuo chin-tai wai-chai, pp. 12–13, 17; CYCT, p. 1872; CWS:KH, 48:19–20, 49:18; David M. Pletcher, The Awkward Years: American Foreign Policy under Garfield and Arthur (Columbia, Mo., 1962), pp. 214–15; James H. Wilson, China: Travels in the Middle Kingdom (New York, 1887), pp. 97–98; Charles C. Tansill, The Foreign Policy of Thomas F. Bayard (Bronx, N.Y., 1940), p. 424; David Healy, US Expansionism: The Imperialist Urge in the 1890s (Madison, Wis., 1970), pp. 74–77; Chang Jo-ku, Ma Hsiang-po hsien-sheng nien-p'u [A chronological biography of Mr. Ma Liang] (Ch'ang-sha, 1939), pp. 162–71, 178; ISHK, 19:7–8; CR: Tientsin (July 23 and 27, 1887); investment proposals in subject file, container 2, Wharton Barker Papers (LC); Foster and Pethick correspondence files, James H. Wilson Papers (LC).

11. The annual export average of $7.6 million was nearly 50 percent over the level of the preceding decade. Imports also expanded steadily, though not so strongly as exports. Overall trade jumped from $19 million in 1890 to $33 million in 1899. Better business brought more American firms to Shanghai, eighty-one by 1900, an increase of forty-nine over the number at the outset of the decade. Despite this upturn, Americans still accounted for only 8 percent of all foreign firms and only 8 percent of all foreign trade. U.S. Bureau of the Census, Historical Statistics, pp. 550, 552; Hsiao, China's Foreign Trade, pp. 22–23, 162; Yang and Hau, Statistics of China's Foreign Trade, pp. 143, 148.

12. Specialized export firms such as Fearon, Daniel and Company, the China and Japan Trading Company, and the American Trading Company purchased from independent mills and shipped the goods to Shanghai for sale to Chinese merchants who carried the goods the last stage of their journey to Chinese customers. Michael

H. Hunt, "Americans in the China Market: Economic Opportunities and Economic Nationalism, 1890s–1931," *Business History Review* (Autumn 1977), 51:285–86.

13. Other incentives to making innovations in marketing were the steady headway of Russian oil on the world market, the need to dispose of a kerosene of lower quality than could be sold domestically, the economies of scale obtained by expanding world trade (by the mid-1890s kerosene was selling off New York piers for one-third its mid-1870s price), and, finally, a hold in China as insurance against any future fall in domestic demand (though to this point the home market was still growing, even during the depression of the 1890s). The treatment of Standard here and below is drawn from Hunt, "Americans in the China Market," pp. 281–82, 284; *FRUS 1888*, part 1, p. 277.

14. The first was the Ch'ien-kang silk filature, established in Shanghai in 1892. The next year the Mercantile Tobacco Company invested $192,000 in a cigarette factory in Shanghai. Though the silk operation soon passed into foreign hands, the cigarette factory lasted out the decade, machine rolling imported American tobacco into a cheaper brand of cigarettes for a growing number of Chinese consumers and a more expensive line for the limited treaty port trade. It closed at last in 1901 after slipping sales put it in financial difficulty. Sun, *Chung-Jih chia-wu*, pp. 25, 51, 81; Sun, *Chung-kuo chin-tai kung-yeh*, pp. 148–52, 246.

15. In 1897 Americans had a hand in the first major project, the million-dollar International (Hung-yüan) Cotton Mill in Shanghai, completed under the new treaty terms, but they eventually sold out to Chinese-Japanese interests. By 1897 Americans were also operating a rice milling plant in Shanghai, capitalized at $140,000. At the tag end of the decade Fearon, Daniel and Company and the American Trading Company each tried to make a go of cotton spinning in China only to fail. Wang Ching-yü, comp., *Chung-kuo chin-tai kung-yeh shih tzu-liao, ti-erh chi, 1895–1914 nien* [Historical materials on modern Chinese industry, second series, 1895–1914] (Peking, 1957), 1:7–11, 180.

16. Wang, *Chung-kuo chin-tai kung-yeh*, 1:12–13.

17. Undaunted, Barker tried again in 1898, once more in 1904–1905, and for the last time in 1918 in an ever more pathetic effort to attract backers with the argument that his contacts with unspecified "Chinese in high position" gave him an inside track on "large and very profitable business." CWS:KH, 99:17–18; CJCC, 5:264; John K. Fairbank et al., eds., *The I.G. in Peking: Letters of Robert Hart, Chinese Maritime Customs, 1868–1907* (Cambridge, Mass., 1975), 1:581; Gresham to Bayard, December 24, 1894, Walter Q. Gresham Papers (LC); China investment materials in subject file, container 21, Wharton Barker Papers; *Hai-fang tang*, "T'ieh-lu," p. 228.

18. Thomas J. McCormick, "The Wilson-McCook Scheme of 1896–1897," *Pacific Historical Review* (February 1967), vol. 36. This scheme might also be interpreted on the basis of the same evidence McCormick uses as primarily an attempt at personal political advancement with investments a decidedly secondary concern.

19. Percy H. Kent, *Railway Enterprise in China* (London, 1907), ch. 12; Lee En-han, "Chung-Mei shou-hui Yüeh-Han lu-ch'üan chiao-she" [Sino-American negotiations over the recovery of rights to the Canton-Hankow railroad], *Chung-yang yen-chiu yüan, chin-tai shih yen-chiu so chi-k'an* (1969), no. 1, pp. 149–54.

20. Hsiao, *China's Foreign Trade*, pp. 22–23, 162; U.S. Bureau of the Census, *Historical Statistics*, pp. 537, 550, 552, 565; Yen, *Chung-kuo chin-tai ching-chi*, pp. 65–66; Remer, *Foreign Investments in China* (Shanghai, 1933), p. 338.

21. DIC (February 2, 1899); Rockhill memo, August 28, 1899, Hay Papers (LC).

22. James B. Richardson, comp., *A Compilation of the Messages and Papers of the Presidents* (Washington, D.C., 1896–1900; supplement, 1903), 10:181, and supplement, p. 64.

23. Milton T. Stauffer, ed., *The Christian Occupation of China* (Shanghai, 1922), pp. 326–27; and Remer, *Foreign Investments*, p. 333.

24. Paul A. Cohen, *China and Christianity: The Missionary Movement and the Growth of Chinese Antiforeignism, 1860–1870* (Cambridge, Mass., 1963), pp. 68–69, 147.

25. Ellsworth C. Carlson, *The Foochow Missionaries, 1847–1880* (Cambridge, Mass., 1974), pp. 77–83, 94–103, 111–112; John K. Fairbank, "Patterns Behind the Tientsin Massacre," *Harvard Journal of Asiatic Studies* (December 1957), 20:492–94.

26. Carlson, *The Foochow Missionaries*, pp. 89–92, 118; Young in *FRUS 1885*, p. 150; examples of missionary interference in MCD (April 26 and November 1, 1897); Chung-yang yen-chiu yüan, chin-tai shih yen-chiu so, comp., *K'uang-wu tang* [Records on mining affairs] (Taipei, 1960), 2:881–99.

27. Fairbank, *The I.G. in Peking*, 2:845.

28. Carlson, *The Foochow Missionaries*, pp. 103–7, 128–32; Ch'ing Ying, comp., *Chung-kuo chin-tai fan-ti fan-feng-chien li-shih ko-yao hsüan* [Selected antiimperialist and antifeudal songs and chants from modern Chinese history] (Peking, 1962), pp. 410–13; SSM, p. 462; Fairbank, "Patterns Behind the Tientsin Massacre"; Carlson, *The Foochow Missionaries*, pp. 128–32; Charles A. Litzinger, "Patterns of Missionary Cases Following the Tientsin Massacre, 1870–1875," *Papers on China* (1970), vol. 23; Philip West, "The Tsinan Property Disputes (1887–1891): Gentry Loss and Missionary 'Victory,' " ibid. (1966), 20:122–23, 138–39; Wilson, *China*, pp. 270–71; Charlton M. Lewis, *Prologue to the Chinese Revolution: The Transformation of Ideas and Institutions in Hunan Province, 1891–1907* (Cambridge, Mass., 1976), ch. 2.

29. *FRUS 1867*, part 1, pp. 488–89; *FRUS 1870*, p. 303; *FRUS 1873*, part 1, p. 119; *FRUS 1875*, part 1, pp. 332–33; *FRUS 1876*, pp. 47–48; MCD (December 5, 1870 and January 18, 1872); DIC (March 2, 1871). British diplomats also saw nothing but vague treaty provisions covering mission work in the interior and indicated that missionaries would have to depend on their own discretion for safety. But British policy, like that of the United States, would eventually move toward a more vigorous defense of the missions. Mary C. Wright, *The Last Stand of Chinese Conservatism: The T'ung-chih Restoration, 1862–1874* (Stanford, Calif., 1957), pp. 261–63; and Edmund S. Wehrle, *Britain, China and the Anti-missionary Riots, 1891–1900* (Minneapolis, Minn., 1966).

30. Thus when the Chinese foreign office, alarmed by the Tientsin massacre, made a major bid to clarify rules applicable to missions, Low would not budge. He pointed to the Catholics as the chief troublemakers and countered that the best solution lay not in cumbersome, inapplicable or restrictive regulations but rather opening troubled areas to the liberalizing influence of foreign commerce and permitting American consuls to reside on the scene to help resolve mission problems that might arise. *FRUS 1871*, pp. 107–10.

31. The Tsinan dispute was resolved in 1892 in favor of the missionaries following a shift in U.S. missionary policy. Nevertheless local opposition remained strong. Miriam Levering, "The Chungking Riot of 1886: Justice and Ideological Diversity," *Papers on China* (May 1969), vol. 22A; CWS:KH, 68:7–8; West, "The Tsinan Property Disputes."

32. Young added hopefully that his policy of protection would require "noth-

ing more than our good will." Shirley W. Smith, *James Burrill Angell* (Ann Arbor, Mich., 1954), pp. 118, 124, 146–47; *FRUS 1882*, pp. 140–41; *FRUS 1885*, pp. 148–51.

33. *FRUS 1886*, pp. 96–100; *FRUS 1888*, part 1, p. 272; *FRUS 1889*, p. 93; *FRUS 1890*, p. 158; *FRUS 1895*, part 1, p. 197; John W. Cassey, "The Mission of Charles Denby and International Rivalries in the Far East, 1885–1898" (Ph.D. diss., University of Southern California, 1959), pp. 136–40.

34. *FRUS 1882*, p. 142; *FRUS 1888*, part 1, p. 266; *FRUS 1890*, pp. 149, 161, 179.

35. Actual American mission losses were slight: damage to a Ch'eng-tu Methodist compound; and at Ku-t'ien one wounded and only minor property losses. Irwin Hyatt, "The Chengtu Riots (1895): Myth and Politics," *Papers on China* (1964), vol. 18; *FRUS 1895*, part 1, pp. 95, 119, 176; CR: Tientsin (July 13, 1895).

36. Justus Doolittle, *Social Life of the Chinese* (New York, 1865); Arthur H. Smith, *Chinese Characteristics* (New York, 1894); Chester Holcombe, *The Real Chinaman* (New York, 1895); Cassey, "The Mission of Charles Denby," pp. 121–27, 130–33; Robert F. McClellan, "Missionary Influence on American Attitudes Toward China at the Turn of the Century," *Church History* (December 1969), 38:476–82; Clifton J. Phillips, "The Student Volunteer Movement and Its Role in China Missions, 1886–1920," in John K. Fairbank, ed., *The Missionary Enterprise in China and America* (Cambridge, Mass., 1974), 101–2, 104–7; Stuart C. Miller, "Ends and Means: Missionary Justification of Force in Nineteenth Century China," in *ibid*, pp. 265–73; W.A.P. Martin, *A Cycle of Cathay* (New York, 1900), p. 406.

37. *North China Daily News*, July 18, 1895, in *FRUS 1895*, part 1, pp. 97–98.

38. *FRUS 1891*, pp. 406–7, 411, 422, 426; *FRUS 1892*, pp. 85, 90; MCD (June 8, 1891); ISHK, 20:14–15.

39. *FRUS 1895*, part 1, pp. 87–88, 94, 96, 117, 126, 137.

40. A. T. Volwiler, ed., *The Correspondence between Benjamin Harrison and James G. Blaine, 1882–1893* (Philadelphia, 1940), pp. 184–85, 257; Lew Wallace, *Life of Gen. Ben Harrison* (Philadelphia, 1888), pp. 305–6; *FRUS 1893*, pp. 230–31, 233–34.

41. Gerald G. Eggert, *Richard Olney: Evolution of a Statesman* (University Park, Pa., 1974); p. 194; CWS:KH, 117:27; *FRUS 1895*, part 1, p. 139.

42. *FRUS 1895*, part 1, pp. 96–97, 109, 125, 134, 139, 142–45, 153, 163, 169–70, 173–74; *FRUS 1896*, pp. 47–48, 53; Eggert, *Richard Olney*, pp. 194–96. The tough stand of the Harrison and Cleveland administrations led the Chinese government to fear some Anglo-American accord on the mission question and elicited concerned inquiries from the Chinese legation. CWS:KH, 85:2–4, 117:2, 27; *FRUS 1895*, part 1, pp. 105, 121–22.

43. *FRUS 1896*, pp. 58–59, 62–64; *FRUS 1897*, pp. 60–69; CWS:KH, 133:1, 134:18–19; Chien Po-tsan et al., eds., *Wu-hsü pien-fa* [The reform movement of 1898] (Shanghai, 1953), 2:35; Chester C. Tan, *The Boxer Catastrophe* (New York, 1955), pp. 58–59; SSM, pp. 462–63.

44. *FRUS 1895*, part 1, pp. 150, 154; MCD (November 14, 1895); *FRUS 1896*, pp. 70–83; *FRUS 1897*, p. 102; CR:Tientsin (June 8, 12, and 16, 1897); *FRUS 1898*, pp. 191–98.

45. *FRUS 1897*, p. 68.

46. *FRUS 1895*, part 1, p. 112.

47. CKS:TC, pp. 1049–51; *K'uang-wu tang*, 2:737, 739, 745–46, 761; Esson M. Gale, "President James Burrill Angell's Diary as United States Commissioner to China,

1880–1881," *Michigan Alumnus Quarterly Review* (May 1943), 49:199; *CWS:KH*, 80:15, 19, 20, 27; *Hai-fang tang*, "Tien-hsien," part 1, p. 379; *SSM*, pp. 303–5, 346–47.

48. Five consulates at that time relied on Chinese whose English proficiency—evident, S. Wells Williams reported, "in such phraseology as, 'You talkee my, my can so fashion talkee he.' "—was as poor as their opportunities to profit from their position were rich. One consulate depended on an American merchant. A physician ill-qualified in Chinese was the interpreter for the busy Shanghai consulate. Only two consulates had Americans who worked full-time and had a knowledge of Chinese, and one of these was Pethick, who was not averse to leaking official documents and confidential information to Li Hung-chang. *FRUS 1864–1865*, part 3, p. 346; Charles Denby, *China and Her People* (Boston, 1906), 2:216–20; *FRUS 1879*, pp. 204–7; Trescott to Evarts, July 23, 1880, William Evarts Papers (Sterling Library, Yale University); MCD (S. W. Williams to Burlingame, September 15, 1862 enclosed in September 16, 1862); CR: Tientsin (October 16, 1872). In general the American diplomatic establishment compared unfavorably with that of even such lesser powers as the Netherlands, Portugal, and Spain.

49. *FRUS 1866*, part 1, pp. 476–77; Lewis C. Arlington, *Through the Dragon's Eye* (London, 1931), p. 121; Denby, *China*, 1:27, 34–45; Tansill, *The Foreign Policy of Thomas F. Bayard*, p. 145; Clive Bingham, *A Year in China, 1899–1900* (London, 1901), pp. 39–41.

50. MCD (Williams to Burlingame, September 15, 1862 enclosed in September 16, 1862); Denby, *China*, 1:33–34.

51. Burlingame had his eye on a $200,000 surplus from an 1858 indemnity (collected for damages done to Americans in Canton) to serve as the school's endowment. Though the surplus was in theory China's to use as she wished, Burlingame warned Washington that it was best to restrict its use to some desirable project such as the school; otherwise the Chinese government "would probably . . . draw upon it in some unexpected way that would be embarrassing." Although endorsed by later ministers and a string of Republican administrations, the Williams-Burlingame proposal finally died in 1885 when Congress returned the surplus unencumbered by noble educational ideals. However, the impulse to educate China's elite in American ways with Chinese money lived on, to shape the terms of the return of the Boxer indemnity surplus. *FRUS 1862*, pp. 843–46; *FRUS 1864–1865*, part 3, pp. 346–48; *FRUS 1871*, pp. 226–27; *FRUS 1872*, pp. 136–38; *FRUS 1885*, pp. 181–82; Paolino, *The Foundations of the American Empire*, pp. 149–50; Ring, "Anson Burlingame, S. Wells Williams and China," pp. 72–73. On the Chinese effort to hasten the return of the money, see *ISHK*, 9:14–15, 14:2–4, 6–7, 15:18–19, 34; *SSM*, pp. 195–200; *CWS:KH*, 31:9–11, 57:9–10.

52. Here too Williams brought matured views of his own that Burlingame drew on. Williams argued for reform and development but also "forbearance and patience while educating a pagan and ignorant people." Though the treaties were "like great charters of civilization and Christianity," they had been imposed by force on a "weak, poor and undecided" China. For the powers to continue to ride roughshod and insist on ever broader treaty rights was "likely to destroy the autonomy of the Government" and its popular support, with unpalatable consequences—continued internal disorder and renewed international rivalry. MCD (June 17, 1862, June 20, 1863, May 21, 1866); Ring, "Anson Burlingame, S. Wells Williams and China," pp. 69–70, 74–75, 80–83, 184–93; Williams to Robert S. Williams, March 24, 1870, Williams Family Papers (Sterling Library, Yale University); *K'uang-wu tang*, 2:757–58.

53. Burlingame speech in New York City, June 23, 1868, reproduced in Frederick Wells Williams, *Anson Burlingame and the First Chinese Mission to Foreign Powers* (New York, 1912), pp. 134–39; Ring, "Anson Burlingame, S. Wells Williams and China," pp. 121, 214–16; Knight Biggerstaff, "The Official Chinese Attitude Toward the Burlingame Mission," *American Historical Review* (July 1936), 41:684–85, 687–90, 694–95; Biggerstaff, "A Translation of Anson Burlingame's Instructions from the Chinese Foreign Office," *Far Eastern Quarterly* (May 1942), vol. 1.

54. Lina F. Browne, ed., *J. Ross Browne: His Letters, Journals and Writings* (Albuquerque, N.M., 1969), pp. 330, 335, 343–48, 352; K'uang-wu tang, 2:852–53; Williams, *Anson Burlingame*, pp. 281–83, 300, 304, 307–8; Ring, "Anson Burlingame, S. Wells Williams and China," pp. 269–87.

55. MCD (January 10, 1871); SSM, pp. 413–14; William H. Morken, "America Looks West: The Search for a China Policy, 1876–1885" (Ph.D. diss., Claremont Graduate School, 1974), pp. 27–32, 70; David L. Anderson, "To the Open Door: America's Search for a Policy in China, 1861–1900" (Ph.D. diss., University of Virginia, 1974), pp. 112–78 passim. The ministers were Frederick Low (1869–1874, given the China post after a failed bid for the U.S. Senate); Benjamin Avery (1874–1875); George Seward (1876–1880, nephew of the former Secretary of State, promoted from his position as the Shanghai consul general); and James B. Angell (who left his position as president of the University of Michigan to serve in 1880 and 1881).

56. Young had also ingratiatingly suggested to Secretary Evarts in the new Hayes administration that he propound an "Evarts doctrine" for East Asia (on a level with the Monroe Doctrine) that would commit the United States as "an almost permanent power" to block European machinations. Offered only a subordinate post in Peking, Young declined and bided his time until Angell's resignation in the fall of 1881. In the meantime, Garfield's assassination and Blaine's departure put appointments in the hands of Arthur and Frelinghuysen, who were glad to gratify the ambitions of a Grant protégé. Young to Evarts, March 4, 25, and 26, 1880, Evarts Papers (LC); and Young to Evarts, May 21, 1880, Evarts Papers (Yale).

57. MCD (October 2 and 9, 1882, August 8, 1883); *FRUS 1883*, pp. 162, 164; Grant to Young, May 18, 1883, John Russell Young Papers (LC).

58. Twenty years of Democratic party service won Denby the post in Peking during the first Cleveland administration. The Republican Harrison kept him on, in part because he was a fellow Hoosier, in part because replacements failed to get Chinese approval. Cleveland's return to the White House and the appointment of Denby's friend Walter Q. Gresham as Secretary of State extended Denby's lease on the legation. Not until after the election of McKinley did party patronage force him out. Denby, *China*, 1:ix–xvi.

59. His opinion of the Empress Dowager was high; he described her in 1889 as "benevolent," "economical," "spotless" in her private conduct, universally esteemed, and sensible in her foreign policy. Li Hung-chang, "the most prominent man in China today" and possessed of "great respect for foreigners," stood even higher. Even after the debacle of the war with Japan, Denby still looked to Li to save a chastened China. The court had only to give him "the power to control and direct material progress," and with the cooperation of the "English speaking people" Li would at last develop China's railways, army and navy, and banking system; pay off the war indemnity; and generally secure the cause of progress in China. *FRUS 1885*, p. 180; *FRUS 1889*, pp. 94–95, 100; MCD (February 26, March 5, 1895).

60. MCD (February 9, 1886); Cassey, "The Mission of Charles Denby," pp. 145–51, 156–59.

61. FRUS 1889, pp. 94–95; MCD (November 16 and 19 and December 8, 1894, April 2, 1897).

62. FRUS 1889, p. 95; MCD (January 17, February 26, March 18, April 8 and 29, 1895, September 2, 1896, April 2, 1897).

63. Tansill, The Foreign Policy of Thomas F. Bayard, p. 424; Cassey, "The Mission of Charles Denby," p. 254; MCD (January 25, 1896, November 5, 1896, May 24, October 20, 1897); Eggert, Richard Olney, pp. 273–74, 276–78; FRUS 1897, pp. 56–58; Hai-fang tang, "T'ieh-lu," pp. 228–29.

64. MCD (December 26, 1894, March 5, 1895, March 15, December 11, 1897, January 31, 1898). Denby may have borrowed from, and possibly even been involved in, one of several schemes then circulating within the Anglo-American community in China. Ting Tse-liang, "Ma-kuan i-ho ch'ien Li-t'i-mo-tai ts'e-tung Li Hung-chang mai-kuo yin-mou ti fa-hsien" [Discovery of Timothy Richard's plot to instigate Li Hung-chang to treason before the Shimonoseki negotiations] in Chung-Jih chia-wu chan-cheng lun-chi [Essays on the Sino-Japanese War of 1894] (Peking, 1954); Foster to Denby, September 26, 1894, Pethick, memo on the political situation in China, September 1894, Pethick to James H. Wilson, September 24 and 29, 1894, and January 7, 1895, and Wilson to Foster, August 10, 1895, Wilson Papers.

65. Conger had made money in farming and banking, served six years in the House, and then resigned in 1890 to spend three uneventful years as minister to Brazil. McKinley's election put a friend back in power. With the additional support of the New Yorkers John W. Foster and William E. Dodge, Conger was returned to Brazil and then in mid-1898, after less than a year, transferred to China.

66. MCD (July 31, 1898, November 3, 1898, December 21, 1899).

67. MCD (August 26, 1898, November 3, 1898, March 1, 1899); FRUS 1899, p. 153.

68. Eggert, Richard Olney, pp. 285–87; John A. Garraty, Henry Cabot Lodge (New York, 1952), pp. 204–5; Rockhill, "The United States and the Future of China," Forum (May 1900), vol. 29; Denby, "Shall We Keep the Philippines?" ibid. (October 1898), vol. 26; Denby, "The Doctrine of Intervention," ibid. (December 1898), vol. 26; Denby, "America's Opportunity in Asia," North American Review (January 1898), vol. 166; Barrett, "Political Possibilities in China," Harper's Weekly (July 7, 1900), vol. 44; Adams, America's Economic Supremacy (New York, 1900), chs. 1, 2, 5; Conant, The United States in the Orient: The Nature of the Economic Problem (Boston, 1900), pp. iii–iv, 29, 32–33, 63, 156–58, 175–76; Strong, Expansion under New World Conditions (New York, 1900), chs. 3, 4, 6, 7; Sheffield, "The Future of the Chinese People," Atlantic Monthly (January 1900), vol. 85; Robert Seager II, Alfred Thayer Mahan: The Man and His Letters (Annapolis, Md., 1977), pp. 459–67.

69. FRUS 1862, p. 839; Olive R. Seward, William H. Seward's Travels Around the World (New York, 1873), p. 216; Williams, Anson Burlingame, pp. 56–57, 211n; Ring, "Anson Burlingame, S. Wells Williams and China," p. 287; Nevins, Fish, 1:249, 2:915–16.

70. DIC (September 8, 1868); Gale, "President James Burrill Angell's Diary," pp. 198–99; Smith, The Life and Letters of James Abram Garfield, 1:488, 592; Tansill, The Foreign Policy of Thomas F. Bayard, pp. 144–45.

71. Gresham to Denby, December 26, 1894, January 4 and April 12, 1895, Gresham Papers (LC); Gresham in CJCC, 5:270.

72. In line with this general faith in free trade, the rhetoric of policy makers

had up until the 1880s been studded with references to the China trade as a source of American prosperity and as a prime "regenerating" force (even more potent than mission work). This China market rhetoric, at least as it was used by policy makers, seems then to have lost currency for a time, perhaps because political prudence in the face of a strong exclusionist campaign dictated minimizing talk about the importance of the China trade with all that it implied about putting exports ahead of the interest of the American worker. As anti-Chinese sentiment subsided in the mid-1890s under a surfeit of restrictive measures, it became easier for policy makers once more to extol the benefits of the China trade. FRUS 1866, part 1, p. 487; Seward, William H. Seward's Travels, pp. 169–70, 201, 265; copies of DIC in Fish Papers (LC); Gale, "President James Burill Angell's Diary," pp. 198–99 (Evarts' views); Theodore C. Smith, The Life and Letters of James Abram Garfield (New Haven, Conn., 1925), 1:488, 592; Richardson, A Compilation, 7:516–19, 569; Charles R. Williams, ed., Diary and Letters of Rutherford Birchard Hayes (Columbus, Ohio, 1922–1926), 3:522–26.

73. Gresham to Denby, April 12, 1895, Gresham Papers; DIC (June 22, 1895); FRUS 1897, p. 56; Eggert, Richard Olney, pp. 273–74, 276–78.

74. Paulsen, "Machinery for the Mills of China," pp. 325–30, 339–41; FRUS 1897, pp. 89, 92.

75. FRUS 1897, p. 59.

76. Charles S. Campbell, Jr., Anglo-American Understanding, 1898–1903 (Baltimore, 1957), pp. 117–20.

77. The strength of the Asiatic squadron had been sapped by the general post–Civil War decline in the navy, falling from thirteen ships in 1867 to three in 1889. But a naval renaissance was by then already under way so that by 1896 the squadron boasted four cruisers (three protected) and three gunboats on the China station. The war with Spain and then the rebellion in the Philippines brought further and substantial additions to the squadron. James M. Merrill, "The Asiatic Squadron: 1835–1907," American Neptune (April 1969), 29:115.

78. Richardson, A Compilation, 10:180–81; William R. Thayer, The Life and Letters of John Hay (London, 1915), 2:241; FRUS 1899, p. 152; John W. Foster, Diplomatic Memoirs (Boston, 1909), 2:257; Hay to Wu T'ing-fang, November 11, 1899, and Hay to Choate, November 13, 1899, Hay Papers.

Part Three: The Patterns Hold

1. Pitkin to Judson Smith, May 14, 1900, ABFM, vol. 28; Isaac C. Ketler, The Tragedy of Paotingfu (New York, 1902), pp. 241–61.

2. Gould to Smith, May 14, 1900, ABFM, vol. 27; Gould letter to parents, May 29, 1900, RG 395, file 906 (enclosed in a letter of August 31, 1900); CR: Tientsin (July 16, 1900); Ketler, The Tragedy of Paotingfu, pp. 143–50, 178–79.

3. Quotes from Robert C. Forsyth, The China Martyrs of 1900 (New York, n.d.), p. 497; and Ketler, The Tragedy of Paotingfu, p. 400. For versions of the Pao-ting massacre, see A. H. S. Landor, China and the Allies (New York, 1901), 1:251, 253; Chien Po-tsan et al., comps, I-ho-t'uan [The Boxers] (Shanghai, 1951), 1:492; U.S. War Department, Annual Reports of the War Department for . . . 1901 (Washington, D.C., 1901), 1:464–65. The account in George Lynch, The War of Civilizations (London, 1901), pp. 204–5, is sensationalistic. Use of the Pao-ting case in missionary literature: Robert E. Speer, A Memorial of Horace Tracy Pitkin (New York, 1903);

Speer, *Young Men Who Overcame* (New York, 1905), ch. 4; and Ketler, *The Tragedy of Paotingfu*. Pitkin's heirs, truer to his spirit, lodged a $100,000 indemnity claim for loss of life. The U.S. government allowed only $12,000 of that and added another $15,000 for household furnishings, described by one neighbor as opulent. MCD (November 8, 1902).

4. Arthur H. Smith, *China in Convulsion* (New York, 1901), 1:232–35, 273–76, 324, 2:365–75, 509; Putnam Weale (pseudonym for Bertram L. Simpson) *Indiscreet Letters from Peking* (New York, 1907), pp. 141–42; Robert Coltman, Jr., *Beleaguered in Peking* (Philadelphia, 1901), pp. 100, 148; *FRUS 1900*, pp. 163–64.

5. John K. Fairbank et al., eds., *The I.G. in Peking: Letters of Robert Hart, Chinese Maritime Customs, 1868–1907* (Cambridge, Mass., 1975), 2:993; Robert Hart, *These from the Land of Sinim* (London, 1903), p. 59.

Chapter 6: China's Defense and the Open Door, 1898–1914

The open door and China's defense—two intimately connected ideas in the early twentieth century—have in the main been divorced one from the other by historians. Raymond A. Esthus has offered the best overview of the development of American policy in "The Changing Concept of the Open Door, 1899–1910," Mississippi Valley Historical Review (December 1959), vol. 46, and a restatement (updated in light of recent scholarship and extended in coverage), "The Open Door and Integrity of China, 1899–1922: Hazy Principles for Changing Policy," in Thomas H. Etzold, ed., Aspects of Sino-American Relations Since 1784 (New York, 1978). Both essays, written from a "realist" perspective, emphasize the tendency of policy makers to embrace a definition of the open door which was beyond their power to implement. Paul A. Varg's The Making of a Myth: The United States and China, 1897–1912 (East Lansing, Mich., 1968), develops the "realist" interpretation in greater detail though with less satisfactory results. My Frontier Defense and the Open Door: Manchuria in Chinese-American Relations, 1895–1911 (New Haven, Conn., 1973), relates one important set of Chinese policy initiatives to the American response to them.

For the encounters between 1898 and 1901, the books by Marilyn Young and Thomas McCormick (both noted in the previous chapter) provide detailed coverage of the American side but from different perspectives. Young is critical of the interventionist, racist, and imperialist strains in turn-of-the-century China policy; McCormick is concerned with tracing Washington's attempt "to colonize China's commerce" (p. 185). McCormick more than Young overestimates the American power to shape events in China in part because he, unlike Young, neglects the Chinese context and tends to take contemporary American self-estimates at face value. William R. Braisted's terse, bare-bones account, "The Open Door and the Boxer Uprising," in Paolo E. Coletta, ed., Threshold to Internationalism: Essays on the Foreign Policies of William McKinley (Jericho, N.Y., 1970), is interesting for its emphasis on McKinley's role in the prolonged China crisis.

Background to Chinese efforts to exploit the American open door pol-

icy at the turn of the century is available through Li Kuo-ch'i, Chang Chih-tung ti wai-chiao cheng-ts'e [Chang Chih-tung's foreign policy] (Taipei, 1970), and Chester Tan, The Boxer Catastrophe (New York, 1955; rev. ed. 1967), still the most comprehensive account on the Boxer crisis, though it is long on diplomacy and short on social history. Victor Purcell's reappraisal, The Boxer Uprising: A Background Study (Cambridge, England, 1963), chs. 8–12, and Wang Shu-hwai's Keng-tzu p'ei-kuan [The Boxer indemnity] (Taipei, 1974) are helpful supplements. The post-Boxer Manchurian crisis gets its due in Masataka Kosaka, "Ch'ing Policy over Manchuria (1900–1903)," Papers on China (1962), vol. 16, and Ian H. Nish, The Anglo-Japanese Alliance: The Diplomacy of Two Island Empires, 1894–1907 (London, 1966).

The battle for Manchuria after the Russo-Japanese War is covered in a variety of works. Together the political biographies by Stephen R. MacKinnon, Power and Politics in Late Imperial China: Yuan Shi-kai in Beijing and Tianjin, 1901–1908 (Berkeley, Calif., 1980); Lee En-han, "T'ang Shao-i yü wan-Ch'ing wai-chiao" [T'ang Shao-i and late Ch'ing diplomacy], Chung-yang yen-chiu yüan, chin-tai shih yen-chiu so chi-k'an (May 1973), no. 4; and Roger V. DesForges, Hsi-liang and the Chinese National Revolution (New Haven, Conn., 1973), provide insight into the domestic basis of Chinese foreign policy. The railway projects which figured prominently in Chinese plans are treated in Robert L. Irick, "The Chinchow-Aigun Railroad and the Knox Neutralization Plan in Ch'ing Diplomacy," Papers on China (1959), vol. 13, and E-tu Zen Sun, Chinese Railways and British Interests, 1898–1911 (New York, 1954), ch. 6. Sun's work, a political and diplomatic but not an economic history, deals in addition with the Hankow-Canton and Hukuang projects, also of concern to Americans, in chs. 3 and 4. On American policy, Charles Vevier's detailed The United States and China, 1906–1913: A Study of Finance and Diplomacy (New Brunswick, N.J., 1955), still deserves attention though it should be read alongside Charles E. Neu, An Uncertain Friendship: Theodore Roosevelt and Japan, 1906–1909 (Cambridge, Mass., 1967); Raymond A. Esthus, Theodore Roosevelt and Japan (Seattle, Wash., 1966); Helen Dodson Kahn, "Willard D. Straight and the Great Game of Empire," in Frank J. Merli and Theodore A. Wilson, eds., The Makers of American Diplomacy (New York, 1974), vol. 2; Walter and Miriam Scholes, The Foreign Policies of the Taft Administration (Columbia, Mo., 1970); Donald F. Anderson, William Howard Taft: A Conservative's Conception of the Presidency (Ithaca, N.Y., 1973), ch. 7; and Hunt, Frontier Defense and the Open Door (noted above).

Wilson's repudiation of "dollar diplomacy" and the subsequent development of his own policy toward the Chinese republic did not receive full historical treatment based on a substantial range of primary sources until 1952 when T'ien-yi Li's Woodrow Wilson's China Policy (Kansas City, Mo.) and Russell H. Fifield's Woodrow Wilson and the Far East: The Diplomacy of the Shantung Question (New York) appeared in print and Roy

W. Curry completed his dissertation, published in 1957 as Woodrow Wilson and Far Eastern Policy (New York). Fifield is short on interpretation, while Li and Curry by contrast highlight what have remained the two most frequently repeated interpretive points—Wilson's moral conception of foreign relations and the difficulties he encountered in translating his ideals into workable policy. Two more recent accounts—Burton F. Beers, Vain Endeavor: Robert Lansing's Attempt to End the American-Japanese Rivalry (Durham, N.C., 1962), and Arthur S. Link, Wilson (Princeton, N.J., 1947–), 2:283–88, 3:192–96, 267–308—provide a fuller sense of the international context Wilson worked in and a more critical, "realist" evaluation of the gap between Wilson's aspirations for China and his achievements. Link judges Wilson's China policy "bankrupt" by 1917 (2:288), while Beers argues that the Wilson administration lost a chance to strike a bargain with Japan and thus avert an ultimately costly showdown over China.

There is only one full-fledged attempt, only partially successful, to reconstruct the place of the United States in Chinese foreign policy in the teens: Patrick J. Scanlan's "No Longer a Treaty Port: Paul S. Reinsch and China, 1913–1919" (Ph.D. diss., University of Wisconsin, 1973). Ernest P. Young's broad interpretive account of The Presidency of Yuan Shih-k'ai: Liberalism and Dictatorship in Early Republican China (Ann Arbor, Mich., 1977) and Madeleine Chi's multiarchival China Diplomacy, 1914–1918 (Cambridge, Mass., 1970) provide the domestic and international setting necessary for a better appreciation of the Scanlan account. Young contends that both Yüan's political drive and his political failures must be understood largely in terms of the pervasive challenge of foreign penetration, and in that sense his book stands as an important rejoinder to Chinese historians—Nationalist as well as Communist—preoccupied with Yüan's "betrayal" of the republic. Jerome Ch'en's Yuan Shih-k'ai, 2d ed. (Stanford, Calif., 1972), once revealingly subtitled Brutus Assumes the Purple, is a notable example of this latter view. Young's work together with MacKinnon's supersede Ch'en's account.

1. Li Kuo-ch'i, Chang Chih-tung ti wai-chiao cheng-ts'e [Chang Chih-tung's foreign policy] (Taipei, 1970), ch. 1.
2. CKCC, 46:22; CWC, 37:36 Hsü T'ung-hsin, Chang Wen-hsiang-kung nien-p'u [A chronological biography of Chang Chih-tung] (Taipei reprint, 1969), p. 96.
3. The doubts about barbarian management that had spread among such notable foreign affairs experts in the late nineteenth century as Feng Kuei-fen, Wang T'ao, and Kuo Sung-tao may account for this caution in drawing policy implications. Wang Chia-chien, Wei Yüan tui hsi-fang ti jen-shih chi ch'i hai-fang ssu-hsiang [Wei Yüan's knowledge of the West and his ideas on maritime defense] (Taipei, 1964), pp. 155–56, 160; Wang Shu-hwai, Wai-jen yü wu-hsü pien-fa [Foreigners and the reform movement of 1898] (Taipei, 1965), pp. 123–24; Chang Tze-mu, Ying-hai lun [A discussion of the circuit of the sea] (circa 1876), part 1, pp. 2–3, 8; Noriko Kamachi, "American Influences on Chinese Reform Thought: Huang Tsun-hsien in California, 1882–1885," Pacific Historical Review (May 1978), 47:242, 245–46, 257;

Tsai-chün, memorial of 1884 in Shao Hsün-cheng et al., comps., *Chung-Fa chang-cheng* [The Sino-French War] (Shanghai, 1955), 6:9; Chung-kuo shih-hsüeh hui, comp., *Yang-wu yün-tung* [The westernization movement] (Shanghai, 1961), 1:238; Hsüeh Fu-ch'eng, diary entry of May 1, 1890, in Ssu-yü Teng and John K. Fairbank, *China's Response to the West: A Documentary Survey, 1839–1923* (Cambridge, Mass., 1961), p. 144; T'ang Ts'ai-ch'ang, 1893 essay in Chien Po-tsan et al., comps., *Wu-hsü pien-fa* [The reform movement of 1898] (Shanghai, 1953), 3:105; anonymous essay (circa 1895–1898) in Wen T'ing-shih et al., *Chung-Jih chia-wu chan-cheng* [The Sino-Japanese War of 1894] (Taipei, 1967), p. 47; Ch'en Chih (1897) in Chao Feng-t'ien, *Wan-Ch'ing wu-shih nien ching-chi ssu-hsiang shih* [Economic thought during the last fifty years of the Ch'ing period] (Peiping, 1933), p. 163.

4. Hsü's mistrust of Britain and Japan and his belief that the rich concessions that Russia had won in Manchuria would lead her to support China all paralleled the views of Li, his early sponsor. Hsü Chüeh, *Fu-an i-chi* [Works of Hsü Chüeh] (Taipei reprint, 1970), 1:13–18, 2:702–3, 707, 753–55; *CJCC*, 5:262, 267–69, 271–72. Hsü's enthusiasm for the United States persisted through the Boxer crisis and the Russo-Japanese War. *Fu-an i-chi*, 1:66, 71ff., 172, 2:776–80, 798ff., 828.

5. The court took Wu's proposal seriously enough to refer it in April to the foreign office for further discussion. Wu T'ing-fang (1895) in Yü Pao-hsüan, comp., *Huang-ch'ao hsü-ai wen-pien* [A collection of essays written under the ruling dynasty] (Taipei reprint, 1965), 3:7; *CWS:KH*, 129:16–17; *SSM*, p. 103; Linda P. Shin, "China in Transition: The Role of Wu T'ing-fang (1842–1922)" (Ph.D. diss., University of California at Los Angeles, 1970).

6. Yung's proposals came to naught in part because of Chang's resistance to railroad plans at odds with his own, and his aversion to foreign loans. Sun Yü-t'ang, "Chung-Jih chia-wu chan-cheng hou Mei-ti-kuo chu-i chüeh-to Lu-Han, Ching-Chen yü Yüeh-Han chu t'ieh-lu ti yin-mou (1895–1898)" [American imperialist plots to seize the Peking-Hankow, Tientsin-Chinkiang and Canton-Hankow railroads after the Sino-Japanese War (1895–1898)], in *Mei-ti-kuo chu-i ching-chi ch'in-Hua shih lun-ts'ung* [Collected articles on the history of American imperialist economic aggression against China] (Peking, 1953), pp. 31–33, 35–37.

Liang Tun-yen, an alumnus of the Yung Wing educational mission to the United States, and a member of Chang Chih-tung's secretariat since 1884, may also have contributed to the interest in the United States, but his views at this time cannot be documented.

7. *CWC*, 45:23–24, 47:4–5, 152:32; Hsü, *Chang Wen-hsiang-kung*, pp. 110–12; Sun, "Chung-Jih chia-wu chang-cheng hou," pp. 38–43; Sheng Hsüan-huai, *Yü-chai ts'un-kao* [Collected papers of Sheng Hsüan-huai] (Taipei reprint, 1968), 2:3–5, 10, 21:10–11, 26:9–10, 12–13, 27:2–3, 29:7, 31:1, 19–21, 23–24; Pei-ching ta-hsüeh li-shih hsi, chin-tai shih chiao-yen shih, comp., *Sheng Hsüan-huai wei-k'an hsin-kao* [Sheng Hsüan-huai's unpublished letters] (Peking, 1960), pp. 40, 57. Sheng was to remain an advocate of reliance on the United States in railway affairs down to 1905, and he argued for securing American assistance in the wake of the Russo-Japanese accords of 1910. *CWS:KH*, 159:2; Sheng, *Yü-chai ts'un-kao*, 66:12–14, 76:6–7.

8. Hay to Wu T'ing-fang, November 11, 1899, and to Choate, November 13, 1899, Hay Papers (LC). See chapter 5 on the contemplated U.S. territorial demands.

9. The Boxer crisis then brewing gave yet another reason to consolidate good relations with the United States and by concluding an agreement forestall new demands that the American China Development Company might be tempted to make. Hsü, *Chang Wen-hsiang-kung*, pp. 130, 135; Sheng, *Yü-chai ts'un-kao*, 7:17–20, 36:4.

10. SL:KH, 453:5–6; Leslie Marchant, "Chinese Anti-Foreignism and the Boxer Uprising" in Leslie Marchant, ed., *The Siege of the Peking Legations* (Nedlands, Western Australia, 1970), pp. 35–36; imperial decree of July 2 in *FRUS 1900*, p. 171.
11. Hsü, *Chang Wen-hsiang-kung*, p. 133; CWS:KH, 143:8, 10–14, 17–20, 144:2–3, 5; *ITS*, pp. 179, 195, 202–3, 212–14, 227–29, 259, 327–28, 339, 344–46, 356–57, 383–84, 392, 445–46, 475; Chien Po-tsan et al., comps., *I-ho-t'uan* [The Boxers] (Shanghai, 1951), 2:497.
12. Hsü, *Chang Wen-hsiang-kung*, pp. 133, 140; *ITS*, pp. 164, 214, 329, 360–61; CWC, 160:22; U.S. Adjutant General, *Correspondence relating to the War with Spain . . . and the China Relief Expedition* (Washington, D.C., 1902), 1:428.
13. *FRUS 1898*, pp. 228–44; *FRUS 1899*, pp. 155–56, 167, 175–77; *FRUS 1900*, pp. 87–88, 93–94, 97, 102–4, 110–14, 117–18, 120–23, 127–29, 132, 139–41, 143, 155; CR:Tientsin (July 16, 1900); Arthur H. Smith, *China in Convulsion* (New York, 1901), 1:214–17; Timothy Richard, *Forty-Five Years in China* (New York, 1916), pp. 295–96.
14. U.S. Adjutant General, *Correspondence*, 1:428; *FRUS 1900*, pp. 263, 287; *I-ho-t'uan*, 1:505–7; Hay to McKinley, July 6, 1900, William McKinley Papers (LC); Michael H. Hunt, "The Forgotten Occupation: Peking, 1900–1901," *Pacific Historical Review* (November 1979), 48:503–5.
15. Hunt, "The Forgotten Occupation," pp. 515, 525–27; MCD (September 21 and October 1, 1900).
16. Charles S. Olcott, *The Life of William McKinley* (Boston, 1916), 2:233; Hay to McKinley, July 6, 1900, McKinley Papers.
17. Louis Morton, "Army and Marines on the China Station: A Study in Military and Political Rivalry," *Pacific Historical Review* (February 1960), 29:52–54; James M. Merrill, "The Asiatic Squadron: 1835–1907," *American Neptune* (April 1969), 29:117; *FRUS 1903*, pp. 85–90; Richard D. Challener, *Admirals, Generals, and American Foreign Policy, 1898–1914* (Princeton, N.J., 1973), pp. 22–23, 286–87; George Dewey to Secretary of the Navy, August 13, 1906, General Board Records (Naval Historical Center, Washington, D.C.).
18. A pied-à-terre on the North China coast—Tientsin was the favored site—was supported by Chaffee and several of his subordinates, by Charles Denby, Jr., by the local naval commander, and by the Tientsin consul. Chaffee wanted an interior base in Peking as well. The navy high command at home wanted to look further south, either to Samsah in Fukien or to the Chusan Islands guarding the approach to Shanghai and the Yangtze Valley. Hay went as far in the immediate aftermath of the Boxer summer as presenting the Chinese with a demand for a tract of land in Tientsin before quietly putting it aside in late 1901 as the prospect of restoring the status quo in China brightened. Hay also inquired about Samsah but encountered Japanese objections. By 1903 the interest of the navy's General Board in a southern base had begun to cool, and by 1906 was at an end for all practical purposes. RG 95, file 329412 (Louis Kempff to Secretary of the Navy, June 22 and 27, 1900); Gardner W. Allen, ed., *Papers of John Davis Long, 1897–1904* (Boston, 1939), p. 334; *FRUS 1901*, pp. 40–59; MCD (November 23 and December 7, 1900); U.S. Adjutant General, *Correspondence*, 1:493; Chaffee to Henry C. Corbin, [Dec. ?] 7, 1900 and April 10, 1901, Corbin Papers (LC); CR: Tientsin (April 22, 1902); George Dewey to Secretary of the Navy, October 10, 1900, General Board Records; DIC (November 16 and 19, 1900); NF 2413/160; Challener, *Admirals, Generals, and American Foreign Policy*, pp. 183–93; Seward W. Livermore, "American Naval-Base Policy in the Far East, 1850–1914," *Pacific Historical Review* (June 1944), vol. 13.

19. Olcott, *The Life of William McKinley*, 2:252, 259–60; James B. Richardson, comp., *A Compilation of the Messages and Papers of the Presidents* (Washington, D.C., 1896–1900; supplement, 1903), supplement, p. 114; MCD (August 9, 14, and 17 and September 2, 1900); DIC (November 23, 1900); *FRUS 1900*, pp. 219, 324; Henry Adams, ed., *Letters of John Hay and Extracts from Diary* (New York reprint, 1969), 3:193, 199–200; Hay to McKinley, July 8, 1900, McKinley Papers; *ITS*, p. 782.

20. *ITS*, pp. 537–42, 582–83, 1142–43; Chung-kuo k'o-hsüeh yüan, li-shih yen-chiu so, ti-san so, comp., *Liu K'un-i i-chi* [Collected works of Liu K'un-i] (Peking, 1959), p. 1438; Alfred Cunningham, *A Chinese Soldier and Other Sketches* (London, 1902), pp. 91–92; Alicia B. Little, *Li Hung-chang: His Life and Times* (London, 1903), pp. 310–11.

21. Germany proposed as a precondition for talks that China agree to hand over guilty officials for Allied punishment. McKinley bluntly replied that punishment was a matter for the Chinese government to attend to (albeit under watchful foreign eyes). McKinley's position prevailed. Washington accommodated the Chinese again later by accepting the court's insistence on making the death penalty optional rather than required for implicated members of the imperial family, by opposing demands in December that China agree unconditionally to the powers' formal peace terms, and by holding out against plans for creating an international fortress in Peking and otherwise enfeebling China militarily (by forbidding the importation of arms, ammunition, or material for their manufacture). *FRUS 1900*, pp. 341–42; DIC (November 27 and December 31, 1900); *FRUS 1901*, Appendix: *Affairs in China*, pp. 82–83, 91, 350; Hsü, *Chang Wen-hsiang-kung*, p. 142; Wang Liang, comp., *Hsi-hsün ta-shih-chi* [Journal of the western inspection trip] (published as a supplement to *CWS*), 2:50–51; *ITS*, pp. 843, 854, 1046–47.

22. *FRUS 1901*, Appendix: *Affairs in China*, pp. 5, 87, 127–28, 141–43, 158–59, 172; Hsü, *Chang Wen-hsiang-kung*, pp. 146–47; DIC (November 20, 1900, January 29, May 10 and 28, 1901); *ITS*, pp. 1046–47, 1063–64, 1069–70, 1075, 1080, 1082–83, 1088–89, 1117–18, 1122–23, 1126, 1142–43, 1150, 1153, 1160–61, 1335–37; *Hsi-hsün ta-shih-chi*, 7:27, 44, 8:18. For evidence on the overcharge, see my "The American Remission of the Boxer Indemnity: A Reappraisal," *Journal of Asian Studies* (May 1972), 31:541–43; and documents in RG 107, file 17712, box 544.

23. Payment in silver, Chinese officials contended, would allow them to circumvent foreign bankers, who manipulated the silver–gold exchange rate to China's disadvantage, and to avoid the uncertainties of shifting exchange rates over the thirty-nine-year term of the indemnity debt. As rates did indeed shift against China in 1902 and 1903, they found to their alarm that they might have to squeeze from their provinces twice the amount originally anticipated. Hsü, *Chang Wen-hsiang-kung*, p. 157; *CWS:KH*, 147:19–20, 152:14–16, 153:15–17, 154:26–28, 157:5–8, 10, 12–15, 158:2–11, 19–20, 159:1–3, 12, 169:3–4, 170:7, 12–13, 15–16, 171:15, 185:1–2, 26–28, 190:7–10; *Hsi-hsün ta-shih-chi*, 11:35–37; *FRUS 1904*, pp. 183–84; *FRUS 1905*, p. 154; Stanley F. Wright, *Hart and the Chinese Customs* (Belfast, 1950), pp. 766–67; *Liu K'un-i i-chi*, pp. 1476–77.

24. Michael H. Hunt, *Frontier Defense and the Open Door: Manchuria in Chinese-American Relations, 1895–1911* (New Haven, Conn., 1973), pp. 55–58, 61–62; *Liu K'un-i i-chi*, p. 1474.

25. This and the paragraph that follows draw on Hunt, *Frontier Defense and the Open Door*, pp. 60–66, 77–81, 84–85; Hay to McKinley, August 20, 1900, McKinley Papers.

26. Under pressure from Britain and Japan as well as the United States, Prince

Ch'ing as head of the foreign office finally agreed in August 1903 to open the ports—but only after restoration of Chinese authority. He sought in addition a quid pro quo in the form of an American commitment to intercede with Russia, in effect drawing the United States closer to the Anglo-Japanese position. But Washington held back in favor of letting the "Chinks," "those poor trembling rabbits" (as Hay liked to call them), take the heat. Hunt, *Frontier Defense and the Open Door*, pp. 69–76, 80–81.

27. Before the Russo-Japanese War Yüan had briefly considered aiding Japan with transport and supplies, but his distrust of Japan, the magnitude of the gamble, and finally the opposition of Britain, the United States, and even Japan to any step that might extend war beyond Manchuria finally decided him on the more conservative course. Chang, despite his preference for Japan over Russia and his hopes for close Sino-Japanese cooperation after the war, sided with Yüan. Hunt, *Frontier Defense and the Open Door*, pp. 85–87, 119–20.

28. Chang Ts'un-wu, *Kuang-hsü sa-i nien Chung-Mei kung-yüeh feng-ch'ao* [The storm in 1905 over the Sino-American labor treaty] (Taipei, 1966), pp. 67–68; Lo Hsiang-lin, *Liang Ch'eng ch'u-shih Mei-kuo* [Liang Ch'eng's ministry to the United States] (Hong Kong, 1977), pp. 284, 287; CWC, 192:42.

29. The Manchurian policy described here and in the following two paragraphs emerges from Robert H. G. Lee, *The Manchurian Frontier in Ch'ing History* (Cambridge, Mass., 1970), pp. 140–55; and Hunt, *Frontier Defense and the Open Door*, pp. 100–6, 127–37, 152–64, 248–49.

30. Having learned in May 1905 that the American Boxer indemnity fund contained a substantial surplus, Yüan had then urged that it be returned to China to finance mining and railway projects. Hsü later embroidered on the idea by suggesting the remitted indemnity either go directly into a development bank or serve as the basis for an American loan of $13 to $20 million for use by the bank.

31. Hunt, *Frontier Defense and the Open Door*, pp. 164–66, 249; Bland to Addis, February 28 and April 29, 1908, Bland Papers (University of Toronto Library).

32. Hermann Hagedorn, ed., *The Works of Theodore Roosevelt* (New York, 1923–1926), 14:141–42, 15:286, 338–39, 18:364–65, 378–81; Roosevelt to Metcalf, June 12, 1905, Theodore Roosevelt Papers (LC); Elting E. Morison et al., eds., *The Letters of Theodore Roosevelt* (Cambridge, Mass., 1951–1954), 2:1329, 1428, 3:112, 172, 709–10, 4:811, 1242, 1310, 1313, 5:206.

33. Morison, *Letters of Roosevelt*, 3:23, 478, 497–98, 500–1, 520, 532, 4:1116, 1230.

34. When the Chinese government complained that the Portsmouth peace settlement had made southern Manchuria into a Japanese sphere of influence without its participation or consent, Roosevelt shot back that China was in no position to "question the efficacy" of the settlement. Privately he expressed doubts that China would ever exercise effective control over Manchuria again. Roosevelt's indifference to Chinese sovereignty in Manchuria is also reflected in his idea, discarded before the 1905 peace settlement, of having Germany designate which Chinese officials would serve in the region. Morison, *Letters of Roosevelt*, 3:105–6, 111–12, 4:830–32, 917, 5:18; Roosevelt to Hay, January 16, 1904, Roosevelt Papers; Cyril Pearl, *Morrison of Peking* (Sydney, Australia, 1967), p. 156; FRUS 1904, pp. 118–25; Charles E. Neu, *An Uncertain Friendship: Theodore Roosevelt and Japan, 1906–1909* (Cambridge, Mass., 1967), pp. 263–64.

35. Hunt, *Frontier Defense and the Open Door*, pp. 170–77.

36. Tsai-tao, "Tsai-feng yü Yüan Shih-k'ai ti mao-tun" [The clash between Tsai-feng and Yüan Shih-k'ai], in Chung-kuo jen-min cheng-chih hsieh-shang hui-i, ch'üan-

kuo wei-yüan hui, wen-shih tzu-liao yen-chiu hui, comp., *Hsin-hai ko-ming hui-i-lu* [Recollections of the 1911 revolution] (Peking, 1961–1963), 6:323–25; Hunt, *Frontier Defense and the Open Door*, pp. 124–26, 190; Stephen R. MacKinnon, *Power and Politics in Late Imperial China: Yuan Shih-kai in Beijing and Tianjin, 1901–1908* (Berkeley, Calif., 1980), ch. 3.

37. Hsi-liang's views, summarized here and in the following paragraph, are drawn from CWS:HT, 3:17–18; Chung-kuo k'o-hsüeh yüan, li-shih yen-chiu so, ti-san so, comp., *Hsi-liang i-kao: tsou-kao* [Collected papers of Hsi-liang: memorials] (Peking, 1959), pp. 1006–9, 1204–6.

38. Hunt, *Frontier Defense and the Open Door*, pp. 195–204, covers Hsi-liang's initial difficulties with Peking (dealt with in this and the following paragraph).

39. Had either Root or Henry Cabot Lodge, chairman of the Senate Foreign Relations Committee, accepted Taft's invitation to serve as Secretary of State, China policy might have remained largely unchanged. They refused and Knox got the job.

40. Taft to Martin Egan, March 25, 1905, to Roosevelt, October 5, 1907, to Rollo Ogden, April 21, 1909, to G. W. Painter, September 6, 1909, all in Taft Papers (LC); Taft cable from Tokyo, October 18, 1907, quoted in Philip Jessup, *Elihu Root* (New York, 1938), 2:27; Taft's October 1907 Shanghai address in his *Present Day Problems* (New York, 1908), pp. 44–48; Taft in *Inaugural Addresses of the Presidents of the United States* (Washington, D.C., 1974), p. 191; Knox quoted in J. B. Osborne memo, September 15, 1909, and "Summary of State Dept Policy and Actions" [Fall 1909], both in Knox Papers (LC); Records of the British Foreign Office (Public Records Office, London), file 371/640 (July 10, 1909).

41. Wilson, an egoistic and abrasive personality, came from a well-to-do Chicago family. After graduating from Yale (1897), he used his family's political connections first to secure an appointment to the Tokyo legation and then in 1906 to get reassigned to Washington as Third Assistant Secretary of State. Though Root had found Wilson obnoxious and his anti-Japanese views objectionable, he did accept part of the young man's reorganization proposal for the department and put him in charge of the new Division of Far Eastern Affairs (his own bureaucratic bailiwick). Straight brought to Knox's State Department a well-developed set of views on the need to stimulate China trade and to work with "progressive" Chinese officials in keeping the door open in Manchuria. Hunt, *Frontier Defense and the Open Door*, pp. 143–49, 182–83, 185–87.

42. Records of the British Foreign Office, file 371/786 (January 5, 1909), 875 (November 11, 1910).

43. See Hunt, *Frontier Defense and the Open Door*, pp. 187–90, 204–10, 212–16, on the beginnings of Knox's policy covered here and in the next two paragraphs.

44. Though Liang Tun-yen in the foreign office and Hsi-liang both denied the existence of an edict approving Hsi-liang's proposed railway, Knox's agents in Peking accepted instead the assurances of Edmund Backhouse. Sir Hugh Trevor-Roper has recently unmasked Backhouse as a fraud who repeatedly deceived credulous concession-hunters and diplomats with his claims to inside information. See Sir Hugh Trevor-Roper, *Hermit of Peking: The Hidden Life of Sir Edmund Backhouse* (New York, 1977).

45. This and the following paragraph summarize Hunt, *Frontier Defense and the Open Door*, pp. 218–23.

46. Knox to Taft, August 20, 1910, quoted in Challener, *Admirals, Generals, and American Foreign Policy*, p. 271; Hunt, *Frontier Defense and the Open Door*, pp. 235–37.

47. The Hukuang line consisted of the Hankow-Canton railway, which China had repurchased from Americans in 1905 with British money, and an extension running from Hankow to Ch'eng-tu in Szechuan province. After evicting the Americans, the Chinese had initially played German against British concessionaires, but by May 1909 the two sets of European rivals had come together with the backing of their governments and brought in a French banking group (also with official support behind it).

48. Knox to Taft, August 20, 1910, quoted in Challener, *Admirals, Generals, and American Foreign Policy,* p. 271; Hunt, *Frontier Defense and the Open Door,* pp. 188–90, 226–27, 236–37.

49. The collapse of Chinese and American Manchurian policies described below is recounted in detail in Hunt, *Frontier Defense and the Open Door,* pp. 230–40 passim, 250–51.

50. *FRUS 1912,* pp. 54–58, 63–64, 74, 79, 102–3, 107, 110, 115–16, 129, 141–42, 147, 149–50, 155; *FRUS 1913,* pp. 166–67.

51. *FRUS 1912,* pp. 71, 81–86; *FRUS 1913,* pp. 89, 91, 163–64, 168; DF 893.00/1105; DF 893.01A/9; DF 893.51/981, 1341–43, Meribeth E. Cameron, "American Recognition Policy toward the Republic of China, 1912–1913," *Pacific Historical Review* (1933), vol. 2.

52. Arthur S. Link et al., eds., *The Papers of Woodrow Wilson* (Princeton, N.J., 1966–), 4:570–71, 11:66, 93, 298–99, 12:11–13, 17–19, 215–16, 223, 14:433, 15:41, 149, 537, 16:120–21, 230, 341, 18:291, 22:159, 23:114.

53. Wilson's campaign talk of "making conquests of the markets of the world" was in any case informed by free trade assumptions which militated against the government serving (as Knox had earlier envisioned) as the "engine" of foreign trade. He advocated a lowering of tariff barriers and ruled out government subsidies or bounties to the merchant marine. After the election Wilson's recommendation to Congress (April 1913) in favor of tariff reduction passed over the problem of surplus product in silence while holding up free trade as the expression of liberty in the economic sphere. Link, *Wilson Papers,* 24:125–26, 194, 208–9, 25:16, 38–39, 101, 27:271, 29:251.

54. Link, *Wilson Papers,* 27:14, 58, 62, 65, 70, 110, 124, 131, 190, 195–99, 210, 259, 492, 28:4, 15, 22–23.

55. Wilson had held back for a time out of a respect for constitutional legitimacy so that Peking could set the machinery of government properly in place. At last in May he pronounced himself satisfied. E. David Cronon, ed., *The Cabinet Diaries of Josephus Daniels, 1913–1921* (Lincoln, Nebr., 1963), pp. 8, 17, 19–22, 25–26, 34, 41; William Jennings Bryan and Mary Baird Bryan, *Memoirs of William Jennings Bryan* (Chicago, 1925), pp. 362–63; David F. Houston, *Eight Years with Wilson's Cabinet* (New York, 1926), 1:44–45, 49, 59–60; *FRUS 1913,* pp. 105, 109–11, 173–74; Link, *Wilson Papers,* 27:192–94, 218, 265–66.

56. T'ang Shao-i (Yüan's first prime minister but out of favor by 1913), Liang Tun-yen, Liang Ju-hao, and Ts'ai T'ing-kan were all Cantonese who had studied in the United States under the auspices of the Yung Wing mission and returned to specialize in foreign affairs. Others educated later in the United States included another Cantonese, Ch'en Chin-t'ao (Columbia and Yale), as well as Chou Tzu-ch'i (Columbia), Shih Chao-chi (Alfred Sze, Cornell) and Ku Wei-chün (Wellington Koo, Columbia). Aside from Liang Ju-hao, T'ang, and Shih, three other officials now serving under Yüan—Chu Ch'i-ch'ien, Hsiung Hsi-ling and Ch'ien Neng-hsün—had held office in Manchuria under Hsü Shih-ch'ang and Hsi-liang and were familiar with the calculations behind the American strategy pursued there.

57. For example, in 1911 two provincial officials independently made the case for trying yet again at turning American resentment of Japan (as well as German jealousy of the Entente powers) into a weapon against the web of special agreements woven about China by Britain, France, Russia, and Japan. The opening of the Panama canal, one noted, would put the United States in an unprecedented position of naval strength in the Pacific. Chang Chien, a prominent entrepreneur and reformer, touched on these same points in an audience with the Regent in the late spring. Though he had supported the 1905 boycott in protest against the "extraordinary illtreatment" visited on Chinese laborers in the United States, he nevertheless later described Sino-American relations as "cordial" and took a keen interest from 1911 on in close cooperation in various mutually advantageous commercial enterprises. Wu Lu-chen, memorial of February 16, 1911, and Shen Jui-lin, July 10, 1911, both cited in Hunt, *Frontier Defense and the Open Door*, p. 248n; Chang Hsiao-jo, comp., *Chang Chi-tzu chiu-lu* [The nine records of Chang Chien] (Shanghai, 1935), "Cheng Wen-lu," 3:9, 37, 4:13.

58. The United States had (so this work asserted) managed to isolate itself from the world, order its internal affairs, and build its industry and commerce to the point that even its home market could not absorb all it produced. China was the only great free market left to American producers. John Hay had begun the task of securing entry. Over the subsequent twenty years his open door policy, which set the United States at odds with powers endangering China, had become as important to Americans as the Monroe doctrine. *Ou-chan hou chih Chung-kuo* [China after the European war] (Taipei reprint, 1966), pp. 102–4. The book was written for Hsü by Huang Fu, a member of the Nationalist party who had resided in California in 1915.

59. Quoted in Ernest P. Young, *The Presidency of Yuan Shih-k'ai: Liberalism and Dictatorship in Early Republican China* (Ann Arbor, Mich., 1977), p. 178.

60. *FRUS 1913*, pp. 82–86, 175; Link, *Wilson Papers*, 27:225, 381–82.

61. The largest of these was a $25 million contract concluded by Liang Tun-yen with Bethlehem Steel in 1911 during his tour of the United States. According to its terms, Bethlehem would build ships and upgrade dockyards and arsenals for the Chinese navy, and the U.S. navy would provide training and technology (including some secret equipment). The second enterprise looked to conservancy of the Huai River, a tributary of the Yangtze. This had been a pet project of Chang Chien since 1903. Disaster relief provided by the Americans during the winter of 1910–1911 and the American Red Cross's subsequent offer to help with long-term flood control had already caught Chang's interest and led to the American organization agreeing to secure a $20 million loan and an American contractor to carry out the conservancy work. The remaining project, a Standard Oil concession to prospect for oil in Shensi province, had also had its inception during the Liang Tun-yen visit. His aide Chou Tzu-ch'i had then broached such a concession with Standard. By 1913 the project had evolved to include a $15 million loan Standard was to raise for the Chinese government in exchange for the concession. If oil were found, the Chinese government would enjoy a share of the profits as a partner in a Standard-dominated development company and at the same time secure an American presence to offset Japanese interest in the province. Like the Huai conservancy project, the Standard concession could be extended to other provinces if it proved mutually advantageous. *FRUS 1913*, pp. 189–91; *FRUS 1914*, pp. 97–98, 103–4; DF 893.00/2050; DF 893.34/109; DF 893.6363/1, 3; DF 893.811/101; Paul S. Reinsch, *An American Diplomat in China* (Garden City, N.Y., 1922), pp. 70–75; Noel H. Pugach, "Standard Oil and Petroleum Development in Early Republican China," *Business History Review* (Winter 1971), 55:454–55, 457; William R. Braisted, "China, the United

States Navy, and the Bethlehem Steel Company, 1909–1929," ibid. (Spring 1968), 42:55–56.

62. The Bethlehem contract was revised to include provisions for construction of a base in Fukien, an area of keen interest to Japan. Standard had refused to provide a loan in return for the concession and in compensation had had to increase China's share in the development company. Bethlehem's work and the Red Cross option on the Huai conservancy still depended on securing financing in the United States.

63. The State Department's guidelines, sent out to the legation in September 1913, indicated that the new administration was "extremely interested" in promoting American enterprise and commerce with China but that it would—in line with Olney's and Root's earlier policy—lend its active support only in response to specific wrongs done Americans, and these it would judge on a case-by-case basis. American businessmen would thus have to "rely primarily upon their own efforts." Standard seems not to have even asked Washington for guarantees for its China project, and Wilson and Bryan offered none. The State Department counselor, John Bassett Moore, also by chance affiliated with the Red Cross, twice pressed for support of the Huai River project and at last won assurances from Wilson of "good offices," coupled with a warning that his administration "would not go to the extent to which some Governments have gone in seeking to enforce the rights of their nationals in the matter of contracts." Wilson refused to honor the navy's earlier commitment to train Chinese sailors and supply up-to-date technology, an important part of the Bethlehem contract. In late 1913, the following May, and again in March 1915 Bryan warned against related plans for a Fukien base as an "unwise" provocation to Japan, and provided Tokyo reassurances on the point.

Washington's reserve alone, however, does not account for the troubles these projects encountered. The financiers selected by the Red Cross to work on the Huai River picked their way forward with great caution until the outbreak of war in Europe undercut all interest in China. The Standard concession came under attack from provincial leaders hostile to Peking's intrusion into local economic affairs and from nationalists opposed to a giveaway of China's economic rights. When Standard failed to find commercially exploitable oil, it let the project lapse. The Peking government's preferential treatment of American enterprise earned it a string of protests from Tokyo coupled with demands for compensation elsewhere. Japanese diplomats announced their opposition to the provision in Bethlehem Steel's contract for construction of a Fukien naval base and demanded a share in any conservancy work extended into adjoining Kiangsi province. Standard's Shensi concession also prompted protests from Japan, this time joined by Britain, and claims by both for similar rights elsewhere.

On the fate of these projects, see Cronon, *Cabinet Diaries*, pp. 40–42; Link, *Wilson Papers*, 28:401; Reinsch, *An American Diplomat*, p. 63; *FRUS 1913*, pp. 183–86; *FRUS 1914*, pp. 99–100, 104–5, 107, 110–19; *FRUS 1915*, pp. 116–17; Braisted, "China, the United States Navy, and the Bethlehem Steel Company," pp. 55–60; Pugach, "Standard Oil," pp. 458–69; DF 893.00/2101; DF 893.6363/3; Samuel C. Chu, *Reformer in Modern China: Chang Chien, 1856–1926* (New York, 1965), p. 154.

64. Cronon, *Cabinet Diaries*, pp. 22–23, 25, 42; Link, *Wilson Papers*, 27:285, 33:51.

65. Wang Yün-sheng, *Liu-shih nien-lai Chung-kuo yü Jih-pen* [China and Japan in the last sixty years] (Tientsin, 1932–1934), 6:46–52; Huang Chia-mo, "Chung-kuo tui Ou-chan ti ch'u-pu fan-ying" [China's initial reaction to the European war], *Chung-*

yang yen-chiu yüan, chin-tai shih yen-chiu so chi-k'an (1969), no. 1; FRUS 1914, Supplement: The World War, pp. 162–63, 183, 187–88; DF 763.72/346; DF 763.72111/490; Reinsch, An American Diplomat, pp. 124–25.

66. Japanese diplomats at the same time upbraided Peking for dragging the Americans into the crisis and for its rumored request for American naval reinforcements. FRUS 1914, Supplement: The World War, pp. 162–63, 172, 190; Wang, Liu-shih nien-lai, 6:46–52; Link, Wilson Papers, 29:362–63, 387, 31:289.

67. As if to confirm this contention, five days after Ts'ao had written, Bryan entered the American claim under the most-favored-nation principle to any generally applicable privileges extracted by Japan from China. Li Yü-shu, Chung-Jih erh-shih-i t'iao chiao-she [The Sino-Japanese negotiations over the twenty-one demands] (Taipei, 1966–), 1:277; Wang, Liu-shih nien-lai, 6:314; FRUS 1915, p. 147.

68. Ting Wen-chiang, comp., Liang Jen-kung hsien-sheng nien-p'u ch'ang-pien ch'u-kao [First draft of a chronological biography of Liang Ch'i-ch'ao] (Taipei, 1958), p. 512; WWP: "K'ang-i an" [File on the protest (against German submarine warfare)], ts'e 1 (February 9, 14, and 19 [three items], 1917) and ts'e 2 (March 8, 1917), "Chüeh-chiao an" [File on breaking relations (with Germany)], ts'e 2 (April 12 and 18, 1917), "Shih-ching–Lan-hsin hsieh-ting hui-i an" [File on the meetings leading to the Lansing-Ishii accord], ts'e 1 (esp. items sent October 6, 13, and 18, and November 5, 6, 8, and 12, 1917).

69. FRUS: The Lansing Papers, 1914–1920, 2:405–6, 413, 415, 422–23; FRUS 1915, p. 108.

70. Reinsch, An American Diplomat, p. 137; FRUS: The Lansing Papers, 1914–1920, 2:407, 409, 416, 426; FRUS 1915, p. 146; Link, Wilson Papers, 32:139, 426, 520, 531, 33:140.

71. FRUS: The Lansing Papers, 1914–1920, 2:407–8; FRUS 1914, Supplement: The World War, p. 190; FRUS 1917, Supplement 1: The World War, pp. 408, 411–12, 419–20; Lansing to Wilson, March 10, 1917, Frank L. Polk Papers (Sterling Library, Yale University), Link, Wilson Papers, 32:120, 319–23.

72. House to Wilson, September 18, 1917, House Papers (Sterling Library, Yale University); Burton F. Beers, "Robert Lansing's Proposed Bargain with Japan," Pacific Historical Review (November 1957), vol. 26; Ray Stannard Baker, Woodrow Wilson: Life and Letters (Garden City, N.Y., 1927–1939), 6:467n, 7:144, 262–63.

73. Reinsch-Wilson memo on Chinese finances, August 14, 1918, House Papers; Warren I. Cohen, The Chinese Connection: Roger S. Greene, Thomas W. Lamont, George E. Sokolsky and American-East Asian Relations (New York, 1978), pp. 49–70.

74. The following account of the last stage of Wilson's China policy is drawn from FRUS: The Paris Peace Conference, 1919, 2:492–94, 508, 524, 528; Madeleine Chi, "China and the Unequal Treaties at the Paris Peace Conference of 1919," Asian Profile (August 1973), 1:60; Patrick J. Scanlan, "No Longer a Treaty Port: Paul S. Reinsch and China, 1913–1919" (Ph.D. diss., University of Wisconsin, 1973), p. 333; Ray Stannard Baker and William E. Dodd, eds., The Public Papers of Woodrow Wilson: War and Peace (New York, 1925–1927), 2:223, 318–19, 342, 362, 407.

Chapter 7: Exclusion Stands the Test, 1898–1914

It took a sharp deterioration in relations between the United States and China, climaxing in the Korean War, to interest historians on either side of the Pacific in the boycott of 1905 and turn-of-the-century American exclusion policy. Howard K. Beale began to search in Theodore Roosevelt's handling of the boycott for the reasons for the failure of the United States "to become the friend and guide of the 'new spirit' in China" (p. 252). His answers supplied in 1956 in Theodore Roosevelt and the Rise of America to World Power (Baltimore), pp. 211–52, covered the American side of the story. Two Harvard seminar papers subsequently did the same for the Chinese side: Margaret Field, "The Chinese Boycott of 1905," Papers on China [1957], vol. 11; and Edward J. M. Rhoads, "Nationalism and Xenophobia in Kwangtung [1905–1906]," ibid. [1962], vol. 16. (Rhoads has since set the boycott in the context of local politics and society in China's Republican Revolution: The Case of Kwangtung, 1895–1913 [Cambridge, Mass., 1975].)

The interest in the boycott was more intense in China where historians held it up as a timely reminder of popular patriotic resistance to American exploitation and abuse. Studies appearing in the journal Chin-tai shih tzu-liao in 1956 and 1958 located the origins of exclusion in a plot by American capitalists, first to exploit Chinese labor and then to divert the nascent American labor movement with the illusory "coolie" issue. These works blamed the collapse of the boycott coalition, in which peasants and laborers ("the masses") allegedly played the most resolute role, on its betrayal by reactionary officials and comprador capitalists dependent politically and economically on the United States. See the anonymous "I-chiu-ling-wu nien fan-Mei ai-kuo yün-tung" [The 1905 anti-American patriotic movement], Chin-tai shih tzu-liao (1956), no. 1; Ting Yu, "1905 nien Kuang-tung fan-Mei yün-tung" [The 1905 anti-American movement in Kwangtung], ibid. (1958), no. 5; and Chin Tsu-hsün, "1905 nien Kuang-tung fan-Mei yün-tung ti p'ien-tuan hui-i" [Fragmentary recollections of the 1905 anti-American movement in Kwangtung], ibid. Chinese scholars interested in the boycott also assembled documentary collections. Chu Shih-chia, a Columbia-trained historian who left his post at the Library of Congress to return to China after the Communist revolution, devoted a section of his Mei-kuo p'o-hai Hua-kung shih-liao [Historical materials on American oppression of Chinese laborers] (Peking, 1958) to boycott materials. Ch'ien Hsing-ts'un, a literary historian writing under the pseudonym A Ying, produced Fan Mei Hua-kung chin-yüeh wen-hsüeh chi [Collected literature on opposition to the American treaty excluding Chinese laborers] (Peking, 1960). Its 698 pages of source materials on the boycott multiplied many-fold what Chu had made available in his slim volume only two years earlier. Ch'ien's interest in fiction connected to the boycott goes back to the 1930s and his work on Wan-Ch'ing hsiao-shuo shih [A history of novels in the late Ch'ing] (Shanghai, 1937), ch. 5.

These works of the late 1950s and early 1960s greatly expanded upon and to some extent superseded the little serious work that had previously been done on the boycott—Mary R. Coolidge, Chinese Immigration (New York, 1909), chs. 14, 16–17; and Charles F. Remer, A Study of Chinese

Boycotts (Baltimore, 1933), ch. 4. But they themselves have in turn been eclipsed by more recent and still fuller scholarship. In 1966 Chang Ts'un-wu, an historian at the Institute of Modern History on Taiwan, gave the boycott its first full dress treatment in his Kuang-hsü sa-i nien Chung-Mei kung-yüeh feng-ch'ao [The storm in 1905 over the Sino-American labor treaty] (Taipei). This work commands the full range of Chinese sources, including the archives of the Chinese foreign office and the documentary materials published in the People's Republic. Some of its findings are repeated and made accessible in English in Tsai Shih-shan, "Reaction to Exclusion: The Boycott of 1905 and Chinese National Awakening," Historian (November 1976), vol. 39. The latest notable addition to the literature is Delber McKee's careful and thoughtful Chinese Exclusion versus the Open Door, 1900–1906: Clashes over China Policy in the Era of Theodore Roosevelt (Detroit, Mich., 1977). In breadth of approach and depth of research, it supplants Beale's treatment in Theodore Roosevelt, and its American focus nicely complements Chang's elaboration of the Chinese perspective. Chang and McKee together with the materials in the Ch'ien Hsing-ts'un and Chu Shih-chia collections and the Theodore Roosevelt Papers are indispensable to understanding the problem of exclusion between 1898 and 1906.

My treatment of exclusion policy after 1906 and the changing political temper of Chinese in the United States draw heavily on neglected State Department files (housed in the Federal Records Center, Suitland, Maryland) and the large literature touching on the role of overseas Chinese in the 1911 revolution. Particularly helpful were Feng Tzu-yu, Ko-ming i-shih [Reminiscences of the revolution] (Shanghai, 1947); Feng, Hua-ch'iao ko-ming k'ai-kuo shih [The overseas Chinese in the revolution and founding of the Republic] (Shanghai, 1947); Harold Z. Schiffrin, Sun Yat-sen and the Origins of the Chinese Revolution (Berkeley, Calif., 1970); and C. Martin Wilbur, Sun Yat-sen: Frustrated Patriot (New York, 1976).

1. For examples of the attacks still suffered by Chinese, see FRUS 1901, pp. 101–28; NF 4767; and DF 311.931 T61.
2. The Bureau was part of the Treasury Department until 1903, when Congress transferred it to the nominal jurisdiction of the newly created Department of Commerce and Labor (after 1913 simply the Department of Commerce). Darrell H. Smith and H. Guy Herring, The Bureau of Immigration (Baltimore, 1924), pp. 19–24.
3. In 1901 Congress allowed $161,000 to fight illegal entries. By 1903 the Bureau was spending $269,000 and thereafter annual expenditures ranged between $500,000 and $600,000 at a time when the number of Chinese arrivals did not usually exceed 2,000 and never 3,000. In San Francisco, the focal point of Bureau activity, the staff grew by 1905 to forty inspectors, who screened arriving Chinese and pursued the many illegal immigrants thought to be hiding in Chinatown. Harley F. MacNair, The Chinese Abroad, Their Position and Protection (Shanghai, 1924), pp. 185–89, 251–53; FRUS 1905, pp. 166–68; Tomas S. Fonacier, "The Chinese Exclusion Policy in the Philippines," Philippine Social Science and Humanities Review

(March 1949), vol. 14; Mary R. Coolidge, *Chinese Immigration* (New York, 1909), ch. 16, are the basis for this and the paragraphs that immediately follow.

4. Angel Island had become the federal inspection center in 1892, replacing a makeshift operation in a large, two-story shed located on the Pacific Mail Company pier. For the Chinese in the United States Angel was to become, as the shed had been, a source of unhappy memories and community lore. Almost certainly among the characters in the tales immigrants told was James R. Dunn, the chief inspector appointed to San Francisco in 1899, a union man fully committed to the Powderly policy and quite literally deaf to the appeals of wronged Chinese. One can imagine Chinese recalling their bewilderment as they played cat and mouse with a stone deaf and ill-tempered Dunn in an interview conducted through an interpreter who, as was sometimes the case, did not speak the same dialect as the man under interrogation. Coolidge, *Chinese Immigration*, pp. 319–22.

5. "Mutual Helpfulness Between China and the United States," *North American Review* (July 1900), vol. 171; *FRUS 1901*, pp. 59–66, 72–97; *FRUS 1902*, pp. 209–21, 236–40, 263–66; CWS:KH, 134:20, 152:19; George E. Paulsen, "The Abrogation of the Gresham-Yang Treaty," *Pacific Historical Review* (November 1971), 40:461–67, 469, 472.

6. Denby, alone among the ranking diplomats in the late nineteenth century, came close to occupying an exclusionist position, and even he had worried over the adverse effects abuse of the Chinese immigrant might have on American interests in China. *FRUS 1868*, part 1, p. 518; *FRUS 1878*, pp. 130–31; petition by Americans in Canton, May 1879 in Grant Papers (LC); Elmer Sandmeyer, *The Anti-Chinese Movement in California* (Urbana, Ill., 1939), p. 93; George Seward, *Chinese Immigration* (New York, 1881); William H. Morken, "America Looks West: The Search for a China Policy, 1876–1885" (Ph.D. diss., Claremont Graduate School, 1974), pp. 79–80; George Seward, "Mongolian Immigration," *North American Review* (June 1882), vol. 134; James G. Angell, "The Diplomatic Relations between the United States and China," *Journal of Social Science* (1883), 17:34–36; MCD (February 9, 1886); John W. Casey, "The Mission of Charles Denby and International Rivalries in the Far East, 1885–1898" (Ph.D. diss., University of Southern California, 1959), pp. 83–88, 91–96, 103–4, 121–24; Denby to Bayard, September 6, 1888, Bayard Papers (LC); *FRUS 1889*, p. 111; John Russell Young, "The Chinese Question Again," *North American Review* (May 1892), vol. 154; Gilbert Reid, "China's View of Chinese Exclusion," *Forum* (June 1893), vol. 15; George E. Paulsen, "The Gresham-Yang Treaty," *Pacific Historical Review* (August 1968), 37:283n, 286; S. L. Baldwin, *Must the Chinese Go?* (pamphlet, 3d ed.; New York, 1890); Charles Denby, *China and Her People* (Boston, 1906), pp. 108–12.

7. James J. Lorence, "Business and Reform: The American Asiatic Association and the Exclusion Laws," *Pacific Historical Review* (November 1970), vol. 39; MCD (The Educational Association of China to Roosevelt, June 12, 1905, enclosed in Rockhill to Hay, June 21, 1905); Taft to Roosevelt, June 25, 1905, and Americans in Canton to Theodore Roosevelt, July 17, 1905, both in Roosevelt Papers (LC); Arthur H. Smith, "A Fool's Paradise," *Outlook* (March 1906), vol. 62; Holcombe, "The Restriction of Chinese Immigration," *Outlook* (April 23, 1904), 76:975; John W. Foster (a former Secretary of State but also an adviser to the Chinese legation), "The Chinese Boycott," *Atlantic Monthly* (January 1906), vol. 97; Seward, *The United States and China* (pamphlet copy of an address delivered in New York, October 28, 1902), p. 8, in Seward Papers (New-York Historical Society); Delber McKee, *Chinese Exclusion versus the Open Door, 1900–1906: Clashes over China Policy in the Era of Theodore Roosevelt* (Detroit, Mich., 1977), pp. 72, 85, 96.

8. McKee, *Chinese Exclusion*, ch. 4, contains the fullest account, but see also Lorence, "Business and Reform."

9. "Life Story of a Chinaman," in Hamilton Holt, ed., *The Life Stories of Undistinguished Americans* (New York, 1906); MacNair, *The Chinese Abroad*, p. xiv; *MPHS* pp. 105–6; Lu Feng-shih, comp., *Hsin-tsuan yüeh-chang ta-ch'üan* [A new compilation of treaty provisions] (Shanghai, 1909), 13:14–16; Ssu-tu Mei-t'ang [Seto Maytong], *Wo t'ung-hen Mei-ti* [I hate American imperialism] (Peking, 1951), p. 22; Nancy Farrar, *The Chinese in El Paso* (El Paso, Tex., 1972), p. 21; A Ying (Ch'ien Hsing-ts'un), comp., *Fan Mei Hua-kung chin-yüeh wen-hsüeh chi* [Collected literature on opposition to the American treaty excluding Chinese laborers] (Peking, 1960), pp. 509–21.

10. *MPHS*, pp. 105–8; Antonio S. Tan, *The Chinese in the Philippines, 1898–1935* (Quezon City, Philippines, 1972), Introduction and pp. 41–65, 98–99.

11. MCD (July 3, 1903 and March 10, 1904); Lu, *Hsin-tsuan yüeh-chang*, 13:47; and McKee, *Chinese Exclusion*, pp. 72–75, 93.

12. Born just outside Canton (Pan-yü district) in 1867, Liang was an experienced diplomat with an American education and a fluent command of English. Though less ardent and controversial than Wu, Liang maintained the struggle against exclusion during his tenure (1903–1907). Walter Muir Whitehall, *Portrait of a Chinese Diplomat: Sir Chentung Liang Cheng* (pamphlet; Boston, 1974); and Lo Hsiang-lin, *Liang Ch'eng ch'u-shih Mei-kuo* [Liang Ch'eng's ministry to the United States] (Hong Kong, 1977).

13. Lo Hsiang-lin, *Liang Ch'eng*, p. 293; Liang Ch'eng to John W. Foster, June 19, 1906, Foster Papers (LC); and Wu T'ing-fang to D.A. Tompkins, August 9, 1905, Tompkins Papers (Southern Historical Collection, Library of the University of North Carolina at Chapel Hill).

14. Quotes from Chinese post office and customs employees' statement, n.d., translation in Roosevelt Papers; and A Ying, *Fan Mei Hua-kung chin-yüeh*, pp. 4–5, 552, 604–8, 610, 648.

15. Charles F. Remer, *A Study of Chinese Boycotts* (Baltimore, 1933), ch. 2; Wang Ching-yü, comp., *Chung-kuo chin-tai kung-yeh shih tzu-liao, ti-erh chi, 1895–1914 nien* [Historical materials on modern Chinese industry, second series, 1895–1914] (Peking, 1957), 2:732–37; Chang Nan and Wang Jen-chih, eds., *Hsin-hai ko-ming ch'ien shih-nien shih-lun hsüan-chi* [Selected articles on current events written during the decade before the 1911 revolution] (Hong Kong, 1962–), 2:3–5.

16. P'eng Tse-i, comp., *Chung-kuo chin-tai shou-kung-yeh shih tzu-liao (1840–1949)* [Historical materials on modern Chinese handicraft industries (1840–1949)] (Peking, 1957), 2:497–500; NF 7385/–; report from Amoy consulate, July 25, 1905 in Roosevelt Papers; *MPHS*, pp. 146–51, 154–62.

17. This and the following paragraph draw on *FRUS 1905*, pp. 112–17; and Edward J. M. Rhoads, *China's Republican Revolution: The Case of Kwangtung, 1895–1913* (Cambridge, Mass., 1975), pp. 63–64, 88–89.

18. Song Ong Siang, *One Hundred Years' History of the Chinese in Singapore* (London, 1923), pp. 374–76; Tan, *The Chinese in the Philippines*, pp. 100–2, 104–5, 179–80; copies of consular reports (under the dates August 8, 10, 22, and 24, and September 21 and 23, 1905) and translation of *Chung Sai Daily*, July 19, 1905, all in Roosevelt Papers; McKee, *Chinese Exclusion*, p. 131; *MPHS*, pp. 162–63.

19. Wu T'ing-fang had returned from the United States in 1902 and entered the foreign office in 1904. T'ang Shao-i had arrived in 1905. From his post in Washington Liang Ch'eng wrote back to the foreign office only days after Shanghai had proposed the boycott endorsing the movement as a spontaneous, popular protest which

the government should allow to run its course. Chang Ts'un-wu, *Kuang-hsü sa-i nien Chung-Mei kung-yüeh feng-ch'ao* [The storm in 1905 over the Sino-American labor treaty] (Taipei, 1966), p. 65; *FRUS 1905*, p. 207; MCD (February 28, 1907); Lin Shao-nien, *Lin Wen-chih-kung tsou-kao* [Memorials of Lin Shao-nien] (Taipei reprint, 1968), pp. 474, 476; Chang and Yüan's message to Roosevelt transmitted in John Barrett to Roosevelt, June 17, 1905, Roosevelt Papers; MPHS, pp. 152–54.

20. *North China Daily News*, July 21, 1905 (in *FRUS 1905*, p. 211).

21. WWP: "Tsai-tse: Ch'u-yang k'ao-ch'a cheng-chih an" [File on Tsai-tse's political investigations abroad] (March 1, 1906); CR: Newchwang (September 12, 1905); Records of the British Foreign Office (Public Records Office, London), file 800/44 (December 14, 1905); Rhoads, *China's Republican Revolution*, p. 88; McKee, *Chinese Exclusion*, p. 156.

22. Yüan argued that China risked losing American diplomatic support during the talks ending the Russo-Japanese War and the ensuing diplomatic adjustments in Manchuria. In late June he took a firm stand within his own jurisdiction, so that even at the height of the agitation American goods traded freely in Tientsin. The balance of the account of the boycott is drawn from Chang, *Kuang-hsü sa-i nien*, pp. 67–68, 153–54; CWS:KH, 190:22–23; *FRUS 1905*, pp. 208, 218; Rhoads, *China's Republican Revolution*, pp. 86–93; Feng Tzu-yu, *Ko-ming i-shih* [Reminiscences of the revolution] (Ch'ang-sha, Chungking and Shanghai, 1939–1947), 1:70, 3:228, 230; NF 785/4a–14, 25–26, 57–63; NF 1879/1–5.

23. Senator Henry Cabot Lodge had previously in an expression of Social Darwinian nonsense deemed exclusion of Chinese laborers from the United States essential to "the survival of the fittest to survive" and supported exclusion in the Philippines as an American duty to the "natives" even if it meant that "their industrial and commercial development should be slow." But by June 1905 Lodge, who also placed considerable value on the China market, wanted exclusion applied with "decency" and "discreetly and reasonably" in order to preserve that market. The detached John Hay, who regarded the Chinese generally with amused contempt, had in March 1904 reacted to Chinese discontent in characteristically ironic and world-weary style. "I am very sorry—but Congress has done its work so well that even Confucius could not be made an American—though he should seek it with prayers and tears." A year later with the boycott in the offing, Hay indicated strong disapprobation of "the barbarous methods of the Immigration Bureau." John W. Foster, *American Diplomacy in the Orient* (Boston, 1903), p. 303; John A. Garraty, *Henry Cabot Lodge: A Biography* (New York, 1953), p. 208; Henry Cabot Lodge, ed., *Selections from the Correspondence of Theodore Roosevelt and Henry Cabot Lodge, 1884–1918* (New York, 1925), 2:15–16, 127, 157; Henry Adams, ed., *Letters of John Hay and Extracts from Diary* (New York reprint, 1969), 3:292; Hay diary entry, June 19, 1905, quoted in McKee, *Chinese Exclusion*, p. 128.

24. Hermann Hagedorn, ed., *The Works of Theodore Roosevelt* (New York, 1923–1926), 14:138–39, 245, 15:270–71, 286, 338–39; James B. Richardson, comp., *A Compilation of the Messages and Papers of the Presidents* (Washington, D.C., 1896–1900; supplement, 1903), 10:426; Elting E. Morison et al., eds., *The Letters of Theodore Roosevelt* (Cambridge, Mass., 1951–1954), 3:249; McKee, *Chinese Exclusion*, p. 60.

25. Morison, *Letters of Roosevelt*, 3:709–10, 4:803.

26. Secretary of War Taft in a speech in Ohio in mid-June reaffirmed the administration's support for American China trade. Taft rhetorically asked his audience if they were willing to give up "one of the great commercial prizes of the world" in order to keep out no more than a couple of hundred coolies by means of

unjust and irritating exclusion measures. Morison, *Letters of Roosevelt*, 5:90–91, 138; W. A. P. Martin, *The Awakening of China* (New York, 1907), p. 251; Howard K. Beale, *Theodore Roosevelt and the Rise of America to World Power* (Baltimore, 1956), p. 220; Lorence, "Business and Reform," p. 427; McKee, *Chinese Exclusion*, p. 127.

27. Morison, *Letters of Roosevelt*, 4:1184, 1235–36, 1240, 1251–52, 1377; Roosevelt to Metcalf, May 16, June 12, 19, and 24, 1905, Roosevelt to L. M. Shaw, August 2, 1905, Roosevelt to Sargent, August 19, 1905, and Roosevelt to President Benjamin Wheeler (University of California, Berkeley), September 8, 1905, all in Roosevelt Papers.

28. Through the American Federation of Labor journal, Gompers charged that the advocates of a modified exclusion policy, a coalition of "sordid profit mongers" and men of the church, really aimed for the ultimate nullification of exclusion. "We make no pretense that the exclusion of Chinese can be defended upon a high ideal. . . . Self-preservation has always been regarded as the first law of nature." Alexander Saxton, *The Indispensable Enemy: Labor and the Anti-Chinese Movement in California* (Berkeley, Calif., 1971), p. 276; Samuel Gompers in *American Federationist* (November 1905), 12:833–34, in ibid. (December 1905), 12:946–48, and in ibid. (February 1906), 13:98–100; McKee, *Chinese Exclusion*, p. 143; Morison, *Letters of Roosevelt*, 4:1327, 1377, 5:90–91, 190n; *FRUS 1905*, pp. xlix–l.

29. Roosevelt to President Edmund J. James (University of Michigan), March 3, 1906, Roosevelt Papers.

30. Rockhill to Roosevelt, August 15, 1905, and to the Secretary of State, August 26, 1905, and Roosevelt to Taft, September 2, 1905, and to Root, November 16, 1905, all in Roosevelt Papers; DIC (August 16 and 25, 1905); *FRUS 1905*, pp. 204–25; Ko Kung-chen, *Chung-kuo pao-hsüeh shih* [A history of Chinese journalism] (Shanghai, 1927), p. 231; Morison, *Letters of Roosevelt*, 4:1310.

31. Merchants were divided over whether "strong steps" or a conciliatory path (such as having the touring Taft assure Canton of liberal treatment for immigrants) was better. Foord, reflecting the latter point of view, cautioned against force. Better that the missionaries "abandon their posts and come home" than precipitate a Sino-American collision even more damaging to commerce than the boycott. Some merchants in China, having given up on Washington, entered direct talks with boycott leaders in a search for some mutually satisfactory solution. Edward J. M. Rhoads, "Nationalism and Xenophobia in Kwangtung [1905–1906]," *Papers on China* (1962), 16:175–78; Senator George Perkins to Roosevelt, August 30, 1905; Portland Chamber of Commerce to Roosevelt, August 31, 1905; Loomis to Loeb, September 1, 1905; J. Parker to Loomis, September 2, 1905; Loomis, memo of telephone conversation, September 2, 1905; Friedlander (secretary, San Francisco Merchants Exchange) to Roosevelt, September 5 and November 6, 1905, all in Roosevelt Papers; Rhoads, *China's Republican Revolution*, p. 89; McKee, *Chinese Exclusion*, pp. 137–39.

32. Morison, *Letters of Roosevelt*, 5:77, 132–33, 205; Roosevelt to Edmund J. James, March 3, 1906, and Roosevelt, draft message to Congress, quoted in Robert W. DeForest to Loeb, November 2, 1905, both letters in Roosevelt Papers; Leonard Wood to Corbin, February 20, 1906, Corbin Papers (LC); Richard D. Challener, *Admirals, Generals, and American Foreign Policy, 1898–1914* (Princeton, N.J., 1973), pp. 216–18; William R. Braisted, *The United States Navy in the Pacific, 1897–1909* (Austin, Tex., 1958), pp. 187–88; Beale, *Theodore Roosevelt*, pp. 239–44.

33. Morison, *Letters of Roosevelt*, 5:165; James B. Reynolds to Roosevelt, May 16, 1905, and to Butler, May 7, 1905, Roosevelt Papers; Coolidge, *Chinese Immigra-*

tion, p. 310; Lorence, "Business and Reform," pp. 431–36; McKee, *Chinese Exclusion,* pp. 172–81.

34. The most astonishing occurred in early 1906 on the arrival of a commission of Chinese officials of the first rank paying a formal visit to the United States. Inspectors first attempted a personal examination of the dignitaries themselves, and finally after much protest settled for taking detailed physical measurements of their staff to guarantee that those who arrived would be the same as those who would eventually leave. Liang Ch'eng to John W. Foster, June 19, 1906, Foster Papers; NF 4739/–; NF 17950 (including Charles Denby, Jr., to Root, January 30, 1906); MCD (February 17, 1906).

35. This and the following paragraphs on the continuing abuses of exclusion and Chinese complaints are based on the documents in NF 785/4a, 8–11; NF 17976/–; NF 18770; NF4467/1–2; NF 13436; NF 17976/1; NF 15479; NF 13596/–8, 11; NF 11250/–3, 5–6; NF 21803/3, 8–9; DF 151.08/1–92 passim; DF 311.93 H29. See also E. T. Williams, memo, n.d. [circa December 1910], Knox Papers (LC); Robert Dollar, *Memoirs of Robert Dollar* (San Francisco, 1917–1925), 1:86, 182–85, 188, 190; MacNair, *The Chinese Abroad,* pp. 190–201; Tan, *The Chinese in the Philippines,* pp. 180–82.

36. NF 17976/–.

37. Edward J. M. Rhoads, "Late Ch'ing Response to Imperialism: The Case of Canton," *Ch'ing-shih wen-t'i* (October 1969), vol. 2.

38. NF 15479/–1.

39. DF 151.08/92.

40. For examples of the Bureau's self-defense, see *FRUS 1902,* pp. 218–20; Metcalf to Roosevelt, June 7 and 24, 1905, and talk by Sargent, June 27, 1905, Roosevelt Papers; NF 11250/5–6; DF 151.08/31, 86. On evasions by the Chinese, see NF 14596, NF 21008, NF 1102, NF 14046, DF 151.06.

41. *FRUS 1908,* p. xlix; Morison, *Letters of Roosevelt,* 6:1241n; Arthur S. Link et al., eds., *The Papers of Woodrow Wilson* (Princeton, N.J., 1966–), 24:241, 243, 382–83; DF 151.08/31.

42. Stewart Culin, "Customs of the Chinese in America," *Journal of American Folklore* (July–September 1890), 3:195; Betty Lee Sung, *The Story of the Chinese in America* (New York, 1971), p. 269; Thomas W. Chinn et al., *A History of the Chinese in California* (San Francisco, 1969), p. 79. For an outsider's impressions of the Chinese community in 1903, see Liang Ch'i-ch'ao, *Hsin-ta-lu yu-chi* [Record of travels in the new world] (Taipei reprint, 1967), parts 37–39.

43. Ivan Light, "From Vice District to Tourist Attraction: The Moral Career of American Chinatowns, 1880–1940," *Pacific Historical Review* (August 1974), vol. 43.

44. A Chinese hospital and Chamber of Commerce were also founded in these years. Liu Pei-chi, *Mei-kuo Hua-ch'iao chiao-yü* [The education of overseas Chinese in the United States] (Taipei, 1957), ch. 2; Chinn, *A History of the Chinese in California,* pp. 65–66.

45. The following treatment of the appeal of reform nationalism is based on Wu Hsien-tzu, *Chung-kuo min-chu hsien-cheng tang tang-shih* [Official history of the Chinese Democratic Constitutionalist party] (San Francisco, 1952), pp. 25–28, 42–43, 60–62; Harold Z. Schiffrin, *Sun Yat-sen and the Origins of the Chinese Revolution* (Berkeley, Calif., 1970), pp. 185–86; Liang in *Hsin-min ts'ung-pao* (1905), no. 20, pp. 75–81; Robert L. Worden, "K'ang Yu-wei, Sun Yat-sen, et al., and the Bureau of Immigration," *Ch'ing-shih wen-t'i* (June 1971), 2:4; Robert L. Worden, "A Chinese

Reformer in Exile: The American Phase of the Travels of K'ang Yu-wei, 1899–1909" (Ph.D. diss., Georgetown University, 1972), pp. 75–77, 81–93, 125–27, 150, 153, 155, 158, 160–62, 170, 172, 201–3, 208–13, 294.

46. Feng Tzu-yu, *Ko-meng i-shih*, 4:4; George T. Yu, *Party Politics in Republican China: The Kuomintang, 1912–1924* (Berkeley, Calif., 1966), pp. 13, 20–21; Sun Yat-sen, *Memoirs of a Chinese Revolutionary*, excerpted in Jean Chesneaux, ed., *Secret Societies in China in the Nineteenth and Twentieth Centuries*, trans. Gillian Nettle (Ann Arbor, Mich., 1971), pp. 145–46.

47. The secret societies retained their preference for the reformers and showed a singular lack of interest in Sun's proposal that secret society members enlist in his party and each pay over to him a two-dollar registration fee. Only the leaders of the San Francisco mother lodge gave any tangible support, replacing one of K'ang's followers with one of Sun's as editor of the society paper. Schiffrin, *Sun Yat-sen and the Origins*, pp. 314–15, 320, 323–24, 326–33; Feng, *Hua-ch'iao ko-ming*, pp. 56–59, 63; Feng, *Ko-ming i-shih*, 1:153–54, 4:20–23; Yu, *Party Politics*, pp. 27–29, 31; Thomas W. Ganschow, "A Study of Sun Yat-sen's Contacts with the United States prior to 1922" (Ph.D. diss., Indiana University, 1971), p. 50n.

48. Sun's fears that the boycott might prove a distraction were realized in Hong Kong, where his supporters joined in the boycott and then split among themselves over how much to demand of the United States. When Sun intervened to restore harmony, he predictably gave his support to the group pushing the more modest demands, presumably because they would be less distracted from the revolution by the peripheral issue of overseas Chinese rights. Feng Tzu-yu, *Chung-kuo ko-ming yün-tung erh-shih-liu nien tsu-chih shih* [A history of China's revolutionary movement through twenty-six years of organizing] (Shanghai, 1948), p. 111; Feng, *Ko-ming i-shih*, 1:70, 3:228, 230.

49. Tsou Jung, *The Revolutionary Army: A Chinese Nationalist Tract of 1903*, trans. John Lust (The Hague, 1968), pp. 73, 108; appeals in DF 151.08/1, 35; recollections in *Hsin-hai ko-ming hui-i-lu* [Recollections of the 1911 revolution] (Peking, 1961–1963), 1:74–84, 559–62, and in Victor G. Nee and Brett DeBary Nee, *Longtime Californ': A Documentary Study of an American Chinatown* (New York, 1974), pp. 73–74, 78–79, 107.

50. Feng, *Ko-ming i-shih*, 1:153–54, 157, 4:135–44, 196–97; Feng, *Hua-ch'iao ko-ming*, pp. 36–38, 64–71; Yu, *Party Politics*, pp. 57, 60; *Hsin-hai ko-ming hui-i-lu*, 1:74–82, 559–62.

51. Earl Swisher, *Chinese Representation in the United States* (Boulder, Colo., 1967), pp. 29–31; Feng, *Hua-ch'iao ko-ming*, p. 36; FRUS 1902, p. 36; Schiffrin, *Sun Yat-sen and the Origins*, p. 104; Lo Hsiang-lin, *Liang Ch'eng*, p. 301.

52. For example, the minister in the United States flashed ahead Sun's departure from New York for London in 1896, thus setting in motion the abduction that was to win Sun fame. K'ang had to sneak into the country posing as an official, and Liang traveled under assumed names to escape detection. Sun, after reorganizing his party in Tokyo in 1905, could not get an organizer past immigration officers into the United States, and when Sun himself came in 1904, he gained entry on the basis of an empty claim to Hawaiian birth, backed with false statements from his brother and friends who resided on the islands. Worden, "K'ang Yu-wei," pp. 7–9; Schiffrin, *Sun Yat-sen and the Origins*, pp. 327–30. The same Chinese diplomats who hounded these political renegades may have had private doubts. They were, like reformers and revolutionaries, attached to the notions of nationalism and political renovation. They too were frustrated over the government's weak defense of overseas Chinese.

And they came in many cases from the same districts as Sun, K'ang, and Liang and had mutual friends and relations.

53. The following two paragraphs draw on FRUS 1901, Appendix: Affairs in China, p. 333; Swisher, Chinese Representation, pp. 33–34; FRUS 1911, pp. 64–66; Chutung Tsai, "The Chinese Nationality Law, 1909," American Journal of International Law (April 1910), vol. 4; NF 21313; Liang Chia-pin, "Chi Ch'ing-chi she-li ch'iao-hsiao yü Chung-Mei wen-chiao chiao-liu" [The establishment of schools for overseas Chinese in the late Ch'ing and Sino-American educational interchange] in Hua-ch'iao wen-t'i lun-wen chi [A collection of essays on the overseas Chinese question] (1954), no. 1; Lo Hsiang-lin, Liang Ch'eng, pp. 188–89, 218–22; July 1905 proclamation by the San Francisco consul general in FRUS 1905, p. 210.

54. FRUS 1912, pp. 50–51; Feng, Ko-ming i-shih, 1:164, 2:316–19; Feng, Hua-ch'iao ko-ming, p. 63; Wu, Chung-kuo min-chu, p. 70; Ssu-t'u, Wo t'ung-hen Mei-ti, pp. 39, 43.

55. Feng, Ko-ming i-shih, 3:388–96, 398; Lin Sen, Ch'ien kuo-min cheng-fu chu-hsi Lin Kung-tzu ch'ao i-chi [Works of Lin Sen, former chairman of the Nationalist government] (Taipei, 1966), pp. 53–55, 260, 459.

56. Feng, Ko-ming i-shih, 3:399; Lin Sen, Ch'ien kuo-min cheng-fu, pp. 56, 58; Yu, Party Politics, pp. 132–35, 138–40; Chün-tu Hsüeh, Huang Hsing and the Chinese Revolution (Stanford, Calif., 1961), pp. 169–74, 177–81.

Chapter 8: American Reform and Chinese Nationalism, 1900–1914

The dialectic between American reformism and Chinese nationalism has largely escaped scholarly treatment. The best available introduction to the Chinese half of the dialectic is Mary C. Wright's masterful survey, "The Rising Tide of Change," in her China in Revolution: The First Phase, 1900–1913 (New Haven, Conn., 1968). Meribeth Cameron, The Reform Movement in China, 1898–1912 (Stanford, Calif., 1931), provides detailed but now somewhat dated coverage of reform nationalism. Akira Iriye, "Public Opinion and Foreign Policy: The Case of Late Ch'ing China," in Albert Feuerwerker et al., eds., Approaches to Modern Chinese History (Berkeley, Calif., 1967), suggests the breadth and potency of nationalist sentiment. None of these works nor any others I know explore in any systematic way the attitude of early twentieth-century nationalists toward the United States, though that topic is the obvious starting point for understanding the new strains already developing between the two countries in these years.

Jerry Israel's Progressivism and the Open Door: America and China, 1905–1921 (Pittsburgh, Pa., 1971), is the most ambitious attempt to date to come to grips with the American fascination with China's development. However, the value of the work is diminished by the author's insistence on engaging in an unrewarding exercise in labeling all American activity in China in terms of competition or cooperation with the other powers (an interpretive theme borrowed from Tyler Dennett) and his unpersuasive argument that these divergent tactics reflected the domestic debate over trust-busting and the relative merits of economic concentration and economic competition.

Chapter 8: Bibliographical Essay

The place to begin on the diplomatic reformers is the biographies of Rockhill by Paul Varg (cited in chapter 5) and by Peter W. Stanley (in Perspectives in American History [1977–1978], vol. 11) and of Paul S. Reinsch: Open Door Diplomat in Action (Millwood, N.Y., 1979) by Noel H. Pugach (though Reinsch's memoir, An American Diplomat in China [Garden City, N.Y., 1922], remains an important source). My account of "The American Remission of the Boxer Indemnity: A Reappraisal," Journal of Asian Studies (May 1972), vol. 31, develops in cameo the conflict between the program of reform laid down by American officials and the divergent goals of Chinese nationalists.

American economic interests in China and the Chinese response to them are treated in survey fashion in my "Americans in the China Market" (cited in chapter 5). It brings together the disparate pieces of the cotton goods export story. The British-American Tobacco Company and its Chinese rivals are treated in full in Sherman G. Cochran's groundbreaking study, Big Business in China: Sino-Foreign Rivalry in the Cigarette Industry, 1890–1930 (Cambridge, Mass., 1980). Standard Oil's operations can be followed through scattered references in the authorized history by Ralph W. Hidy and Muriel E. Hidy, Pioneering in Big Business, 1882–1911 (New York, 1955).

The Hankow-Canton railway concession has been studied from a variety of perspectives beginning with Percy H. Kent's solid contemporary account, Railway Enterprise in China (London, 1903), ch. 12. Some five decades later William R. Braisted's essay (cited in Chapter 5) added details on the American side drawn chiefly from the State Department archives and the Roosevelt Papers. Several accounts from the Chinese perspective have since followed. Sun Yü-t'ang et al., Mei-ti-kuo chu-i ching-chi ch'in-Hua shih lun-ts'ung [Collected articles on the history of American imperialist economic aggression against China] (Peking, 1953), contains four accounts of American involvement in the Hankow-Canton and Hukuang projects. They are unusual among works on Sino-American relations which appeared during the Korean War for their heavy reliance on printed primary sources in Chinese, though not for their conventional indictment of American imperialism and its Chinese collaborators. Lee En-han's valuable "Chung-Mei shou-hui Yüeh-Han lu-ch'üan chiao-she" [Sino-American negotiations over the recovery of rights over the Canton-Hankow railway], Chung-yang yen-chiu yüan, chin-tai shih yen-chiu so chi-k'an (1969), no. 1, and E-tu Zen Sun, Chinese Railways and British Interests, 1898–1911 (New York, 1954), chs. 3 and 4, have further developed the Chinese side of the story. Lee's work has recently been summarized in Daniel H. Bays, China Enters the Twentieth Century: Chang Chih-tung and the Issues of a New Age, 1895–1909 (Ann Arbor, Mich., 1978), ch. 8, and rendered in English by Lee himself in a slightly revised version in China's Quest for Railway Autonomy, 1904–1911: A Study of the Chinese Railway-Rights Recovery Movement (Singapore, 1977), ch. 2.

The most recent installment of the Hankow-Canton railroad story, G. Kurgan-van Hentenryk's Léopold II et les groupes financiers belges en Chine:

Chapter 8: Bibliographical Essay

La politique royale et ses prolongements (1895–1914) (Brussels, 1972), has added some fascinating new insights from Belgian sources on the backstage financial maneuvers involving American investors. The international investment activity of American financiers such as Morgan and Harriman has aroused considerable historical attention and controversy, but not until the opening of more sources comparable to those made available for Léopold II will the generalizations move beyond mere surmise supported by highly circumstantial evidence.

Crucial to understanding mission work and cultural nationalism is Jessie G. Lutz's *China and the Christian Colleges, 1850–1950* (Ithaca, N.Y., 1971), admirable for its breadth of coverage. Two Harvard-sponsored collections of essays—*American Missionaries in China* (Cambridge, Mass., 1966), edited by Kwang-Ching Liu, and *The American Missionary Enterprise in China and America* (Cambridge, Mass., 1974), edited by John K. Fairbank—contain some helpful insights on the multiple facets of mission work in China in the late nineteenth and early twentieth centuries. Four monographs, also of Harvard origin, provide valuable in-depth treatment. Irwin T. Hyatt, Jr., *Our Ordered Lives Confess: Three Nineteenth-Century American Missionaries in East Shantung* (Cambridge, Mass., 1976), is a finely drawn portrait of these missionaries and their very different responses to the frustrations of evangelization. It should be supplemented by a reading of Sidney A. Forsythe's *An American Missionary Community in China, 1895–1905* (Cambridge, Mass., 1971), which catches in microcosm a mission world isolated from the Chinese and at times deeply contemptuous of them. Shirley Garrett, *Social Reformers in Urban China: The Chinese Y.M.C.A., 1895–1926* (Cambridge, Mass., 1970), is an institutional study which is quite revealing about the changing missionary approach to China at the turn of the century. Valentin H. Rabe, *The Home Base of American China Mission, 1880–1920* (Cambridge, Mass., 1978), lays the groundwork for a better understanding of the expansion and institutionalization of foreign mission work at home, but is disappointingly weak in interpretation.

A new generation of scholarship represented by these works has largely superseded the pioneering survey by Paul A. Varg, *Missionaries, Chinese and Diplomats: The American Protestant Mission Movement in China, 1890–1952* (Princeton, N.J., 1958), and the still older chronicle by Kenneth Scott Latourette, *A History of Christian Missions in China* (New York, 1929). Missing from the new literature as well as the old is a sense of the Chinese church—its origins and outlook, its institutional framework, and its ties to local society. Nor has the role of women in the mission movement received its due, an omission that Jane Hunter's "Imperial Evangelism: American Women Missionaries in Turn-of-the-Century China" (Ph.D. diss., Yale University, 1981) should help correct.

1. For some helpful discussion of the various strains of early Chinese nationalism, see Mary B. Rankin, *Early Chinese Revolutionaries: Radical Intellectuals in*

Shanghai and Chekiang, 1902–1911 (Cambridge, Mass., 1971), pp. 8–12, 25–33; Chang Nan and Wang Jen-chih, eds., *Hsin-hai ko-ming ch'ien shih-nien shih-lun hsüan-chi* [Selected articles on current events written during the decade before the 1911 revolution] (Hongkong, 1962–), vol. 2, pp. 10–13 of the preface; Harold Z. Schiffrin, *Sun Yat-sen and the Origins of the Chinese Revolution* (Berkeley, Calif., 1968), ch. 10; and Ernest P. Young, *The Presidency of Yuan Shih-k'ai: Liberalism and Dictatorship in Early Republican China* (Ann Arbor, Mich., 1977), introduction and ch. 1.

2. "Lun Mei-kuo tsai Chung-kuo chih chü-tung" [On the behavior of the United States in China], *Ching-chung jih-pao* [The tocsin], September 25, 1904. On the persistence of interest in George Washington as a figure of exemplary virtue and the American revolution as a major event in world history, see Michael Gasster, *Chinese Intellectuals and the Revolution of 1911: The Birth of Modern Chinese Radicalism* (Seattle, Wash., 1969), pp. 39–40, 104, 196; Schiffrin, *Sun Yat-sen and the Origins,* pp. 15, 162; Liang Ch'i-ch'ao in *Ch'ing-i pao* [Journal of pure discussion] (August 5, 1900), no. 53, p. 2; Yüan Shih-k'ai in Pai Chiao, *Yüan Shih-k'ai yü Chung-Hua min-kuo* [Yüan Shih-k'ai and the Chinese republic] (Taipei reprint, 1962), p. 81; and Mao Tse-tung's recollections in Edgar Snow, *Red Star Over China* (New York, 1961), p. 134.

3. Sun's view persisted well into the teens. In 1912 he held up the United States, which had built 200,000 miles of track with foreign funds and become strong, as an example for China to copy. Five years later Sun was still carrying the conventional view of the United States as a friendly country whose technical expertise and wealth could be studied and exploited to China's benefit. But he now also harbored reservations about her usefulness as a diplomatic ally. A country that discriminated against Chinese immigrants was also unlikely to defend a non-white country against other major powers. To underline his point, Sun cited the American failure to support Li Hung-chang's Korean policy. Thomas W. Ganschow, "A Study of Sun Yat-sen's Contacts with the United States prior to 1922" (Ph.D. diss., Indiana University, 1971), pp. 8–11, 33–41, 47–48, 50n, 72–74, 169–78; Schiffrin, *Sun Yat-sen and the Origins,* pp. 334–37, 360; C. Martin Wilbur, *Sun Yat-sen: Frustrated Patriot* (New York, 1976), p. 24.

4. Revolutionaries took some comfort in the fact that even after his split with the Americans, Aguinaldo and his small and poorly armed band of followers still managed to resist crack American troops for several years, suggesting the difficulty intervening powers would have in subduing a fully aroused Chinese people. "Lun Mei-kuo tsai Chung-kuo" (cited above); Chang and Wang, *Hsin-hai ko-ming ch'ien,* 1:81–82, 2:458–59, 465–66, 468; "Lun T'ai-p'ing-yang lieh-ch'iang chih shih-li" [On the influence of the powers in the Pacific], *Ching-chung, jih-pao* (August 19, 1904); Tsou Jung, *The Revolutionary Army: A Chinese Nationalist Tract of 1903,* trans. by John Lust (The Hague, 1968), passim.

5. See chapter 6 for a full development of these points.

6. Chang and Wang, *Hsin-hai ko-ming ch'ien,* 1:55, 2:67; articles from *Shang-wu jih-pao, Chung-wai jih-pao, Hsin-wen pao,* and *Su pao,* all reproduced in *Hsin-min ts'ung-pao* [New citizen] (February 22, 1902), 1(2):83–88, and (March 24, 1902), 1(4):81–90; "Lun Jih-O i-ho hou chih Chung-kuo wai-chiao" [On Chinese diplomacy after the Russo-Japanese peace negotiations], *Shih pao* [Times] (November 4, 1905); Hsü Nai-lin, *Ch'ou-pien ch'u-yen* [Plain talk on managing the frontier] (Taipei reprint, 1969; written sometime after 1903), 4:29; "Mei-kuo nüeh-tai Fei-lieh-pin t'u-jen chi-shih" [A record of American abuse of Filipinos], *Tung-fang tsa-chih* [Far

eastern magazine] (July 12, 1909), 6(6):33–38; Hunan anti-American publication on the Philippines (April 1914) in DF 893.00/2124, 2181; Schiffrin, *Sun Yat-sen and the Origins*, pp. 287, 289.

7. Liang's conception of the passing of a golden age of American virtue and promise was anticipated by Huang Tsun-hsien and by the traditionalist Ku Hung-ming (an interpreter for Chang Chih-tung between 1885 and 1905 and later a member of the foreign ofice). Ku Hung-ming, *Papers from a Viceroy's Yamen* (Shanghai, 1901), pp. 146–74; R. David Arkush, "Ku Hung-ming (1857–1928)," *Papers on China*, (1965), vol. 19.

8. Philip C. Huang, *Liang Ch'i-ch'ao and Modern Chinese Liberalism* (Seattle, Wash., 1972), p. 12; Chen Chi-yun, "Liang Ch'i-ch'ao's 'Missionary Education,'" *Papers on China* (1962), 16:104–5; Hao Chang, *Liang Ch'i-ch'ao and Intellectual Transition in China, 1890–1907* (Cambridge, Mass., 1971), p. 161; Liang Ch'i-ch'ao "Lun Mei-Fei, Ying-Tu chih chan-shih kuan-hsi yü Chung-kuo" [The relation to China of the Filipino-American and Boer Wars], in his *Yin-ping-shih ho-chi: wen-chi* [Collected works from the ice drinker's studio: essays] (Taipei reprint, 1960), vol. 4, part 11. In 1906 Liang returned to the argument that instability could convert the commercial powers to a policy of partition to preserve at least a portion of their market. Chang and Wang, *Hsin-hai ko-ming ch'ien*, 2:287–89.

9. Liang had been impressed to learn during his American tour that eight of some forty-odd major southern cotton mills had failed in 1900 when the Boxer crisis temporarily shut off their North China market. He concluded that the United States simply could not afford the economic disruption at home that the permanent loss of even a portion of the China market would entail. Liang was also struck during his visit by the inequalities of wealth and the racial injustices that existed within the context of unparalleled prosperity, and contact with work-a-day American politics shook some of his democratic idealism. A reading of Edward Bellamy's *Looking Backward* in 1896 may have prepared the grounds for these views. Chang, *Liang Ch'i-ch'ao*, pp. 238–41, 243–45; Liang Ch'i-ch'ao, "Erh-shih shih-chi chih chu-ling to-la-ssu" [The great spirit of the twentieth century—the trust], *Yin-ping-shih*, vol. 5, part 14, pp. 60–61; Liang Ch'i-ch'ao *Hsin-ta-lu yu-chi* [Record of travels in the new world] (Taipei reprint, 1967), pp. 307–11; Chang and Wang, *Hsin-hai ko-ming ch'ien*, 2:287; Martin Bernal, *Chinese Socialism to 1907* (Ithaca, N.Y., 1976), p. 25.

Liang's views corresponded with those that his mentor K'ang Yu-wei had developed in the 1890s. K'ang insisted that foreign commercial penetration was as potentially dangerous to China's independence as territorial aggression. He also respected the United States ("Of all the countries on earth, none is as prosperous. . . .") for its command of advanced science and technology. Liang's contribution was an analysis which related economic imperialism, which he and K'ang both feared, to the very features of material mastery and abundance, which they both admired. Hsiao Kung-chuan, *A Modern China and a New World: K'ang Yu-wei, Reformer and Utopian, 1858–1927* (Seattle, Wash., 1975), p. 209; Hsü Hsü-tien, "Wu-hsü pien-fa yün-tung yü wei-hsin p'ai ti tui-wai t'ai-tu" [The 1898 reform movement and reformers' attitude toward foreign affairs] in Wu Yü-chang et al., *Wu-hsü pien-fa liu-shih chou-nien chi-nien lun-wen-chi* [Essays commemorating the sixtieth anniversary of the 1898 reform movement] (Peking, 1958), p. 72.

10. Liang Ch'i-ch'ao, "Chung-kuo huo-pi wen-t'i" [China's currency question] in his *Yin-ping-shih ho-chi: wen-chi*, vol. 6, part 16, pp. 121, 123; Liang Ch'i-ch'ao, "Tu 'Chin-hou chih Man-chou' shu hou" [After reading "Manchuria in the immediate future"], *Hsin-min ts'ung-pao* (1905), 3(20); Liang in various issues of *Kuo-*

feng pao: (February 20, 1910) 1(1) [unpaginated article entitled "Lun ko-kuo kan-she . . ."]; (March 11, 1910), 1(3):30–39, 42–43; (April 20, 1910), 1(7):32–33; (October 3, 1910), 1(24):9–21; (October 13, 1910), 1(25):5–32; (November 12, 1910), 1(28):51–58; (March 21, 1911), 2(5):27–30; (April 19, 1911), 2(8):53–61. For commentary by others, see *ibid.* (March 11, 1910), 1(3):71–72, 80, 84, 87, 92; (September 24, 1910), 1(23):21; (December 2, 1910), 1(30):17–29; and summaries from other reform periodicals in Archives of the French Ministry of Foreign Affairs (Quai d'Orsay, Paris), Chine, NS 178 (August 4, 1909).

11. An amateur, ill-schooled diplomatic service played some role in the problems American officials had in coming to grips with Chinese nationalism. The American legation and consulates still lagged embarrassingly behind the other powers both with respect to the size of their staffs and the number and quality of their linguists. The single effort to train a small corps of young men in the language and culture fell victim to the very forces that impeded improvement of the foreign service generally—congressional indifference, party politics, and a lack of funds. The Roosevelt and Taft years saw some successes in upgrading the consular service and also the rise of a Division of Far Eastern Affairs within the State Department, though whatever potential the Division had as a source of intelligence and analysis was nullified by the generally low quality of the information from the legation that it worked with, by the shortage of competent staff, and by its negligible influence over key policy decisions. To secure essential expertise, the foreign service resorted again to an infusion from the ranks of missionaries, an expedient made possible by the post-Boxer cooling of the mission question. The two most prominent missionary recruits, E. T. Williams and Charles D. Tenney, had come to China in the 1880s and later abandoned evangelism for diplomacy. Williams started in the Shanghai consulate in 1896, and later served alternately in the legation (1901–1908, 1911–1913) and in the Division of Far Eastern Affairs (1909–1910, 1914–1918). Tenney joined the legation in 1908 and was a mainstay there for over a decade. CR: Tientsin (October 5, 1903); CR: Canton (July 9 and November 19, 1904); Cheshire to Rockhill, July 5 and November 6, 1904, Rockhill Papers (Houghton Library, Harvard University); NF 9880/37; H. H. D. Peirce, *Report . . . Upon a Tour of Consular Inspection in Asia* (Washington, D.C., 1904), pp. 7–15, 24–25; Roosevelt to F. B. Loomis, January 16, 1905 and to John Hay, January 16, 1905, Roosevelt Papers (LC); Elting E. Morison et al., eds., *The Letters of Theodore Roosevelt* (Cambridge, Mass., 1951–1954), 6:1496; Michael H. Hunt, *Frontier Defense and the Open Door: Manchuria in Chinese-American Relations, 1895–1911* (New Haven, Conn., 1973), pp. 66–67, 235–36; Richard H. Werking, *The Master Architects: Building the United States Foreign Service, 1890–1913* (Lexington, Ky., 1977), pp. 130–34, 151, 163–64.

12. During his early years in China Rockhill developed a specialty—more Sinological than diplomatic—in China's inner Asian frontier. After several years out of the diplomatic service, Rockhill went to Washington to serve as the State Department's first resident China expert between 1893 and 1897 and Hay's chief guide to Chinese affairs at the time of the writing of the first open door note in 1899 and thereafter. Hunt, *Frontier Defense and the Open Door*, pp. 58–59; Rockhill to Hippisley, October 30, 1894, Rockhill Papers.

13. Rockhill to Hippisley, July 6, 1901, Rockhill Papers; MCD (September 21 and October 1, 1900); *FRUS 1901*, Appendix: *Affairs in China*, p. 114.

14. Rockhill also pressed for revisions in the imperial audience ceremony, long considered demeaning by him and other diplomats, and for the posthumous rehabilitation of officials executed for offending the pro-Boxer faction at court. On all

these points the final agreement gave full satisfaction. W. W. Rockhill, "Diplomatic Missions to the Court of China: The Kowtow Question," *American Historical Review* (1896/1897), 2:638–43; *FRUS 1901*, Appendix: *Affairs in China*, pp. 120–22, 349, 356; *ITS*, pp. 1260–62.

15. Rockhill also wanted improved opportunity for foreign mining investments, strict adherence to the principle of nondiscrimination in trade and investment, and improvement of the waterways serving Tientsin, Shanghai, and Newchwang. The powers could not agree among themselves on how to fund China's new indemnity, and that dispute in turn stood in the way of a settlement of all these commercial questions except for river conservancy work for Tientsin and Shanghai. However, by conceding a nominal customs increase, the United States as well as Japan and Britain secured in return a promise from the Chinese to negotiate separate commercial treaties. DIC (December 29, 1900, April 11 and 29, 1901, January 25, 1902); *FRUS 1901*, Appendix: *Affairs in China*, pp. 170–73, 227.

16. Accordingly, Rockhill demanded—and got—punishment of pro-Boxer officials; indemnity for mission losses; acts of contrition, including expiatory monuments in desecrated foreign cemeteries; and guarantees for the future, mainly in the form of imperial pronouncements holding officials responsible for protecting foreigners, dealing fairly with converts, and preventing the renewal of Boxer propaganda. For good measure Rockhill secured suspension of official exams in those areas where foreigners had been attacked.

17. Hunt, *Frontier Defense and the Open Door*, pp. 59, 95, 97. Rockhill's pessimism over China's future was subsequently compounded by the outcome of the 1903 treaty talks, where Chinese concessions on paper in regard to abolishing likin, revising mining regulations to encourage foreign investment, protecting trademarks, patents, and copyrights, and opening new Manchurian ports to American trade all proved worthless. *CWC*, 187:31–32; *FRUS 1906*, pp. 228–61; *FRUS 1907*, pp. 249–54; *FRUS 1910*, p. 200; *FRUS 1912*, p. 176. Currency reform, which Rockhill and Hay expected to stimulate trade, also proved a disappointment. At China's request, Congress had appointed an advisory commission. Its most active member, Professor Jeremiah W. Jenks of Cornell, submitted in early 1904 detailed recommendations for introducing the gold standard under the direction of a foreign expert (he hoped himself?) armed with broad powers. Before Jenks even got away from Peking, however, Chang Chih-tung had scuttled the plan by condemning it as an intolerable interference in a vital area of national life and unsuitable in any case to Chinese social and economic conditions. Yang Tuan-liu, *Ch'ing-tai huo-pi chin-yung shih-kao* [A draft history of money and finance under the Ch'ing dynasty] (Peking, 1962), pp. 325–30.

18. These reforms looked to the promotion of Chinese commerce and industry, the recovery of foreign economic concessions, and an end to extraterritoriality (thereby bringing foreigners under Chinese jurisdiction). The reform of provincial administrations began in 1905 with Manchuria as the model, and the classical literary exam fell in favor of modern education to be pursued abroad as well as in China. Sweeping political reforms soon followed, including plans for provincial assemblies and a national parliament.

19. Hunt, *Frontier Defense and the Open Door*, pp. 95–96; *FRUS 1905*, p. 182; *FRUS 1906*, pp. 275–76, 349; *FRUS 1907*, pp. 179–80; *FRUS 1908*, p. 190.

20. *FRUS 1904*, pp. 192–93; *FRUS 1905*, pp. 185, 205; *FRUS 1909*, pp. 70–92, *FRUS 1910*, pp. 353–60; Rockhill to Hippisley, March 29, 1906, Rockhill Papers; MCD (August 26, 1905); Rockhill to Roosevelt, August 15, 1905, and January 11,

1906, and to the Secretary of State, August 26, 1905, Roosevelt Papers; NF 788/193; NF 1518/–2, 27; NF 1787/10–12; NF 2321/–; NF 3543/1–2; NF 10251/29–35.

21. Rockhill to Roosevelt, October 30, 1905, Rockhill Papers; MCD (July 25, 1905), Rockhill to Roosevelt, July 7 and August 17, 1905, Roosevelt Papers.

22. Ch'ing officials in their increasing concern with reclaiming sovereign rights began to seek to narrow the geographical extent of old treaty-opened ports (where foreigners could operate free of Chinese taxation and regulation) and to insure that in the ports thereafter opened by China herself the Chinese government would retain the right to define the terms and extent of foreign residence. Yüan Shih-k'ai and Chang Chih-tung had applied this approach to the new Manchurian ports in 1903. Conger and Hay, anxious to spare their countrymen the evils of Chinese administration, would only agree that the ports be formally called self-opened while delaying the real issue of what regulations would prevail for negotiations to follow the end of the Russo-Japanese War. Hunt, *Frontier Defense and the Open Door*, pp. 72–76, 113–14.

23. Roosevelt's secretary to Rockhill, August 14, 1905, Roosevelt Papers; Morison, *Letters of Roosevelt*, 4:1326–27, 5:29–30.

24. Root therefore instructed the legation to continue to insist that open ports, whether in Manchuria or elsewhere, be broadly defined. Where other nationalities gained special advantages, he wanted Americans to share them too. DIC (February 26, 1906); FRUS 1906, p. 399; FRUS 1907, pp. 234–35; FRUS 1908, pp. 122–23, 138–39.

25. MCD (January 16, 1906 for Denby, Jr.'s comments on Rockhill's dispatch of February 26, 1906); NF 785/1; NF 8372/25; NF 13556/– (memo, circa May 14, 1908); J. B. Loomis to Roosevelt's secretary, August 26, 1905, Roosevelt Papers. Rockhill seems to have fallen afoul of Taft precisely because of this apparent lack of zeal. Once installed in the White House, Taft had Rockhill recalled on the grounds that he was a "dilettante" and temperamentally unsuited to furthering American trade and investment. Taft to Rollo Ogden, April 21, 1909, Taft Papers (LC).

26. NF 5971/10–14, 18; MCD (January 20, 1905; Denby, Jr., memo of April 2, 1906); NF 3543/–; Morison, *Letters of Roosevelt*, 5:809. How could the younger generation of "modern" treaty port Chinese—with their demonstrated entrepreneurial skills, their ties to the United States, their Christian faith, and their Western education—so easily fall under the sway of old antiforeign biases? The question came up most insistently during the boycott. The foreign service found the most satisfactory answer in the supposed immaturity and impracticality of students and intellectuals. Time and tutelage would set them straight.

27. After 1905 American educational influence in China did in fact increase: mission schools expanded; Cornell, Yale, and Wellesley created special scholarships; and the excess Boxer indemnity was used to establish an American-run preparatory school (to become Tsinghua College) and to send Chinese (by 1914 nearly 900) for advanced studies in the United States. NF 379/–; NF 10530/3; William Martin (consul, Nanking) to Rockhill, July 9, 1901, Rockhill Papers; Y. C. Wang, *Chinese Intellectuals and the West, 1872–1949* (Chapel Hill, N.C., 1966), pp. 71–72, 158, 510.

28. Hunt, *Frontier Defense and the Open Door*, pp. 163, 171–73.

29. NF 774/8–9; FRUS 1906, p. 353; FRUS 1907, p. 147. Washington waited some nine months after issuing invitations to the other powers before drawing Peking into the conference preparations, and then insisted that it accept as binding any conclusions drawn at the international gathering. Apprehensive over some new form of foreign interference, Peking refused to cooperate until the State Department

withdrew its demands. NF 774/2; *FRUS 1906*, pp. 353, 360–61; *FRUS 1907*, pp. 158–59, 163–66; *FRUS 1908*, pp. 76–78, 87, 92–93, 97, 108, 111. Official American doubts about the sincerity of the Chinese program of suppression were misplaced but also largely irrelevant to the ultimate success of the antiopium campaign. That depended essentially on rigorous Chinese efforts and the cooperation of the British, whose nationals were the major importers of opium. The Chinese and British governments already had talks under way well before the United States proposed a conference, and had come to a crucial agreement in January 1908, over a year before that conference opened in Shanghai. Meribeth E. Cameron, *The Reform Movement in China, 1898–1912* (Stanford, Calif., 1931), ch. 7.

30. *FRUS 1903*, p. 79; *FRUS 1904*, pp. 118–25, 203; *FRUS 1909*, p. 206; *FRUS 1911*, p. 55; NF 24074/12, 16, 19, 21; DF 893.00/439.

31. Yüan had crushed the Boxers in Shantung and later helped commit the central government to wide-ranging reforms. In Chihli he had created a model province where opium had been suppressed, educational changes carried out, modern armies trained, and a degree of public political participation encouraged. His credentials in foreign affairs were impeccable. He had suppressed the offensive boycott agitation in his jurisdiction in 1905 and opposed it elsewhere, and kept foreigners and foreign-educated Chinese advisers by his side. Cameron, *The Reform Movement*, pp. 73–74, 89–91, 103, 111, 144, 147, 171; NF 1518/181A, 194, 195; NF 9414/–; NF 12280/–.

32. DF 893.00/1628, 2049, 2100, 2152; DF 893.51/843; *FRUS 1912*, pp. 102, 105; *FRUS 1914*, pp. 51–52; Paul S. Reinsch, *An American Diplomat in China* (Garden City, N.Y. 1922), pp. 3, 5–6. Yüan would have no serious rivals for the affection of the foreign service. Chang Chih-tung had against him his conservative educational philosophy, his view of Japan as a model for China, and his leading role in the recovery of the Hankow-Canton railway. The leaders of the successful revolution of 1911 and the abortive one of 1913 were suspect in the eyes of the foreign service because of their resort to intrigue and violence (even though they went out of their way to protect foreign interests). Even Sun Yat-sen, a "modern man" with his training in Western medicine, his Christian background, and his residence abroad, appeared in foreign service reports in 1911 and 1912 as an impractical dreamer without the strength to hold China together. By 1914 those reports, taking note of Sun's unyielding opposition to Yüan and his criticism of the unequal treaties, had made him into a dangerous fanatic, an ideologue, even a traitor depriving his country of peace and unity. DF 893.77/1152–53; DF 893.51/385A; *FRUS 1912*, pp. 102, 105; Roger Greene to Nelson T. Johnson, March 29, 1912, Greene Papers (Houghton Library, Harvard University); Ganschow, "A Study of Sun Yat-sen's Contacts," pp. 59–63, 167.

33. *FRUS 1910*, pp. 357–58; *FRUS 1911*, pp. 72–73; *FRUS 1914*, pp. 98, 103–4, 107, 111–22; DF 393.115 St2/57; DF 693.116/55; Samuel C. Chu, *Reformer in Modern China: Chang Chien, 1853–1926* (New York, 1965), pp. 149–51, 153–54, 157; Reinsch, *An American Diplomat*, pp. xi–xii, 42, 60, 65, 80–81, 106, 176; Noel H. Pugach, *Paul S. Reinsch: Open Door Diplomat in Action* (Millwood, N.Y., 1979), pp. 104, 206, 261.

34. This continued support for the opium suppression effort was inspired by the hope of not only liberating China's latent energy and relieving her poverty but also checking the alarming spread of drug use to the Philippines and the United States, a development blamed on Chinese immigrants. *FRUS 1909*, pp. 107–11; *FRUS 1910*, pp. 292–93; *FRUS 1912*, pp. 183–84, 188, 206–7, 220; *FRUS 1913*, pp. 215,

218–19, 249, 269–70, 273–74; FRUS 1914, pp. 930, 932–34; NF 774/–2; DF893.00/565A. The international opium conference in Shanghai in 1909 and The Hague in 1913 and 1914 accomplished little, however. The outbreak of the World War ended the conference movement and diverted American narcotics control efforts homeward. Arnold H. Taylor, "Opium and the Open Door," South Atlantic Quarterly (Winter 1970), 69:86–94.

35. William Howard Taft, Present Day Problems (New York, 1908), pp. 48–49; Hunt, Frontier Defense and the Open Door, pp. 144, 148, 168, 152; NF 5315/727A; FRUS 1911, pp. 74–75, 83; FRUS 1914, pp. 122–27; FRUS 1915, pp. 217–22, 226–27, 230; DF 693.116/55.

36. Exports—the overwhelming proportion of them cotton goods, kerosene, and tobacco—went from $10 to $21 million, and imports increased from $18 to $39 million. The American share of China trade remained roughly unchanged (8 percent in 1900, 7 percent in 1913). But in the commercial standings the United States had already by 1900 fallen from second behind Britain to third behind Japan. (By 1913 the Japanese share of the China trade would be more than twice that of the Americans.) In relation to total U.S. exports, exports to China remained stuck at about 1 percent. Investments told the same disappointing story, accounting for no more than 2 to 3 percent of total foreign investment in China and little more than 1 percent of U.S. investment abroad during this decade and a half, and trailing behind the share of China investments held by all the other major powers. Liang-lin Hsiao, China's Foreign Trade Statistics, 1864–1949 (Cambridge, Mass., 1974), pp. 162–63; Yen Chung-p'ing et al., comps., Chung-kuo chin-tai ching-chi shih t'ung-chi tzu-liao hsüan-chi [Selected statistical materials on modern Chinese economic history] (Peking, 1955), p. 65; U.S. Bureau of the Census, Historical Statistics of the United States, Colonial Times to 1957 (Washington, D.C., 1960), pp. 537, 550, 552, 565; Charles F. Remer, Foreign Investments in China (New York, 1933), pp. 332–33, 338; Hsü I-sheng, Chung-kuo chin-tai wai-chai shih t'ung-chi tzu-liao, 1853–1927 [Statistical materials on the history of foreign investment in China in modern times, 1853–1927] (Peking, 1962), p. 93; and Shu-lun Pan, The Trade of the United States with China (New York, 1924), pp. 59–60, 110–11.

37. Michael H. Hunt, "Americans in the China Market: Economic Opportunities and Economic Nationalism, 1890s–1931," Business History Review (Autumn 1977), 51:285–86.

38. William R. Braisted, "The United States and the American China Development Company, 1895–1911," Far Eastern Quarterly (February 1952), vol. 11; George Kennan, E. H. Harriman's Far Eastern Plans (Garden City, N.Y., 1917); Yen, Chung-kuo chin-tai ching-chi, pp. 173, 180; Wang Ching-yü, Chung-kuo chin-tai kung-yeh shih tzu-liao, ti-erh chi, 1895–1914 nien [Historical materials on modern Chinese industry, second series, 1895–1914] (Peking, 1957), 1:140–47.

39. RG 94, file 329412 (Foord to Hay, August 21, 1900); Foord, "China in Regeneration," Engineering Magazine (August 1900), vol. 19; Chien Po-tsan et al., comps., I-ho-t'uan [The Boxers] (Shanghai, 1951), 3:241; FRUS 1901, Appendix: Affairs in China, p. 217; Shanghai merchants to Rockhill, October 26, 1900, McKinley Papers (LC); James J. Lorence, "Coordinating Business Interests and the Open Door Policy: The American Asiatic Association, 1898–1904," in Jerry Israel, ed., Building the Organizational Society (New York, 1972), pp. 130–41; Herbert Ershkowitz, The Attitude of Business Toward American Foreign Policy, 1900–1916 (University Park, Pa., 1967), pp. 10–20; Hunt, Frontier Defense and the Open Door, pp. 60, 80, 82–83.

40. Foord tried to keep textile producers interested in China, at least as a future

market of importance; however, neither he nor his successor Willard Straight could check the declining interest in China or the Association's loss of influence. James J. Lorence, "The American Asiatic Association, 1898–1925: Organized Business and the Myth of the China Market" (Ph.D. diss., University of Wisconsin, 1970), chs. 8–10; Ershkowitz, *The Attitude of Business*, pp. 21–28.

41. Hunt, *Frontier Defense and the Open Door*, pp. 145, 149.

42. Annual messages to Congress, 1901–1907, in Hermann Hagedorn, ed., *The Works of Theodore Roosevelt* (New York, 1923–1926), vol. 17. Hay, guided by Rockhill, did not worry about Russian trade discrimination in Manchuria until late 1903, but by then Roosevelt's decision not to press Russia on the occupation issue precluded action. After 1905 consular charges of Japanese discrimination were undercut by skeptical reports by Rockhill, several consuls in Manchuria, the Tokyo legation, a committee of American Shanghai businessmen who made an on-the-scene investigation in 1906, trade specialists in the Department of Commerce and Labor and in the State Department, and British officials in China. Root accordingly moved warily. He refused just after the Russo-Japanese War to do more than prod the Chinese and Japanese into arranging an early and unrestricted opening of Manchuria to foreign trade, and in the spring of 1908 he rejected Huntington Wilson's elaborate indictment of Japan's proceedings in Manchuria. Hunt, *Frontier Defense and the Open Door*, pp. 59, 65n, 83–84, 108–13, 147–51.

43. DF 893.77/1217. This failure of understanding is striking, for these economic expansionists prided themselves on being practical men of affairs and have recently been depicted as the adept servants of corporate interests.

44. The following discussion of cotton goods draws on my own "Americans in the China Market," pp. 286–88.

45. While the Japanese (to take 1910–1913 as an example) sold 20 percent of their total production (1.0 billion square yards of cloth) abroad, the American industry had immediately at hand a market that was large, still growing, protected by a substantial tariff, and could be counted on to absorb all but 6 percent of total production (6.2 billion square yards).

46. The reserve of the financiers is reflected in the tendency of American bankers to stay clear of long-term commitments and instead devote themselves primarily to short-term financing of trade, dealings in foreign exchange, and accepting deposits from American citizens, firms, and government agencies in China. The American bank with the greatest staying power was the International Banking Corporation, a latecomer by European standards when it set up shop in Shanghai in 1902. Later taken over by the National City Bank, it elaborated a string of eight branch offices. A later, postwar flowering of American-connected banks proved short-lived. Hunt, "Americans in the China Market," pp. 299–300.

47. This and the following two paragraphs draw on G. Kurgan-van Hentenryk, *Léopold II et les groupes financiers belges en Chine: La politique royale et ses prolongements (1895–1914)* (Brussels, 1972), pp. 226–54, 436–541, 549; MCD (January 19, 1901); Rockhill to Hay, January 29, 1901, Hay Papers (LC); *FRUS 1905*, pp. 127–28, 130; John W. Foster, *Diplomatic Memoirs* (Boston, 1909), 2:298–300; memo to the Chinese minister on the terms of cancellation, May 29, 1905, Roosevelt Papers; and Morison, *Letters of Roosevelt*, 4:1109, 1278–79, 1303, 5:15, 29–30.

48. Hunt, *Frontier Defense and the Open Door*, pp. 153–57, 189, 206, 211.

49. Taft quoted in Henry F. Pringle, *The Life and Times of William Howard Taft* (New York, 1939), 2:655; Hunt, *Frontier Defense and the Open Door*, pp. 189–90, 219–20, 242.

50. The Standard Oil Company refused to make a political loan in order to obtain an oil concession in Shensi. The financiers drawn in to underwrite the Huai River conservancy work shared none of Reinsch's sense of urgency; the outbreak of war in Europe got them off the hook. Even a loan tied to a ship contract secured by Bethlehem Steel was never consummated. Pugach, *Paul S. Reinsch*, ch. 5; *FRUS 1916*, p. 126; Reinsch, *An American Diplomat*, pp. 91–94, 217–18, 221–22, 227, 240–41.

51. The boycott of 1905 offers a striking example of the diplomatists' failure of vision on this count. To Rockhill, Conger, and several consuls on the scene it was inconceivable that self-interested Chinese merchants would join the boycott at the expense of their profitable trade in American goods. Aside from misjudging the intensity of resentment in the treaty ports over U.S. policy, they simply failed to count on the possibility that merchants could make money dealing in substitute goods of non-American origin and some even of Chinese manufacture. *FRUS 1905*, p. 211; Delber McKee, *Chinese Exclusion versus the Open Door, 1900–1906: Clashes over China Policy in the Era of Theodore Roosevelt* (Detroit, Mich., 1977), pp. 109, 160.

52. Among the central government's principal initiatives were the creation of a ministry devoted to the promotion of industry and commerce, the compilation of commercial law codes and mining regulations, the pursuit of tariff revision, and the establishment of state banks. Working with provincial authorities, it sponsored technical schools, special official bureaus, commercial expositions, and merchant associations. Kung Chün, *Chung-kuo hsin kung-yeh fa-chan shih ta-kang* [An historical outline of the development of China's modern industry] (Shanghai, preface 1931), pp. 67–69; Wang Ching-yü, *Chung-kuo chin-tai kung-yeh*, 1:637; Yang, *Ch'ing-tai huo-pi*, pp. 366–69, 374–79.

53. Hunt, "Americans in the China Market," p. 296.

54. Lee En-han, "China's Response to Foreign Investment in Her Mining Industry (1902–1911)," *Journal of Asian Studies* (November 1968), vol. 28; Wang, *Chung-kuo chin-tai kung-yeh*, 1:26–28; Chi-ming Hou, *Foreign Investment and Economic Development in China, 1840–1937* (Cambridge, Mass., 1965), pp. 109–11; G. C. Allen and Audrey D. Donnithorne, *Western Enterprise in Far Eastern Economic Development: China and Japan* (London, 1954), p. 152; Chung-yang yen-chiu yüan, chin-tai shih yen-chiu so, comp., *K'uang-wu tang* [Records on mining affairs] (Taipei, 1960), 1:86–90, 93–94, 100–9, 4:2396–97, 2414–15.

55. *K'uang-wu tang*, 1:91, 4:2388–89, 2403, 2410, 2412; *FRUS 1904*, pp. 153–54, 161–67; *FRUS 1905*, pp. 234–36, 261–73; *FRUS 1908*, pp. 152, 175; NF 2648/10–13; *FRUS 1914*, pp. 133–35.

56. The opposition of the Fengtien military governor in 1906 scared off American participants in two joint ventures in coal mining in that province. One project was capitalized in 1904 at $60,000. The other was to include a factory for mining machinery; it depended on an American investment of $320,000. In other cases where Americans failed to go ahead, the deterrent effect of the regulations can only be inferred. An American involved in 1896 with English and German associates in a $70,000 Chihli coal mine subsequently dropped out. An investment of nearly a million dollars was planned in 1901 for Kirin silver and copper mines, but the American participant, a Shanghai resident with some backing from Guggenheim interests, would not go forward, arousing in turn local criticism and demands that his contract be annulled. By 1910 he was effectively out of the picture. A Sino-American partnership, formed about 1903 to operate silver and gold mines in Jehol, fell through. Finally, American support for a gold mining operation on the Fengtien-Kirin border

evaporated, leaving the Chinese and British partners to supply the $660,000 in capital needed to carry on. K'uang-wu tang, 6:3869, 3874–75, 3885–90, 3927–47, 7:4131–47, 4158–59; NF 2648/2–5, 27; Wang, Chung-kuo chin-tai kung-yeh, 1:12–13, 132–34, 140–47, 149, 155–59; Lu Feng-shih, comp., Hsin-tsuan yüeh-chang ta-ch'üan [A new compilation of treaty terms] (Shanghai, 1909), 13:40–41.

57. The Americans had fallen far behind in their construction schedule and run far over even the revised cost estimate. The chief American agent had embezzled funds, neglected his job, and antagonized his Chinese counterparts. His subordinates had behaved in a generally disreputable fashion. Lee En-han, "Chung-Mei shou-hui Yüeh-Han lu-ch'üan chiao-she" [Sino-American negotiations over the recovery of rights over the Canton-Hankow railroad], Chung-yang yen-chiu yüan, chin-tai shih yen-chiu so chi-k'an (1969), no. 1, pp. 161–85, 192–94; Kurgan-van Hentenryk, Léopold II, pp. 484–86; Edward Rhoads, China's Republican Revolution: The Case of Kwangtung, 1895–1913 (Cambridge, Mass., 1975), p. 63; CWS:KH, 159:2, 171:14, 182:14, 186:4, 190:1; CWC, 189:23–26.

58. Chang was back to where he had been in 1898—with a strategic railway still to build and funds to build it in short supply in China. When Chang reached out for a foreign loan in 1906, the line became the focus of domestic and international controversy even greater than before. CWS:KH, 159:2, 191:7, 10, 192:25, 197:2, 201:18; CWS:HT, 3:40, 13:1, 15:45, 16:39, 17:37, 18:9, 20:22, 27, 36, 21:15, 29; FRUS 1905, pp. 124–25; FRUS 1909, pp. 177–78, 192, 197; FRUS 1910, p. 291; CWC, 65:21–29.

59. NF 5315/412; Chün-tu Hsüeh, Huang Hsing and the Chinese Revolution (Stanford, Calif., 1961), pp. 85–86; Ch'en Chen et al., comps., Chung-kuo chin-tai kung-yeh shih tzu-liao, ti-erh chi [Historical materials on modern Chinese industry, second series] (Peking, 1958), pp. 327–31; DF 893.00/2124.

60. Other American manufacturers had only limited success. The Singer Company, already well on its way as the first American-based multinational to become entrenched in the European market, turned to China in the 1880s. Disappointing sales and a discouraging market survey led the company to turn its business over to a commission house. In 1904 Singer tried direct sales again, but China took perhaps no more than 1 percent of U.S. sewing machine exports in the following decade. Baldwin, a manufacturer of railway equipment, secured orders intermittently through the 1890s and 1900s, but fragmentary figures suggest that even in good years China claimed less than 1 percent of its locomotive exports. American manufacturers of textile machinery made no headway against the preference of China's developing textile mills for British-made machinery. Only during World War I with British machinery unavailable did American manufacturers enjoy temporary favor. Robert B. Davies, Peacefully Working to Conquer the World: Singer Sewing Machines in Foreign Markets, 1854–1920 (New York, 1976), ch. 7; History of the Baldwin Locomotive Works, 1831–1923 (Philadelphia [circa 1924?]), pp. 93, 106, 146, 157, 182, 184; H. D. Fong, Cotton Industry and Trade in China (Tientsin, 1932), 1:79–81, 84.

61. Hunt, "Americans in the China Market," pp. 281–84. After the breakup of the Standard Oil trust in 1911, its former export arm, Standard Oil of New York, continued operations in China.

62. This and following treatment of BAT comes from Sherman G. Cochran, Big Business in China: Sino-Foreign Rivalry in the Cigarette Industry, 1890–1930 (Cambridge, Mass., 1980), esp. pp. 10–40, 225; Y. C. Wang, "Free Enterprise in China: The Case of a Cigarette Concern, 1905–1953," Pacific Historical Review (1960), vol. 29; and NF 14732.

63. For example, in 1887 Chang Chih-tung, then governor general of Kwangtung and Kwangsi, requested a ban on kerosene imports on the grounds that they were hurting the peasants who grew peanuts, the processors who extracted oil from them, and the merchants who sold it. "The livelihood of our people is at stake and we are indeed obliged to prohibit [kerosene]." Ultimately, growing demand overseas for Chinese vegetable oils not only drove up the price but weakened guild opposition to kerosene. Hunt, "Americans in the China Market," pp. 291–92.

64. Standard sometimes paid a "voluntary tax," generally in the form of a contribution to local charity, thus avoiding a local confrontation without formally abandoning treaty rights. Diplomats deprecated Standard's policy of temporization because they felt that it was necessary "to take a strong stand so as not to encourage the Chinese." But they could not stop Standard from carrying out its own policy—especially one so eminently successful in economic terms. Hunt, "Americans in the China Market," pp. 292–95.

65. DF 393.115 St2/70; DF 893.00/795.

66. Indeed, the federal government during the Roosevelt and Taft years made its major impact on Standard Oil and BAT quite inadvertantly—by prosecuting trusts at home and thus endangering the dominant position in the domestic market enjoyed by Standard and Duke's American Tobacco Company. The domestic market was important to these firms above all as a source for the funds necessary to carry overseas operations through the difficult early experimental stage. The domestic market also provided initial economies of scale (as with Standard's refineries) and experience with production and advertising methods that could later be adapted to foreign markets (as with BAT). Trust-busting also served as an unintended incentive to Duke (and Harriman as well) to turn energies abroad.

67. Loss of life and property damage were particularly heavy in Shansi, Chihli, and Shantung provinces, where most missions were simply obliterated. (The American Board alone lost $346,000 in property there.) But destruction also occurred elsewhere—for example in Kiangsi (five American mission churches destroyed) and in Kwangtung and Kwangsi (total losses by American missionaries and their converts amounting to $113,000). Arthur H. Smith, *China in Convulsion* (New York, 1901), 2:497–98, 613–16, 643–49, 665–712; A. H. S. Landor, *China and the Allies* (New York, 1901), 1:250–69; Robert C. Forsyth, *The China Martyrs of 1900* (New York, n.d.), pp. 353–59; FRUS 1900, pp. 193–94; RG 94, file 329412, box 2184 (Reverend H. D. Porter's account of the Shansi massacres); RG 395, file 906, box 5 (Karnes' report of March 20, 1901); Peking station report for 1900–1901 and Luella Miner to Pao-ting-fu friends, March 28, 1901, in ABFM, vols. 24 and 27 respectively; ITS, pp. 281, 825–27, 1238–41; Judson Smith to Hay, November 26, 1900, and acting Secretary of State Hill to Cortelyou, February 6, 1901, McKinley Papers.

68. Chinese officials got into the act more and more, with the result that some locales which had already paid once to missionaries had to pay again. Confused accounts of this crazy patchwork of payments make difficult a precise estimate of what converts gained, but in the Peking area, to take one example, they received at least $86,000 in recompense (a figure that may take account of none of the early informal collections). Ament to Judson Smith, November 13, 1900, Dr. Peck to Smith, March 14, 1901, and Tewksbury to Smith, February 18, 1901, ABFM, vols. 25, 27, and 30 respectively; Smith, *China in Convulsion*, 2:501, 729–30; W. A. P. Martin, *The Siege in Peking: China Against the World* (New York, 1900), p. 136; ITS, pp. 1203, 1209–13, 1238–41; CWS:KH, 150:9–10; Lu, *Hsin-tsuan yüeh-chang*, 13:40; Michael H. Hunt, "The Forgotten Occupation: Peking, 1900–1901," *Pacific Historical*

Review (November 1979), 48:526–27; Wilbur J. Chamberlin, Ordered to China (London, 1904), pp. 98, 127, 156, 190, 353–54.

69. "Memorandum on Reform," August 15, 1900, letter to McKinley, March 1901, Porter et al., cable, October 4, 1900, program addressed to American and British diplomatic representatives (Fall 1900?), all in ABFM, vol. 23; Chauncey Goodrich to Judson Smith, August 24 and December 4, 1900, and Luella Miner to Smith, November 8, 1900, ABFM, vol. 26; H. P. Perkins to Smith, August 6 and 18, 1900, H. D. Porter to Smith, September 8, 1900, Sheffield to Smith, March 26, 1901, ABFM, vol. 28; Sheffield to Smith, July 16, August 19, November 15, 1900, ABFM, vol. 29; Martin, The Siege in Peking, pp. 155–57; and Chien, I-ho-t'uan, 3:229.

70. Prudence, however, dictated that missionaries keep their criticisms of Hay and McKinley to themselves and direct their public fire instead at Wu T'ing-fang, who had allegedly saved the dynasty with his lies, and at unsympathetic American officials in China, particularly Rockhill and Chaffee. A. H. Smith and Tewksbury to Board, 1901, ABFM, vol. 23; Ament to Smith, November 13, 1900, and July 13, 1901, ABFM, vol. 25; H. D. Porter to Smith, August 26, October 10 and 28, 1900, H. P. Perkins to Smith, November 30, 1900, ABFM, vol. 28; Sheffield to Smith, February 26 and March 26, 1901, ABFM, vol. 29; Elwood Tewksbury to "Dear Ones," January 13, 1901, and to Smith, February 18, 1901, ABFM, vol. 30.

71. Chamberlin, Ordered to China, pp. 146, 212–13, 253–54; Albert B. Paine, Mark Twain: A Biography (New York, 1912), 3:1121; Mark Twain, "To the Person Sitting in Darkness," and his "To My Missionary Critics," North American Review (February and April 1901), vol. 172; Gilbert Reid, "The Ethics of Loot," and his "The Ethics of the Last China War," Forum (July 1901), 31:586, and (December 1901), 32:449–50; Stuart C. Miller, "Ends and Means: Missionary Justification of Force in Nineteenth Century China," in John K. Fairbank, ed., The Missionary Enterprise in China and America (Cambridge, Mass., 1974), pp. 273–80.

72. The board leaders thus rejected the missionary call for payment of "remote" indemnity and counseled against missionaries taking upon themselves the collection of any indemnity. A majority agreed it was best to let the "civil authorities" decide whether to keep the army in China and how to punish Chinese officials. Martin, The Siege in Peking, pp. 162, 175–85; Landor, China and the Allies, 2:192–93; Hay to McKinley, July 6, 1900; McKinley Papers; FRUS 1900, p. 195. Mission apologists set to work about this time to blunt charges of missionary misconduct and provocation. They traced the recent turmoil instead to the aggressiveness of some powers, the inroads made by foreign commerce, the feebleness of the Chinese government in the face of popular unrest, crop failures, race hatred, and aggressive Catholic missionaries. Forsyth, The China Martyrs, pp. 1–8; Charles Denby, China and Her People (Boston, 1909), 1:225, 2:19, 71, 171–73; Ada Haven Mateer, Siege Days (New York, 1903), pp. 23–24; Luella Miner, China's Book of Martyrs (Philadelphia, 1903), pp. 17–18; Reverend Z. Charles Beals, China and the Boxers (New York, 1901), chs. 2 and 13.

73. Minutes of the North China Mission, January 24 and April 30, 1901, and A. H. Smith and Sheffield, cable, received March 20, 1901, ABFM, vol. 23; Ament to Smith, June 18, 1901, ABFM, vol. 25. The recurrence of missionary incidents subsequently helped sustain the new outlook. In 1902 bands employing Boxer slogans attacked converts in the Ch'eng-tu area of Szechuan. A number of incidents followed in 1905–1907 suggesting that the far southern coastal region might explode as easily as the North. The most serious of the cases (discussed below) occurred in 1905 at Lien-chou in Kwangtung. FRUS 1902, pp. 170–80; FRUS 1903, p. 79; MCD (November 7 and 11, 1905); NF 10251/3–7.

74. W. A. P. Martin, *A Cycle of Cathay* (New York, 1900), pp. 233–35, 293–327; Peter Duus, "Science and Salvation in China: The Life and Work of W. A. P. Martin (1827–1916)," in Kwang-Ching Liu, ed., *American Missionaries in China* (Cambridge, Mass., 1966). Charles D. Tenney was another of the missionaries working with the elite. For two decades he took a leading role in the schools of Chihli province, first under Li Hung-chang and after him Yüan Shih-k'ai. The Americans Gilbert Reid and Young J. Allen were also active, helping the Englishman Timothy Richard with the main propaganda organ for the Society for the Diffusion of Christian and General Knowledge, which had its origins in the Anglo-American mission community in Canton in the 1830s and hit its peak of influence among Chinese intellectuals in the 1890s. Their papers, published and confidential, argued for social, economic, and educational reform; development of China's resources and the building of railways; and dependence on the financial and diplomatic strength of the two most advanced countries in the world, the United States and Britain. Paul A. Cohen, "Missionary Approaches: Hudson Taylor and Timothy Richard," *Papers on China* (1957), vol. 11; Wang Shu-hwai, *Wai-jen yü wu-hsü pien-fa* [Foreigners and the 1898 reforms] (Taipei, 1965), pp. 129–32; Gilbert Reid to Prince Kung, January 7, 1896, *Hai-fang tang*, "T'ieh-lu," pp. 299–306; CR: Tientsin (April 11, 1905).

75. Isaac Taylor Headland, "The Reform Movement in China," *Outlook* (June 30, 1900), 65:494–97; Clifton J. Phillips, "The Student Volunteer Movement and Its Role in China Missions, 1886–1920," in Fairbank, *The Missionary Enterprise*, p. 106.

76. Conversion was a matter of faith, not knowledge, the evangelists in these debates contended, and education was limited to the secondary function of sustaining Christian values. To purvey Western science, philosophy, or languages was to risk inculcating skepticism or losing graduates to lucrative middleman jobs in the treaty ports. Consequently, mission schools in the latter part of the nineteenth century concentrated on the training of students, generally poor and from Christian families, as elements in a native army that missionary generals would lead in the final assault on heathendom (to use the military metaphor of Judson Smith, head of the American Board). By one estimate, one graduate in eight from mission colleges went into the ministry, and one-half found employment in mission schools and hospitals. Irwin T. Hyatt, Jr., "Protestant Missions in China (1877–1890): The Institutionalization of Good Works," and Roberto Paterno, "Devello Z. Sheffield and the Founding of the North China College," in Liu, *American Missionaries*; Phillips, "The Student Volunteer Movement," p. 104; Irwin T. Hyatt, Jr., *Our Ordered Lives Confess: Three Nineteenth-Century American Missionaries in East Shantung* (Cambridge, Mass., 1976), pp. 176–82; Ellsworth Carlson, *The Foochow Missionaries, 1847–1880* (Cambridge, Mass., 1974), pp. 86–87.

77. Out of an evolution from primary to secondary schools to junior colleges, a handful of institutions of higher learning had emerged at the old centers of mission effort. The first was Shantung Christian University, founded by Mateer in Teng-chou in 1864 (and later relocated to Tientsin), followed by St. John's University in Shanghai (1879), Canton Christian College (1888), the University of Nanking (1888), and Peking University (1890), which was to combine with three other institutions in the late teens to make up Yenching University, the premier Christian college. The post-1900 blossoming, which continued down to 1917, extended the school system to central and western China (Ginling College, Hangchow Christian College, Hsiang-ya in Ch'ang-sha, Fukien Christian University, and West China Union University), and it added two more institutions to the coast (Soochow and Shanghai Universities) and one to the Peking area (North China Union College). Medical missionaries, thanks to their enhanced professional status at home, had already in the late nineteenth

century begun to slough off the evangelical work that boards favored, and now finally came into their own with the establishment of medical schools, preeminently the Yale-sponsored Hsiang-ya and the Peking Union Medical College, as well as a network of hospitals that grew from 60 in 1900 to 200 in 1914. William R. Hutchinson, "Modernism and Missions: The Liberal Search for an Exportable Christianity, 1875–1935," in Fairbank, *The Missionary Enterprise*; Jessie G. Lutz, *China and the Christian Colleges, 1850–1950* (Ithaca, N.Y., 1971), ch. 4; William Reeves, Jr., "Sino-American Cooperation in Medicine: The Origins of Hsiang-Ya (1902–1914)," in Liu, *American Missionaries*; Charles F. Remer, *American Investment in China* (Honolulu, 1929), pp. 26–27. For a sketch of three careers illustrating the tendency of education to eclipse evangelism in this period, see Philip West, *Yenching University and Sino-Western Relations, 1916–1952* (Cambridge, Mass., 1976), pp. 23–24.

78. Phillips, "The Student Volunteer Movement," p. 104; Shirley S. Garrett, *Social Reformers in Urban China: The Chinese Y.M.C.A., 1895–1926* (Cambridge, Mass., 1970).

79. Even during the most serious of these incidents, at Lien-chou in November 1905, local officials first offered sanctuary to the missionaries (who instead fled—some into the hands of their tormentors) and then got the survivors safely to Canton. The provincial government at once cashiered the local magistrate and army commander and in line with American demands razed the local temple, erected a memorial tablet, executed one of the rioters, and after some negotiation paid a $25,000 indemnity. Worried by this outbreak and a string of others reaching back over several years, the central government issued an imperial edict in 1907 warning officials to be vigilant against unrest and tolerant of missions and converts. *FRUS 1903*, p. 80; *FRUS 1906*, pp. 308–24; *FRUS 1907*, pp. 176–77, 215–18; MCD (November 7 and 11, 1905).

80. Lutz, *China and the Christian Colleges*, pp. 78, 503; Harlan P. Beach and Burton St. John, eds., *World Statistics of Christian Missions* (New York, 1916), pp. 78–79; Cameron, *The Reform Movement*, pp. 62, 178; Hilary Beattie, "Protestant Missions and Opium in China, 1858–1895," *Papers on China* (May 1969), vol. 22A; Taylor, "Opium and the Open Door," pp. 85–86.

81. Smith, draft of reply to Twain, March 1901, ABFM, vol. 23; A. H. Smith to Judson Smith, September 23, 1900, and January 31, 1901, ABFM, vol. 29; the series of articles by Smith that appeared in *Outlook* in 1901 (they were brought together later that year in *China in Convulsion*); Arthur H. Smith, *China and America To-day* (New York, 1907), pp. 115–18, 123–45, 158–83, 220; Arthur H. Smith, *The Uplift of China* (New York, 1907), pp. 38–48, 110, 158, 178, 225–27. Smith also penned a pioneering sociology, *Village Life In China* (New York, 1899), based on firsthand experience in Shantung, and numerous articles in *Outlook* and mission journals.

Other missionary leaders also recognized the force of reform nationalism. Arthur J. Brown in *New Forces in Old China* (New York, 1904) anticipated Smith by three years. His sense of trepidation reflected in that work's subtitle, *An Unwelcome but Inevitable Awakening*, was balanced by the faith he reposed in Yüan Shih-k'ai to control the "fires . . . smouldering beneath" the surface of imperial China (pp. 338–45). W. A. P Martin appears to have gone through the same shift in views that Smith did. Compare Martin's *The Siege in Peking* (1900) with his *The Awakening of China* (New York, 1907), especially the preface and ch. 29.

82. Wilson's reversal of Taft administration policy in 1913 earned him missionary accolades for his "clearness of vision, breadth of statesmanship, and high moral

tone." But by then, missionary enthusiasm was already on the decline, blunted by growing fears "that liberty may run to license." Increasingly, missionaries lined up with diplomats in denigrating Sun the agitator and emphasizing their dependence on Yüan not so much as the exponent of progress, but as the guarantor of order and security essential to the mission movement. *Missionary Review of the World* (February 1913), new series, 26:139; Arthur J. Brown, *The Chinese Revolution* (New York, 1912), pp. 158, 177, 180–81; DF 893.00/634 (James W. Bashford to Woodrow Wilson, November 28, 1912, and to Huntington Wilson, January 20, 1913); Arthur S. Link et al., eds., *The Papers of Woodrow Wilson* (Princeton, N.J., 1966–), 27:27–28; Eugene P. Trani, "Woodrow Wilson, China, and the Missionaries, 1913–1921," *Journal of Presbyterian History* (1971), 49:333; Michael V. Metallo, "American Missionaries, Sun Yat-sen, and the Chinese Revolution," *Pacific Historical Review* (May 1978), vol. 47.

83. American missionaries were still overwhelmingly middle class but recruited in larger numbers from the upper Midwest as the yield from the old New England and upstate New York stronghold declined. Milton T. Stauffer, ed., *The Christian Occupation of China* (Shanghai, 1922), pp. 326–27, 346–47; Beach and St. John, *World Statistics of Christian Missions*, p. 63; Remer, *American Investment in China*, pp. 26–27; Valentin H. Rabe, *The Home Base of American China Missions, 1880–1920* (Cambridge, Mass., 1978), chs. 1 and 2; Phillips, "The Student Volunteer Movement," p. 105.

84. Arthur J. Brown, "Future Missionary Policy," in Martin, *The Siege in Peking*, p. 182; Brown, *New Forces*, pp. 233–34, 241–43; NF 10155/1 (Brown to Root and Roosevelt, November 25 and 26, 1907); Brown to Presbyterian missions in North China, July 1, 1909, Brown Papers (Yale Divinity School Library).

85. Brown, "Future Missionary Policy," p. 182; China Centenary Missionary Conference, *Records: Report of the Great Conference held at Shanghai . . . 1907* (Shanghai, 1907), pp. 340–41, 344, 743. On the continuing lack of agreement in mission ranks, see Bashford, "The True Policy of the United States in the Far East," October 1910, in DF 711.93/17; Bashford to Taft, December 12, 1910 (enclosed in Taft to Knox, January 12, 1911), in DF 711.933/–; and Reverend B. H. Niebel to Knox, May 31, 1910, in DF 893.00/397.

86. The YMCA's promotion of Chinese leadership was a major exception. Even so, it remained identified as a foreign institution. Garrett, *Social Reformers*, pp. 65–67, 81, 131, 175. Stauffer, *The Christian Occupation*, p. 384, offers statistics on the Chinese church's limited degree of self-support.

87. The treaty of 1903 recognized the right of mission societies to buy and rent property in the interior (a demand of some forty years' standing) as long as deeds and contracts were submitted in advance for the approval of local officials. The State Department and its agents in China refused to defend mission property deals involving deceit, saw no way that missionaries as individuals could hold land in the interior under the treaty, and opposed trying to secure property in the face of local hostility, but characteristically refused to take a public stand that might weaken "any equitable or quasi legal [sic] rights which may have arisen from custom." *FRUS 1902*, pp. 131–38; *FRUS 1906*, pp. 277–78; *FRUS 1907*, pp. 206–9; Lu, *Hsin-tsuan yüeh-chang*, 13:31–36; NF 5944/–; NF 20005/31; DF 393.116/2–3, 7–8, 29; DF 893.52/21.

88. The following discussion of the Lien-chou case draws on Arthur J. Brown, *The Lien-Chou Martyrdom* (New York, 1905), pp. 1–15; NF 167/17, 29, 98–100; *FRUS 1906*, pp. 308–24; and *FRUS 1907*, pp. 215–18.

89. This and the following paragraph on Chinese mission policy is based on

Mary C. Wright, *The Last Stand of Chinese Conservatism: The T'ung-chih Restoration, 1862–1874* (Stanford, Calif., 1957), pp. 276–77, 296–98; Paul A. Cohen, *China and Christianity: The Missionary Movement and the Growth of Chinese Antiforeignism, 1860–1870* (Cambridge, Mass., 1963), pp. 73–74, 123–26, 149–54, 250–55; IWSM:TC, 54:20–21; Chih Chün, comp., *Ch'i-pu-fu-chai cheng-shu* [Official papers of Chou Chia-mei] (Taipei reprint, 1973), 2:42–44; ISHK, 20:14–15; FRUS 1891, pp. 410–11; FRUS 1894, pp. 143, 152–60; FRUS 1897, pp. 102–5; Chien, I-ho-t'uan, 4:212–15; Chung-kuo k'o-hsüeh yüan, li-shih yen-chiu so, ti-san so, comp., *Keng-tzu chi-shih* [A record of events in 1900] (Peking, 1959), p. 232; ITS, pp. 846–47, 854–55, 1119–22, 1275–76; FRUS 1903, pp. 202–5; NF 13957/1; FRUS 1907, pp. 176–77; CWS:KH, 77:2, 85:2–4, 118:25–27, 120:22–27, 129:13–16, 163:15, 26–27; and CWS:HT, 19:6.

90. However, as long as they could not directly control missionaries and prevent the incidents their activities gave rise to, Chinese officials sought some way to minimize the diplomatic consequences of those incidents. Li Hung-chang, for example, proposed the appointment of a special foreign envoy to attend to Protestant and Catholic affairs and to negotiate settlements directly with the concerned church organizations, thus avoiding meddling by diplomats of the various powers. Another device proposed to prevent diplomatic interference was to promise the automatic grant of an indemnity when incidents occurred through no fault of the missionary.

91. For example, Tenney resigned as president of Chihli province's university after a fight with the Chinese directors, and the government eased W. A. P. Martin out as head of the Peking Imperial University and at the same time got rid of his staff, which was predominantly made up of Chinese Christians.

92. The mission medical schools, by contrast, enjoyed Chinese support. Lutz, *China and the Christian Colleges*, pp. 205–7; CR: Tientsin (March 22, 1906); FRUS 1906, pp. 342–45; Cameron, *The Reform Movement*, pp. 66, 69, 71–72, 74, 76–78, 84–85.

93. Garrett, *Social Reformers*, pp. 172–74, 178–83; Philip West, "Christianity and Nationalism: The Career of Wu Lei-ch'uan at Yenching University," and Garrett, "Why They Stayed: American Church Politics and Chinese Nationalism in the Twenties," both in Fairbank, *The Missionary Enterprise*; Lutz, *China and the Christian Colleges*, pp. 170–74, 193–97; Edward J. M. Rhoads, "Lingnan's Response to the Rise of Chinese Nationalism: The Shakee Incident (1925)," in Liu, *American Missionaries*.

Epilogue: The Special Relationship in Historical Perspective

1. Michael H. Hunt, "Pearl Buck—Popular Expert on China, 1931–1949," *Modern China* (January 1977), vol. 3; John K. Fairbank, *The United States and China*, rev. ed. (Cambridge, Mass., 1958), pp. 1, 7, 261, 318, 319.

2. Acheson in U.S. Department of State, *United States Relations with China with Special Reference to the Period, 1944–1949* (Washington, D.C., 1949), pp. iii, xv–xvii, and in U.S. Senate, Committees on the Armed Services and Foreign Relations, *The Military Situation in the Far East* (Washington, D.C., 1951), Appendix II, p. 1868; Dulles in *Department of State Bulletin* (July 15, 1957), 37:95; and Rusk in ibid. (May 2, 1966), 54:693.

3. Wu T'ing-fang, for example, had tried to stir up the interest of Wall Street investors in a railway loan in the late 1890s only to be told that even in the midst

of a depression "their money could be just as easily, and just as profitably, invested in their own country, and with better security than was obtainable in China." His successor in the Washington legation made the same discovery. Wu T'ing-fang, *America Through the Spectacles of an Oriental Diplomat* (New York, 1914), pp. 71–72; Lo Hsiang-lin, *Liang Ch'eng ch'u-shih Mei-kuo* [Liang Ch'eng's ministry to the United States] (Hong Kong, 1977), p. 343.

4. Sun Yat-sen could be added to the list of antiimperialists who in the 1920s saw no distinction between the United States and the other powers. All were economic and political oppressors. C. Martin Wilbur, *Sun Yat-sen: Frustrated Patriot* (New York, 1976), pp. 199–200; T'ang preface, p. 9 of Ch'i Shu-fen, *Ti-kuo chu-i t'ieh-t'i hsia ti Chung-kuo* [China under the imperialist heel] (Shanghai, 1925); Chiang quoted in Harold R. Isaacs, *Images of Asia: American Views of China and India* (New York, 1962), p. 202; P. Cavendish, "Anti-imperialism in the Kuomintang, 1923–28," in Jerome Ch'en and Nicholas Tarling, eds., *Studies in the Social History of China and South-east Asia* (Cambridge, England, 1970); Stuart Schram, ed., *The Political Thought of Mao Tse-tung* (New York, 1963), pp. 143, 266; Edgar Snow, *Red Star Over China* (New York, 1961), p. 154.

5. MacMurray was chief, Division of Far Eastern Affairs (1919–1924), Assistant Secretary of State (1924–1925), and Minister to China (1925–1929). Johnson served as chief, Division of Far Eastern Affairs (1925–1927), Assistant Secretary of State (1927–1929), and Ambassador to China (1929–1935). Hornbeck was chief, Division of Far Eastern Affairs (1928–1937), and departmental adviser on political relations (1937–1944). Others who had seen service in China before 1914 and enjoyed prominence thereafter include William Phillips (Third Assistant Secretary of State, 1914–1917; Assistant Secretary of State, 1917–1920; and Undersecretary of State, 1922–1924), Clarence E. Gauss (Counselor of the embassy in China, 1933–1935; and Ambassador to China, 1941–1944), and Willys R. Peck (Counselor of the embassy in China, 1935–1940).

On official views on China in the interwar period, see *FRUS 1925*, 1:799–802; *FRUS 1926*, 1:899; MacMurray memo for the State Department, November 1, 1935, MacMurray Papers (microfilm copy, Sterling Library, Yale University); Russell D. Buhite, *Nelson T. Johnson and American Policy Toward China, 1925–1941* (East Lansing, Mich., 1968), pp. 47, 66, 87–89; Buhite, "The Open Door in Perspective: Stanley K. Hornbeck and American Far Eastern Policy," in Frank J. Merli and Theodore Wilson, eds., *Makers of American Diplomacy* (New York, 1974), vol. 2; James C. Thomson, Jr., "The Role of the State Department," in Dorothy Borg and Shumpei Okamoto, eds., *Pearl Harbor as History: Japanese-American Relations, 1931–1941* (New York, 1973), pp. 88–94; and Christopher Thorne, *The Limits of Foreign Policy: The West, the League, and the Far Eastern Crisis of 1931–1933* (New York, 1972), pp. 51–52, 56, 227, 236, 301.

6. *FRUS 1925*, 1:826; *FRUS 1927*, 2:350–53; Dorothy Borg, *American Policy and the Chinese Revolution, 1925–1928*, expanded ed. (Hamden, Conn., 1968).

7. Dorothy Borg, *The United States and the Far Eastern Crisis of 1933–1938* (Cambridge, Mass., 1964); George E. Gallup, *The Gallup Poll: Public Opinion, 1935–1971* (New York, 1972), 1:72–73, 159–60.

8. Liang Chin-tung, *General Stilwell in China, 1942–1944: The Full Story* (New York, 1972); Frank Ninkovich, "Cultural Relations and American China Policy, 1942–1945," *Pacific Historical Review* (August 1980), vol. 49; Robert Dallek, *Franklin D. Roosevelt and American Foreign Policy, 1932–1945* (New York, 1979), pp. 329–30, 388–91, 427–29, 499, 517–19.

9. James Reardon-Anderson, *Yenan and the Great Powers: The Origins of Chinese Communist Foreign Policy, 1944–1946* (New York, 1980).
10. *FRUS 1945*, 7:762, 768; *FRUS 1946*, 9:937, 945; *New York Times*, March 5, 1946, p. 1; Steven I. Levine, "A New Look at American Mediation in the Chinese Civil War: The Marshall Mission and Manchuria," *Diplomatic History* (Fall 1979), vol. 3.
11. *FRUS 1946*, 9:935; RG 59, Records of the Office of Chinese Affairs, Box 14, file 050.014 (1949; comments by Philip Sprouse, Livingston Merchant, Philip Jessup, and Max Bishop).
12. Michael H. Hunt, "Mao Tse-tung and the Issue of Accommodation with the United States, 1948–1950," in Dorothy Borg and Waldo Heinrichs, eds., *Uncertain Years: Chinese-American Relations, 1947–1950* (New York, 1980). For a different appraisal emphasizing ideological inflexibility in Chinese Communist foreign policy, see Steven M. Goldstein, "Chinese Communist Policy Toward the United States: Opportunities and Constraints, 1944–1950," in *ibid.*; and Okabe Tatsumi, "The Cold War and China," in Yōnosuke Nagai and Akira Iriye, eds., *The Origins of the Cold War in Asia* (New York, 1977).
13. Betty Lee Sung, *The Story of the Chinese in America* (New York, 1971), pp. 77–86, 92–94.
14. Nancy B. Tucker, "An Unlikely Peace: American Missionaries and the Chinese Communists, 1948–1950," *Pacific Historical Review* (February 1976), vol. 45; Philip West, *Yenching University and Sino-Western Relations, 1916–1952* (Cambridge, Mass., 1976), ch. 7; Mary E. Ferguson, *China Medical Board and Peking Union Medical College* (New York, 1970), chs. 9 and 10; Warren W. Tozer, "Last Bridge to China: The Shanghai Power Company, the Truman Administration and the Chinese Communists," *Diplomatic History* (Winter 1977), vol. 1.
15. U.S. Department of State, *United States Relations with China*, pp. iii–iv, xiv–xvii; and *Department of State Bulletin* (January 16, 1950), 22:113–15.
16. *Ibid.* (July 15, 1957), 37:91–95, and (May 2, 1966), 54:686–95.
17. Robert G. Sutter, *China-Watch: Toward Sino-American Reconciliation* (Baltimore, 1978), chs. 5–7; Linda D. Dillon et al., "Who Was the Principal Enemy?: Shifts in Official Chinese Perception of the Two Superpowers, 1968–1969," *Asian Survey* (May 1977), vol. 17.
18. *New York Times*, March 12, 1972, p. 5, and December 16, 1978, p. 8. In China the idea of the special relationship has undergone a dramatic transformation. After several decades of denouncing the myth of friendship, Chinese policy makers with a keen new interest in the United States as a makeweight in the Asian power balance began reflexively to garnish their public commentary with references to the old ties of amity. No sooner were diplomatic relations restored than some Chinese historians also started to make claims for past friendly relations. For a discussion of this reinterpretation and the historical controversy it set off, see the essays by Michael H. Hunt, Luo Rongqu, and Jiang Xiangze in Warren Cohen, ed., *New Frontiers in American–East Asian Relations* (New York, 1983).
19. Fairbank, *The United States and China*, p. 9.

Index

Abeel, David, 47
Acheson, Dean, 299–300, 310–11
Adams, Brooks, 177
Adams, Charles Francis, 82
Adams, John, 12
Adams, John Quincy, 15, 34
Allen, Horace N., 132, 349n36
Allen, Young J., 399n74
Ament, W. S., 287–88
American Asiatic Association, 248; organized, 152; opposition to exclusion, 231, 245, 248; and the Russian occupation of Manchuria, 275
American Baptist Board of Foreign Missions, 25
American Board of Commissioners for Foreign Missions, 25, 26, 31, 185, 286, 397n67
American China Development Company, 181, 350n46, 367n9; acquires railway concession, 150–51, 192; supplemental railway agreement, 193, 367n9; loss of concession, 274, 277–78, 396n57
American economic enterprise in China: profitability, 2, 11–12, 24, 145, 283; under Canton system, 5–15 passim; mercantile capitalism, 6–7, 22–24, 144–45, 151, 355n2; China trade goods, 7–8, 10, 21–22, 145; Chinese associates and intermediaries, 8–11 passim, 22, 38, 149, 282–85 passim, 302, 323n29, 355n2; trade and investment statistics, 8, 22–23, 143–44, 149–52 passim, 274, 276, 282–83, 321n10, 354n1, 355n4, 356n11, 393n36; and government support, 12–15 passim, 145, 151–52, 267, 277–85 passim, 390nn15,17, 397nn64,66; reserve of investors, 24, 144, 145, 356n7, 394n46, 402n3; early views of China, 35–39 passim; rise of multinationals and specialized import-export firms, 144, 151–52, 282–85, 356n12, 396n60, 397n66;
shipping, 144–46, 321n10, 324nn31,32, 355nn5,6; investment projects, 146–51 passim, 355nn2,7, 356n9, 357nn14,15,17,18, 373n61, 374nn62,63, 395n56; mining, 149, 274, 280–81, 395n56; and immigration question, 231, 240, 245–48 passim, 381n31; historical literature on, 318–19, 352–53, 385–86; see also American China Development Company; American Group; Cotton goods trade; Opium trade; Textile mills in China
American educational influence, 391n27; see also Chinese Educational Mission
American Federation of Labor, 232, 249, 381n28
American Group: creation of, 211; and relationship with United States government, 212, 213, 216, 279; unites with European bankers, 213, 215; and currency reform loan, 214; and Woodrow Wilson, 218, 225
American investment, see American economic enterprise in China
American Red Cross, 220, 373n61, 374nn62,63
Americans in Eastern Asia (Tyler Dennett), ix
Americans in Korea, 131–32, 140, 349n36
American Tobacco Company, 397n66
American trade, see American economic enterprise in China
American Trading Company, 356n12, 357n15
American view of China: Enlightenment influence, 32–33; popular, 33–34, 40, 309, 312; Chinese compared with blacks, 93, 337n57; parallels with Chinese view of foreigners, 301; historical literature on, 320; see also Sinophobia
Anderson, Rufus, 28

Angel Island, 228, 378n4
Angell, James B.: and immigration treaty, 88, 100, 101; and mission movement, 160; as minister, 361n55
Anglo-Japanese alliance, 201–2, 222
Anti-American boycott (1905–1906), 284, 373n57, 392n31; evolution in China, 233–41, 383n48, 395n51; American reaction to, 241–47, 268, 269; historical literature on, 375–77
Anti-Christian movement: roots of, 156–58; resort to violence, 160, 162, 167–68, 295–96, 359n35, 398n73; impact of nationalism on, 297–98; central preoccupations, 301; historical literature on, 353–54; impact on missionary educators, 402n91; see also Boxer Uprising; Chinese foreign policy
Arena, 287
Armstrong, James, 323n27
Arthur, Chester A., 90, 181, 361n56
Associated Chambers of Commerce of the Pacific, 248
Astor, John Jacob, 8, 11, 322n14
Avery, Benjamin, 162, 343n39, 361n55; and Taiwan dispute, 59, 60; and missionary question, 158–59

Backhouse, Edmund, 371n44
Baldwin (manufacturer of railroad equipment), 396n60
Barker, Wharton, 147, 149, 357n17
Barrett, John, 177
Bayard, Thomas, 91, 92, 105, 132, 161, 179, 180
Bethlehem Steel, 220, 373n61, 374nn62,63, 395n50
Blaine, James G., 93–94, 129, 161–62
Board of Rites, 346n5, 348n23
Boxer indemnity: collected from China, 200, 369n23, 397n68; surplus returned by the United States, 207, 213, 270, 360n51, 370n30, 391n27
Boxer Uprising, 185–87
Brice, Calvin, 150, 277–78
Bridgman, Elijah C., 25, 26, 27, 36; and diplomacy, 31; view of China, 35, 38, 326n51; literary work, 44
British-American Tobacco Company, 282, 283–85, 397n66

British policy toward China: overthrow of Canton system, 14; and mission movement, 358n29
Brown, Arthur J., 293, 294, 296, 400n81
Browne, J. Ross, 172, 173, 178, 343n39
Browns of Providence (China traders), 8, 11–12
Bryan, William Jennings, 273; appointed Secretary of State, 218; view on economic foreign policy, 218, 374n63; on cooperation with China, 221, 222; and Japanese claims in China, 223, 374n63, 375n67
Buchanan, James, 19, 21
Buck, Pearl, 299
Burgevine, Henry, 60
Burlingame, Anson, 86, 179, 307; as envoy for China, 59, 60, 172–73; and mission question, 156, 158–59; views as American minister, 172, 360nn51,52
Burlingame mission, 98

Calhoun, William J., 271–72, 276
Canton Christian College, 399n77
Canton delta: map, 62; and exclusion movement, 102–3, 105, 235–41 passim, 247
Canton Self-Government Society, 247
Canton system: collapse of, 3, 14; described, 9–10; American view of, 37–39; control of foreigners under, 318n4
Carnegie Steel, 150
Carter, Jimmy, 312
Catholic missionaries, 324n37
Central Pacific Railroad, 336n51
Chae Chan Ping v. the United States, 344n46
Chaffee, Adna, 196, 368n18, 398n70
Chang Chien, 373nn57,61
Chang Chih-tung, 134, 196, 208, 262, 367n6; and immigration question, 103, 104, 105, 238; background, 190; and interest in the United States, 192–202 passim; and railway development, 192–93, 281, 367n6, 396n58; and Boxer Uprising, 194; American view of, 198, 392n32; aligns with Yüan Shih-k'ai on Manchurian policy, 203, 370n27,

391n22; as economic nationalist, 390n17, 397n63
Chang Yin-huan, 140; and immigration question, 104–5, 111–12, 113, 344nn49,50; career, 342n36
Chao Wan-sheng, 252
Chase National Bank, 150
Ch'en Chin-t'ao, 372n56
Cheng Tsao-ju, 103–4, 105, 111–12
Ch'eng-tu riot (1895), 162, 165, 166, 359n35
Ch'en Lan-pin, 99
Chiang Kai-shek, 304, 305, 306
Ch'ien Neng-hsün, 372n56
China and America To-day (Arthur H. Smith), 292
China and Japan Trading Company, 356n12
China market, see American economic enterprise in China
China Reform Association, 252
Chinese American Citizens Alliance, 250
Chinese antiimperialism, see Chinese nationalism
Chinese Characteristics (Arthur H. Smith), 163, 166, 273
Chinese community in the United States: motives and methods of emigration, 61–66, 69, 335n40; prominent role of merchants in, 65–66; social organization of, 65, 66–69, 250; lineage organizations in, 66–67, 111; place associations in, 67; occupational guilds in, 67–68, 111; secret societies in, 68, 69, 111, 252, 253, 254, 383n47; economic conditions in, 68–72 passim, 109, 250, 335n42, 336n51; intermediaries with white community, 69, 110, 301, 344n46; as a bachelor society, 70–71, 250; population of, 71, 72, 76, 108, 249–50, 310, 337n56; movement out of the West, 71, 108; second generation in, 71, 109–10, 250, 309–10; prostitution in, 71, 336n53, 344n49; social tensions within, 72–73, 110–11, 256; as an asset to China, 99, 336n49; estrangement from imperial government, 99–100, 251–57 passim; struggle against exclusion, 105, 107, 108, 110, 232–33, 235, 237, 247, 344n46, 382n44; impact of exclusion on, 108–14 passim, 249–51, 300, 309; education in, 250–51; historical literature on, 329–30, 331, 377

Chinese Educational Mission: described, 60; influential alumni of, 204, 214, 367n6, 372n56; historical literature on, 329

Chinese foreign office, 146; and immigration question, 100–1, 104, 105, 121, 238, 240; role in policymaking, 116; foreign view of, 116, 346n2; and defense of tribute states, 124, 125, 128, 135, 137, 139, 348nn20,23; American effort at reform of, 267, 268

Chinese foreign policy: resistance to treaty system, 14; militant outlook, 54–55, 115, 134, 140, 190, 193–94, 351n52; and tribute states, 83, 346–47n5; and Chinese overseas, 96–99, 255; decisionmaking in, 116; and loss of Korea, 125–32 passim, 136–40 passim; and Vietnam crisis, 133–36; and war with Japan, 136–40; and mission movement, 167, 291, 296–98, 358n30, 359n42, 400n79, 402nn90,91; and Boxer crisis, 194; and defense of Manchuria, 200–9 passim, 214–16, 369n26, 370n34; and international financial consortium, 215, 216, 220; under early Republic, 219–25 passim; position of foreign experts in, 301–2; historical literature on, 328–29, 338–39, 344–46, 364–66, 385; critics of barbarian management, 333n24, 366n3; restricting foreign rights, 391n22, 395n56; *see also* Canton system; Chinese policy toward the United States

Chinese foreign service in the United States, 96, 97–98, 255–56

Chinese historians, writings on Sino-American relations, 317n1, 318, 328, 344–45, 353–54, 376, 385, 404n18

Chinese legation in the United States: establishment, 98–99; control over Chinese community, 99–100, 111–14, 255–56, 344n50; and exclusion movement, 100, 103–8 passim, 230–31; influence of nationalism in, 230, 383n52

Chinese merchants: in cohong, 9–10;

Chinese merchants (Continued)
capital entrusted to Americans, 10, 145, 146, 147, 355n7, 356n9; and immigration question, 235–41 passim, 247, 395n51; see also American economic enterprise in China
"Chinese museums," 34
Chinese nationalism: and boycott, 234–36, 251–52, 261, 263; and foreign penetration, 258–64 passim, 374n63; view of the United States, 259–66 passim, 304; impact on Chinese government policy, 262, 280–81, 292, 303–4, 308, 390n18, 395n52; and American economic enterprise, 280–82, 283–84; and mission movement, 297–98; historical literature on, 384
Chinese Nationalist League, 254, 257
Chinese overseas: process of emigration, 61, 64, 97; relation to host societies, 73; as an asset to China, 97, 98; protest against exclusion, 233, 235, 237, 247; historical literature on, 330–31
Chinese policy toward the United States: search for diplomatic and financial support, 52–59 passim, 117–42 passim, 147, 149, 195, 198–202, 212, 214–15, 222–23, 302–6 passim, 311–12, 372n56, 373nn57,61; and aid against the Taiping, 53, 55; fear of American collusion with Japan, 60, 128; interest in American education and technology, 60, 334n37, 373n61; immigration question, 98–108 passim, 121, 233–34, 238, 239–40, 247, 343n39; historical literature on, 328–29, 339, 344–46, 364–66, 385; see also Chang Chih-tung; Hsi-liang; Li Hung-chang; Yüan Shih-k'ai
Chinese Repository, 27, 31, 34, 38, 40
Chinese students: and boycott, 236, 239, 240–41; and agitation in mission schools, 298
Chinese view of the United States: scholar-officials, 3, 41–51, 190–92, 219, 222–23; impact of open door constituency on, 44, 47; as a developmental model, 45–50 passim, 191, 208, 259, 260, 387n3, 388n9; admiration for George Washington and the American revolution, 46–50 passim, 191, 259,

260, 262; economic expansion and the open door policy, 208, 259–65 passim, 373n58, 388n9; annexation of the Philippines, 260–64 passim, 387n4; parallels with American view of Chinese, 301; during Korean War, 309; historical literature on, 327–28, 384; see also Chinese nationalism; Chinese policy toward the United States
Ch'i-ying, 43–44, 53, 55, 57–58
Choate, Joseph, 344n46
Chou En-lai, 312
Chou Fu, 129, 241
Chou Tzu-ch'i, 220, 372n56, 373n61
Chu Ch'i-ch'ien, 372n56
Chungking riot (1886), 160
Cleveland, Grover, 141; and immigration issue, 91–92, 94, 105; and protection of missionaries, 166; and appointments to Peking legation, 343n39
Cohong, 9–11, 22, 323n29
Collins, Perry McDonough, 146
Compradors, 144, 323n29; see also Chinese merchants
Conant, Charles, 177, 178
Conger, E. H., 185, 193, 232, 237, 277, 278; views as minister, 176–77, 195, 196, 274, 391n22, 395n51; background, 362n65
Coolie trade, 23, 64, 97, 324n31, 335n42
Cotton Goods Export Association, 248
Cotton goods trade: American role in, 22–23, 148, 274, 276–77, 356n12, 393n40, 394n45; British competition, 148; Japanese competition, 276–77, 394n45; Chinese competition, 280
Cowpland (captain of the Emily), 1
Cultural intermediaries, 301–2
Cushing, Caleb, 18–19, 31, 327n61
Cushing, John Perkins, 11, 18
Cushing, Jonathan, 1
Cushing mission, 15, 18–19, 53

Danish Great Northern (telegraph company), 356n8
Denby, Charles, Jr., 169; and outbreak of Sino-Japanese War, 138, 350n47; view on China policy, 246, 270, 271, 368n18; career, 350n46
Denby, Charles, Sr., 132, 141, 162, 177,

179, 180, 181, 277; and Sino-Japanese War, 138, 139; views on China and United States policy, 138, 139, 168, 174–76, 178, 274, 361n59, 362n64, 378n6; and mission movement, 161, 165, 167; appointment as minister, 361n58
Dennett, Tyler, ix
Denny, Owen N., 132, 349n36
Dodge, William E., 362n65
Dollar, Robert, 247, 355n6
Duke, James B., 283
Dulles, John Foster, 300, 311
Dunn, James R., 378n4
Dunn, Nathan, 34

Eddy, Sherwood, 293
Einstein, Lewis, 213
Elements of International Law (Henry Wheaton), 289
Eliot, Charles W., 218
Empress Dowager Tz'u-hsi, 191, 203, 207; political role, 116, 134, 136, 140, 194, 207; American view of, 267, 361n59
Empress of China (merchant ship), 2
Episcopal Mission, 25
Evarts, William M., 87, 127, 128, 179, 348n17
Everett, Alexander, 19
Ever-Victorious Army, 59–60
Exclusion movement: resort to violence, 75–76, 81, 102, 227; success at the state and local level, 76–77, 78, 247, 337n63; influence in Congress, 77, 86–95 passim, 105, 232, 234, 243, 246; impact on national party politics, 78, 85–86, 87, 91–92, 93, 242, 339n1; domestic opposition to, 78–79, 87–94 passim, 95, 231, 240, 245–48 passim, 340n6; appeal to labor, 81; defended by United States government, 106; reaches to overseas territories, 229–30, 232; historical literature on, 329, 338, 339; *see also* American Federation of Labor; Anti-American boycott; Immigration treaties; Knights of Labor; Sinophobia; United States Bureau of Immigration

Fairbank, John K., 299, 313
Fearon, Daniel and Company, 356n12, 357n15

Feng Hsia-wei, 236, 241
Feng Kuei-fen, 366n3
Ferry, Jules, 133, 350n39
Fifteen Passenger Bill (1879), 86, 87, 89, 95
First National Bank, 211
Fish, Hamilton, 178
Fong Yue Ting v. the United States, 344n46
Foord, John: and the threat of partition, 152, 275; and the immigration question, 231, 232, 245, 381n31; and the cotton textile trade, 393n40
Forbes, John M., 24, 326n55
Forbes, Robert Bennett, 24
Foster, John W., 140, 141, 149, 351n51, 362n65
France, and the Vietnam crisis, 133–36 passim
Franklin, Benjamin, 32, 33, 34
Frazar and Company, 146
Frelinghuysen, Frederick, 161, 179, 181, 361n56
Fukien Christian College, 399n77

"Gang of four," 312
Garfield, James A., 89–90, 123, 179
Gauss, Clarence, 403n5
Geary Act (1892), 94, 105, 344n46
Ginling College, 399n77
Girard of Philadelphia (China trader), 8
Gompers, Samuel, 242, 244, 381n28
Goodnow, Frank J., 220
Gould, Annie, 185–86, 286
Grand Council, 116
Grant, Ulysses S., 128, 141, 179, 349n36; stand on exclusion, 86, 120, 121, 339n4; bargain with Li Hung-chang on immigration restriction, 88, 120–21, 347n12, 348n17; and Liuchiu dispute, 119–24 passim; political aspirations, 119–20, 123, 348n16; views on East Asia and United States Asian policy, 121–23, 178; and a policy of coercion, 178
Gresham, Walter Q., 141, 361n58; and immigration question, 106–7; and Sino-Japanese War, 137–38, 140, 350n45; cautious China policy of, 165–66, 179–80
Guggenheim mining interests, 395n56

Hai-kuo t'u-chih (Wei Yüan), 45–46, 333n20
Hakkas, 337n56; origins in China, 62, 63; social solidarity in the United States, 65, 67, 69, 72
Hamilton, Alexander, 12
Hangchow Christian College, 399n77
Hankow-Canton railway, 237, 241, 244, 281; see also American China Development Company; Chang Chih-tung
Harriman, E. H., 146, 397n66; and Hankow-Canton railway, 150, 279; and Manchurian railways, 204–5, 274; and American Group, 279
Harrison, Benjamin, 92–93, 165, 343n39, 361n58
Harris, Townsend, 37
Hart, Robert, 136, 187, 323n30
Hay, John, 237, 267, 307, 373n58, 390n17, 398n70; and first open door notes (1899), 153, 154; and alliance with Britain, 181–82, 199; on territorial acquisition, 182, 193, 368n18; and Boxer crisis and Russian occupation of Manchuria, 195–202 passim, 394n42; and the American China Development Company, 278; on exclusion, 380n23; and opening new treaty ports, 391n22
Hayes, Rutherford B., 87–88, 89, 348n17
Headland, Isaac Taylor, 289
Heard and Company, 172; operations in China, 23, 24, 144; view of partners, 35, 36–37, 327n61
Heard, Augustine, Jr., 35
Heng-fu, 58
"Highbinders," 69, 111
Hippisley, Alfred, 153
Ho Ju-chang, 118, 348n23
Ho Kuei-ch'ing, 55–56, 333n28
Holcombe, Chester, 147; diplomatic activities, 130, 169; as investment promoter, 149; and mission movement, 160; Li Hung-chang's dislike of, 169; and exclusion, 232
Hornbeck, Stanley K., 305, 403n5
Ho Ts'ai-yen, 237
House, Edward M., 224, 225
Howqua, 10
Hsiang-shan district, 337n56; as source of Chinese in the United States, 61–62; as basis of social solidarity in the United States, 67, 68
Hsiang-ya, 399n77
Hsien-feng Emperor, 55, 56, 57
Hsi-liang, 262, 371n44, 372n56; background, 208; view of the United States, 208; Manchurian policy, 208, 209, 214, 215–16
Hsiung Hsi-ling, 220, 372n56
Hsü Chi-yü, 55, 189, 191, 208, 259, 264, 303, 304; view of the United States, 46–50, 55; career of, 46, 334n30; implications of his writings for Chinese policy, 51
Hsüeh Chüeh, 112–13, 191, 367n4
Hsü Shih-ch'ang, 221, 372n56; role in Manchuria, 204–5, 208, 370n30; view of the United States, 219, 373n58
Huai River conservancy project, 373n61, 374nn62,63, 395n50
Huang En-t'ung, 50
Huang Fu, 373n58
Huang Hsi-ch'üan, 82
Huang Hsing, 257, 282
Huang Tsun-hsien, 50, 388n7
Hughes, Charles Evans, 305
Hukuang railway, 213, 215, 282, 372n47
Hung Hsiu-ch'üan, 326n56
Hung Jen-k'an, 50, 332n12

I-hsin, see Prince Kung I-hsin
I-huan, see Prince Ch'un I-huan
I-k'uang, see Prince Ch'ing I-k'uang
Immigration treaties: 1868 treaty (Seward-Burlingame), 77, 86, 87, 88, 98, 102, 173, 339n4; 1880 treaty (Angell treaty), 87–88, 100–2; aborted 1888 treaty, 91, 104–5; 1894 treaty (Gresham-Yang), 94, 106–8
Imperial Maritime Chinese Customs, 22, 323n30
Imperial Tobacco Company, 283
International Banking Corporation, 394n46
Ishii Kikujiro, 225

"Jackal diplomacy," 17
Japan: and the Liuchiu dispute, 118, 121, 123, 124; and Korea, 126; and war with China, 136, 137, 139; opposition to

American influence in China, 212, 374n63, 375n66; see also Twenty-one demands
Jay, John, 12
Jefferson, Thomas, 12, 13, 33, 34, 322n14
Jenks, Jeremiah, 220, 245, 390n17
Johnson, Nelson T., 305, 403n5
Juan Yüan, 1, 2, 43, 52
Jung Hung, see Yung Wing

K'ang Yu-wei, 255, 384n52; and exclusion, 251–52; political activities in the United States, 252, 383n52; view on the United States and the foreign threat, 388n9
Kearney, Denis, 76, 337n62
Kellogg, Frank B., 305
Kerosene, exported by the United States to China, 148–49; see also Standard Oil
King, Rufus, 11
Kissinger, Henry, 312
Knights of Labor, 81, 227, 232
Knox, Philander C., 275, 277; pursuit of dollar diplomacy in China, 209–17 passim, 272; legalistic style of, 210–11; controversy over his policy, 212–13, 216–17; appointment as Secretary of State, 371n39
Korea: seclusion policy under pressure, 125–27; resistance to Chinese direction, 129, 130, 131–32
Korean-American treaty (1882), 128–31
Kuei-liang, 57–58, 333n28
Kuhn, Loeb and Company, 150, 211
Ku Hung-ming, 388n7
Kuo-feng pao, 265
Kuo Sung-tao, 366n3
Ku-t'ien riot (1895), 162, 166
Ku Wei-chün, 223, 372n56
Kwang-hsü Emperor, 136, 289

Lansing-Ishii agreement (1917), 223, 225
Lansing, Robert, 224, 225
Laymen's Missionary Movement, 293
LeGendre, Charles, 60
Leopold II (King of Belgium), 277–78
Liang Ch'eng, 234, 255, 379nn12, 19
Liang Ch'i-ch'ao, 220, 255, 383n52; political activities in the United States, 251, 252, 383n52; view of the United States, 263–65, 281–82, 388n7; view of the foreign threat, 264, 388nn8,9
Liang Ju-hao, 372n56
Liang T'ing-nan, 44
Liang Tun-yen, 221, 371n44, 372n56; mission to the United States, 214, 215, 373n61; view of the United States, 367n6
Libby, William H., 149
Lien-chou massacre, 245, 295–96, 400n79
Li Hung-chang, 83, 150, 189, 191, 196, 307, 356n7; early interest in the United States, 59–60; and protection of Chinese overseas, 99, 100, 104, 105–6, 121, 124–25; background, 115–16; American view of, 116, 122, 130–31, 174–75, 198, 346n3, 361n59; estimate of his policy, 116–17, 141–42; view of the United States, 117, 128, 132–33, 137, 139, 199; and the Liuchiu dispute, 118–24; view of Ulysses S. Grant, 119, 123, 124; and Korean policy, 126–32 passim, 136–40 passim, 348n23, 349n36; and the Vietnam crisis, 133–36, 350nn39,41; and the Sino-Japanese War, 137, 350n47; visit to the United States (1896), 141; and Sino-American bank proposal, 147–48; and Boxer crisis and the Russian occupation of Manchuria, 194, 199–201; historical literature on his foreign policy, 344–46; on the mission question, 402n90
Li Hung-tsao, 100, 101
Lin Shao-nien, 238
Lin Tse-hsü, 44–45, 52
Lodge, Henry Cabot, 177, 178, 371n39, 380n23
Looking Backward (Edward Bellamy), 388n9
Low, Frederick, 127, 158–59, 343n39, 358n30, 361n55

MacArthur, Douglas, 309
McCook, John J., 150
Ma Chien-chung, 129, 130, 350n41
McKinley, William, 259, 275, 398n70; and threat of China's partition, 181, 182; and Boxer crisis, 195–96, 198, 369n21

McLane, Robert, 20–21, 37, 54, 327n61, 333n23
MacMurray, John V. A., 305, 308, 403n5
Madison, James, 12, 33
Mahan, Alfred Thayer, 177
Mao Tse-tung, and the United States, 304–9 passim
Marshall, Humphrey, 20, 37, 54, 322n25, 325n46, 327n61
Martin, W. A. P., 160, 217, 288; and China policy, 32, 164, 327n61; view of China, 36, 400n81; as missionary-educator, 289, 402n91
Mateer, Calvin, 290, 294, 399n77
Mattingly, Garrett, quoted, x
Mei-li-ko ho-sheng kuo chih-lüeh (Elijah C. Bridgman), 44, 47
Mencius, quoted, 54
Mercantile Tobacco Company, 357n14
Merchant and Chemical Bank (New York), 149
Metcalf, Victor, 245
Methodist missions, 25, 27
The Middle Kingdom (S. Wells Williams), 27, 40
Mission movement: established in Canton and the treaty ports, 25–30; statistics on, 25, 26, 154, 291–92, 293; reliance on evangelism, 26–27, 28–30, 32, 154–56; and literary work, 27, 28, 29, 44; and medical work, 27–28, 29, 399n77, 402n92; commitment to education, 28, 29, 289–98 passim, 399nn74,76,77; and Chinese associates, 29, 155–56, 286, 293–95, 302; and diplomatic support, 30, 155, 158, 162–64, 267, 286–87, 289, 294–98, 390n16, 398n70, 401n87; involvement in diplomacy, 30–32, 169, 389n11; view of China, 35–36, 38, 39, 162–64, 286–87, 292–98, 400nn81,82; and exclusion movement, 78, 231, 248, 292; impact of Boxer experience on, 286–90, 397n67, 398n72, 401n83; historical literature on, 319, 353–54, 386
Mitkiewicz, Count (representative of Wharton Barker), 147–48
Monroe, James, 33
Moore, John Bassett, 374n63
Morgan, J. P., 149, 150, 211, 264, 278

Morison, Samuel Eliot, quoted, 6
Morton, Levi, 150
Mott, John R., 217, 218, 293

Nanyang Brothers Tobacco Company, 283–84
Nation, 287
National Association of Manufacturers, 152, 246
National City Bank, 150, 211, 394n46
Native Sons of the Golden State, 250
Nativism, see Anti-Christian movement; Boxer Uprising; Exclusion movement; Sinophobia
Na-t'ung, 208–9, 212–16 passim
New York Chamber of Commerce, 248
New York Herald, 119
New York Sun, 287
New York Times, 287
Ng Yup (five districts), 334n39
Nixon, Richard, 312
North American Review, 34
North China Union College, 399n77
Nye, Gideon, Jr., 21

O'Keefe, Daniel J., 249
Olney, Richard, 141, 177; and the immigration question, 107; and China policy, 166–67, 178, 180, 181
Olyphant and Company, 8, 144, 172, 355n2
Olyphant, D. W. C., 1, 326n55
Open door constituency: defined, xi, 5; under the Canton system, 2; early view of China, 35–40; internal relations, 37, 326n55, 327n61; impact in the United States, 39–40, 143; and exclusion policy, 231–32, 234, 245–46, 247–48; fate in the twentieth century, 300–1, 310, 312; see also American economic enterprise in China; Mission movement; United States foreign service in China; United States legation in China
Open door ideology: rise to prominence, 177–78, 258, 302; and Chinese nationalism, 187–88
Open door policy, see United States China policy
Opium trade: American role in, 7–8, 9, 11, 22, 23–24; Chinese campaigns against,

14, 392n29; American view of, 39; American support for suppression of, 292, 391n29, 392n34
Overseas Chinese, see Chinese community in the United States; Chinese overseas

Pacific Mail Company, 145–46, 279
Pan Shih-ch'eng, 52, 333n17
Pao-yün, 100
Parker, Peter, 302, 326n55; involvement in diplomacy, 19–20, 21, 31, 323n27; as medical missionary, 27–28; view of China, 35, 36; Chinese view of, 56, 325n46
Peck, Willys R., 403n5
Pei-yang kuan-pao, 233
Peking Union Medical College, 400n77
Peking University, 399n77
Perkins and Company, 7, 8, 322n14
Perkins, Thomas H., 15
Perry, Matthew, 21, 37, 120
Pethick, William, 118–19, 140, 169, 360n48
Philadelphia Bible Society, 25
Phillips, William, 403n5
Pierce, Franklin, 20
Pitkin, Horace T., 185, 186, 286, 291, 364n3
Powderly, Terence V., 227–28
Presbyterian missions, 294, 296
Prince Ch'ing I-k'uang, 202, 207, 208, 238, 369n26
Prince Ch'un I-huan, 134
Prince Ch'un Tsai-feng, 208, 209, 214–15
Prince Kung I-hsin, 59, 116, 119, 120, 134
Protect the Emperor Society, 252

The Real Chinaman (Chester Holcombe), 166
Red Cross, see American Red Cross
Reed, William B., 21, 56, 57, 327n61
Reid, Gilbert, 160, 161, 288, 290, 399n74
Reinsch, Paul S., 222, 224, 225, 275, 276, 277, 307; appointment as minister, 218; as advocate of support for China, 220, 225, 272, 279–80; and exclusion policy, 248
The Revolutionary Army (pamphlet by Tsou Jung), 254

Revolutionary Party, 257
Reynolds, James B., 245
Richard, Timothy, 399n74
Roberts, Edmund, 31, 38, 39
Roberts, Issachar J., 36, 326n56
Rockhill, W. W., 169, 177, 196, 200, 205, 220, 237, 278, 299, 302, 394n42; emergence as a China expert, 152, 153, 195, 266–67, 389n12; and mission movement, 167, 296, 298, 398n70; and immigration issue, 232, 242, 244, 268, 269, 395n51; as advocate of reform, 267, 270, 271, 389n14, 390nn15,16,17; reaction to reform nationalism, 268–69; influence in Washington, 269–71, 273, 391n25
Rock Springs massacre, 81–82, 102, 103
Roosevelt, Franklin D., 306
Roosevelt, Theodore, 251, 264, 275, 277; and Korea, 140; and Japan, 206–7; and China, 206–7, 270, 271, 370n34; and economic foreign policy, 212, 269, 275, 278, 394n42; and immigration question, 241–45, 249, 269, 275
Root, Elihu, 196, 371n39; and mission of T'ang Shao-i, 207, 270; defense of treaty rights, 269, 391n24; and Japanese trade discrimination, 394n42
Root-Takahira agreement (1908), 207
Ruschenberger, W. S. W., 38–39
Rush, Benjamin, 33
Rusk, Dean, 300, 311
Russell and Company, 147, 172, 332n19; origins and activities, 8–9, 12, 16, 23, 24, 135, 355n2; decision to close down, 144–45
Russell, Samuel, 1
Russo-Japanese agreement on Manchuria (1910), 212

St. John's University, 399n77
Sam Yup (three districts), 337n56; source of immigrants to the United States, 61, 63, 334n39; basis of social solidarity in the United States, 65–73 passim, 110–11
San Francisco Call, 287
San Francisco Chamber of Commerce, 248
Sargent, Aaron A., 77, 82, 127, 128, 141

Sargent, Frank P., 242, 243, 249
Schiff, Jacob, 150
Scott Act (1888), 92, 95, 105, 125, 344n46
Scott, William L., 92
Sec, Fong F., 232
Senggerinchin (Seng-ko-lin-ch'in), 58–59
Seward, George F., 87–88, 100, 232, 361n55
Seward William, 86, 127, 146, 178, 179
Shanghai Chamber of Commerce, 235
Shanghai Steam Navigation Company, 145
Shanghai University, 399n77
Shantung Christian University, 399n77
Sheffield, Devello Z., 177, 286–87
Sheng Hsüan-huai, 150, 192, 367n7
Sheng-wu chi (Wei Yüan), 45
Shen Kuei-fen, 342n28
Sherman, John, 152, 181
Shih Chao-chi, 372n56
Shufeldt, Robert, 141; mission to Korea, 126–31 passim; view on Asians, 127, 130–31; career, 127, 349n32; view on United States role in Asia, 129
Singer Company, 396n60
Sino-Japanese War, 136–40
Sinophobia, 73–76, 78, 94–96, 301; see also Exclusion movement
Six Companies, 112, 250; composition, 68–69; reaction to exclusion, 110, 246–47; social controls, 336n49
Smith, Arthur H., 292, 299
Smith, Judson, 288, 399n76
Social Life of the Chinese (Justus Doolittle), 163
Society for the Diffusion of Christian and General Knowledge, 399n74
Society for the Diffusion of Useful Knowledge, 47
Society to Revive China, 253
Soochow University, 399n77
Southern Pacific Railroad, 146
Special Sino-American relationship: defined, x, 300; promoted by the United States legation in China, 170–71; persistence of an idea, 299–300, 310–11, 312–13; discussed in China, 404n18
Ssu-chou chih (Lin Tse-hsü), 44–45
Stalin, Joseph V., 306
Standard Oil, 397n66; marketing strategy in China, 148–49, 282–83, 284–85, 357n13, 397n64; Shensi concession, 220, 282, 373n61, 374nn62,63, 395n50, 397n63
Stilwell, Joseph, 306
Stimson, Henry L., 305–6
Straight, Willard, 205, 208, 394n40; in Manchuria, 209, 275; in Knox's State Department, 210, 371n41
Straus, Oscar, 245
Strong, Josiah, 177
Student Volunteer Movement, 154, 163, 291
Sun Pao-ch'i, 220
Sun Yat-sen, 218, 255, 256, 293, 384n52; political activities in the United States, 252–55, 256–57, 383nn47,52; and immigration question, 253–54, 383n48; view of the United States, 260, 387n3, 403n4; American view of, 392n32, 401n82
Swift, John F., 88, 343n39
Sze Yup (four districts), 337n56; source of immigrants to the United States, 61, 63, 334n39; basis of social solidarity in the United States, 66, 67, 68, 72–73, 110–11; boycott activity in, 237

Taft, William Howard, 205, 244; and China policy, 209–10, 273; and Japan, 210; and immigration issue, 249, 380n26; and economic foreign policy, 279, 380n26; search for a Secretary of State, 371n39; and W. W. Rockhill, 391n25
Taiping Rebellion, American reaction to, 20, 36–37
T'ang Ch'iung-ch'ang, 252
T'ang Shao-i, 216, 268, 372n56, 379n19; service in Manchuria, 204–8 passim; view of the United States, 204, 304; American view of, 270
T'an T'ing-hsiang, 57
Tao-kuang Emperor, 54
Tattnal, Josiah, 323n27
Teng Hsiao-p'ing, 312
Tenney, Charles D., 389n11, 399n74, 402n91
Terranova affair, 1–2
Terranova, Francesco, 1, 2

Index

Tewksbury, Elwood, 287
Textile mills in China: American investment in, 146–47, 149; export of American machinery to, 396n60
Thomas, James A., 283
Thomson, Charles, 33
Treaty of Shimonoseki (1895), 149
Treaty of Wanghsia (1844), 19
Trescott, William H., 88
Triads, 68, 252
Tsai-feng, see Prince Ch'un Tsai-feng
Ts'ai T'ing-kan, 372n56
Tsai-tse, 208–9, 212, 214
Ts'ao Ju-lin, 222
Ts'en Ch'un-hsüan, 238, 240–41
Tseng Chu, 236, 244
Tseng Kuo-fan, 59, 60, 115, 146
Tsinan missionary dispute, 160, 358n31
Tsinghua College, 391n27
Tso-chuan, quoted, 42
Tso Tsung-t'ang, 134
Tsui Kuo-yin, 107, 113–14
Tung-fang tsa-chih, 263
Twain, Mark (Samuel L. Clemens), 288
Twenty-one demands, 222, 223, 224, 257
Tyler, John, 15
Tz'u-hsi, see Empress Dowager Tz'u-hsi

Union Pacific Railroad, 81, 82
The United States and China (John K. Fairbank), 299
United States Army, 196, 197–98, 245
United States Bureau of Immigration, 255, 377n2, 380n23; extends restrictions against Chinese, 227–29, 233, 246–47, 378n4, 382n34; attempted reform under Theodore Roosevelt, 242, 243, 245, 246, 249; defends procedures, 248–49; staff and funding, 377n3
United States China policy: promotion of economic interests, 11–17 passim, 146–47, 152–54, 180–81, 209–11, 273–85 passim, 362n72, 372n53, 374n63, 394n42, 397n66; cooperation with Britain, 16, 17, 181–82; interest in territorial acquisitions, 17, 120, 182, 193, 198, 368n18; reaction to foreign threats to China, 17, 135–38 passim, 179–83 passim, 201–2, 206–14 passim, 218–26 passim, 306–7, 350n45; and the

issue of coercion, 19, 20, 21, 178–79, 197–98, 244–45; clash between open door and exclusion policies, 82–83, 85, 87, 90, 93, 161–62, 341n16, 362n72, 381n31; critical of tribute relations, 131, 141–42; open door notes (1899 and 1900), 153–54, 183, 193, 197; response to Chinese nationalism, 153, 239, 240, 244–45, 266–70, 273, 280, 281, 305; and support for mission movement, 161–62, 165–67, 168, 179, 295–96, 401n87; and the Boxer crisis, 195–200, 267, 369n21, 389n14, 390nn15,16; view of Chinese leaders, 198, 207, 307–8, 311; relations with Communists, 307–9, 311, 312; historical literature on, 319–20, 345–46, 351–53, 354, 364–66, 384
United States Department of Commerce and Labor, 377n2
United States Department of State: and immigration question, 106, 249, 344n50; China experts in, 210, 269, 270, 273, 371n41, 389n11; Division of Far Eastern Affairs, 389n11
United States foreign service in China: takes root, 13, 15–16; views on China and China policy, 17, 35–39 passim, 274–75, 304–5, 391n26, 392n32, 395n51, 397n64, 403n5; China experts in, 169, 302; amateur character of, 169–70, 327n61, 360n48, 389n11; promotion of American educational influence by, 172, 270, 360n51; and exclusion, 231–32, 248; historical literature on, 385
United States Korean policy, 125, 126–28, 131, 132, 137–38, 140
United States legation in China, 343n39, 361n55; guidelines for promoting economic interests, 153, 180, 275, 374n63; and protection of missionaries, 158–61, 162, 165, 167–68; views on China and China policy, 170–77, 267–72 passim; historical literature on, 354
United States Navy: presence in Chinese waters, 13, 15, 16, 165, 168, 198, 205, 245, 363n77; and use of force, 17–18, 323n27, 333n25; incidents created by

United States Navy (Continued)
  sailors, 237, 241; interest in a China base, 368n18; training and technology for the Chinese navy, 373n61, 374n63
United States Supreme Court, 106, 107
United States Treasury Department, 106, 377n2
University of Nanking, 399n77
The Uplift of China (Arthur H. Smith), 292

Walker, Robert J., 20
Wang Mao-yin, 333n20
Wang T'ao, 50, 366n3
Wang Wen-shao, 342n28
Ward, Frederick Townsend, 59
Ward, John E., 21–22, 58, 333n28
Washington, George, 33
Webster, Daniel, 15, 18–19
Wei Yüan, 189, 191, 208, 259, 303; view of the United States, 45–46; policy implications of his writings, 51; career, 331n7
West China Union University, 399n77
Western Union Company, 146
Wetmore and Company, 8
Wetmore, W. S., 146–47, 181
Whittier, Charles A., 278
Williams, Edward T., 220, 272, 273, 389n11
Williams, S. Wells, 160, 170, 217, 299, 302, 326n55; involvement in diplomacy, 21, 31, 32, 169, 360n52; mission work, 27, 40; views on China, 35, 39; quoted on the foreign service, 170, 360n48; promotion of American educational influence, 172, 360n51; quoted on exclusion, 231
Wilson, Huntington, 275; service in the State Department, 210, 211, 270, 394n42; background, 371n41
Wilson, James H., 132, 147, 149–50
Wilson, Woodrow: stand on economic foreign policy, 217, 218, 221, 225, 279, 372n53, 374n63; policy toward China, 217–26 passim, 372n55; and exclusion, 249
Workingmen's party, 76, 337n62

Wu Chien-chang, 53–54, 332n19
Wu Ch'ung-yüeh, 10
Wu Ping-chien, 10
Wu T'ing-fang, 268, 379n19; activities as minister to the United States, 182, 192, 193, 195, 230–31, 402n3; background, 191; promotes interest in the United States, 191–92, 367n5; American view of, 270, 398n70; on control of missionaries, 297
Wu Yüan-hua, 10

Xenophobia, see Anti-Christian movement; Boxer Uprising; Exclusion movement; Sinophobia

Yale-in-China, 186
Yang-ju, 107–8, 112, 113, 344nn49,50
Yeh Ming-ch'en, 55, 56–57, 333n23
Yenching University, 399n77
Ying-huan chih-lüeh (Hsü Chi-yü), 46–50, 334n30
Young, John Russell, 141, 179, 349n36; as protégé of Ulysses S. Grant, 119–20, 348n17; as minister to China, 133–36, 147, 173–74, 349n38, 361n56; and mission movement, 156, 160–61, 358n32; views on East Asian policy, 173–74, 361n56
Young Men's Christian Association (YMCA), 218, 291, 401n86
Young People's Mission, 291
Yüan Shih-k'ai, 256, 257, 262, 270, 307; background, 202–3; view of the United States, 203; and Manchurian policy, 203, 205, 370nn27,30, 380n22, 391n22; and immigration controversy, 203, 238, 239–40, 380n22; political eclipse and return to power, 207–8, 216; and foreign policy during the early Republic, 219–22; American view of, 271–72, 293, 392n31, 400n81
Yü-chien, 52
Yü Ch'ien-yao, 128
Yung-cheng Emperor, 96
Yung Wing (Jung Hung), 60, 99, 192, 334n34, 367n6